Application-Driven Quantum and Statistical Physics

A Short Course for Future Scientists and Engineers

Volume 1: Foundations

Essential Textbooks in Physics

ISSN: 2059-7630

The *Essential Textbooks in Physics* explores the most important topics in Physics that all Physical Sciences students need to know to pass their undergraduate exams (years 1, 2 and 3 of the BSc). Some topics are run-of-the-mill topics, others introduce students to more applied areas (e.g. Quantum Optics, Microfluidics…).

Written by senior academics as well lecturers recognised for their teaching skills, they offer in around 200 to 250 pages a theoretical overview of fundamental concepts backed by problems and worked solutions at the end of each chapter.

Their lively style, focused scope and pedagogical material make them ideal learning tools at a very affordable price.

Most authors are based at prestigious universities: Imperial College London, Oxford, UCL, Ecole Polytechnique.

Published

Application-Driven Quantum and Statistical Physics: A Short Course for Future Scientists and Engineers
Volume 1: Foundations
by Jean-Michel Gillet

A Guide to Mathematical Methods for Physicists: With Problems and Solutions
by Michela Petrini, Gianfranco Pradisi and Alberto Zaffaroni

Introduction to General Relativity and Cosmology
by Christian G. Böhmer

Newtonian Mechanics for Undergraduates
by Vijay Tymms

Essential Textbooks in Physics

Application-Driven Quantum and Statistical Physics

A Short Course for Future Scientists and Engineers

Volume 1: Foundations

Jean-Michel Gillet
Professor at CentraleSupélec (Paris-Saclay University, France)
Professor at Ecole Centrale Pékin (Beihang University, China)

NEW JERSEY • LONDON • SINGAPORE • BEIJING • SHANGHAI • HONG KONG • TAIPEI • CHENNAI • TOKYO

Published by

World Scientific Publishing Europe Ltd.
57 Shelton Street, Covent Garden, London WC2H 9HE
Head office: 5 Toh Tuck Link, Singapore 596224
USA office: 27 Warren Street, Suite 401-402, Hackensack, NJ 07601

Library of Congress Cataloging-in-Publication Data
Names: Gillet, Jean-Michel, author.
Title: Application-driven quantum and statistical physics : a short course for future scientists and engineers / Jean-Michel Gillet (CentraleSupâelec, France).
Other titles: Essential textbooks in physics.
Description: Singapore ; Hackensack, NJ : World Scientific Publishing Co. Pte. Ltd., [2018] | Series: Essential textbooks in physics Contents: volume 1: Foundations -- volume 2: Equilibrium -- volume 3: Transition.
Identifiers: LCCN 2018023720| ISBN 9781786345547 (hc ; v.1 ; alk. paper) | ISBN 1786345544 (hc ; v.1 ; alk. paper) | ISBN 9781786345578 (hc ; v.2 ; alk. paper) | ISBN 1786345579 (hc ; v.2 ; alk. paper)
Subjects: LCSH: Quantum theory--Textbooks. | Statistical physics--Textbooks.
Classification: LCC QC174.12 .G54 2018 | DDC 530.12--dc23
LC record available at https://lccn.loc.gov/2018023720

British Library Cataloguing-in-Publication Data
A catalogue record for this book is available from the British Library.

First published 2019 (Hardcover)
Reprinted 2019 (in paperback edition)
ISBN 9781786346902 (pbk)

For any available supplementary material, please visit
https://www.worldscientific.com/worldscibooks/10.1142/Q0164#t=suppl

Desk Editors: V. Vishnu Mohan/Jennifer Brough/Koe Shi Ying

Typeset by Stallion Press
Email: enquiries@stallionpress.com

Preface

Le concret c'est de l'abstrait rendu familier par l'usage.[1]

Paul Langevin

"La notion de corpuscules et d'atomes"

16th October 1933

This first volume of a triptych sets the general quantum landscape. It is built upon three connected parts that can be considered as a general introduction to quantum physics.

Classical physics had shown all its power and ability to accompany the industrial revolution with the tremendous success of — in particular — thermodynamics, with the development of more efficient heat engines or electromagnetism, with the birth of modern radio-communications. Nevertheless some scientists at the beginning of the twentieth century were still dissatisfied with unexplained experimental results. It is thus the purpose of the first part to give the general ambiance that led a well-established physical paradigm to progressively crumble. It then shows how many daring scientists came to invent the necessity of a new universal physical constant — the

[1]The concrete is the abstract rendered familiar through use.

celebrated Planck constant — that will prove to be the cornerstone of an alternative way to interpret the physical world.

In the second part, we introduce new rules according to which one believes the quantum game is played. They are presented as mere postulates that have — so far — proved to be corroborated by every single experiment and at any level of accessible precision. Most of these rules have lifted former obstacles in explaining physical observations, while some of them have led to develop new methods of investigation at the microscopic scale or even new technologies. It has been stated that a rough 25% of the USA gross national product is the result of inventions based on quantum physics. It is not to be overoptimistic to foresee an increasing input of quantum consequences in the coming years.

Finally, the third part of this volume presents connections between quantum and classical descriptions. While the quantum paradigm solves many difficult problems, it also introduces a new level of complexity that is both cumbersome and unnecessary in most of daily life macroscopic challenges. We then raise a quite legitimate question: when do we really need to resort to a quantum description of our physical world? The links between those two points of view turn out to be multifaceted and require a careful inspection.

Three excellent references, which are not mentioned explicity in the main text but have guided me throughout my study years and are still of great inspiration for my daily teaching, are worth being cited here. *Quantum Mechanics* by C. Cohen-Tanoudji, B. Diu and F. Laloe (Wiley, 1991), *Physique Statistique* by B. Diu, C. Guthmann, D. Lederer, B. Roulet (Hermann, 1989), and *Quantics: Rudiments of Quantum Physics* by J.-M. Levy-Leblond and F. Balibard (North-Holland, 1990). I have been lucky enough to enjoy the very inspiring teaching of Françoise Balibard and Bernard Diu who gave me the taste for learning, then transmitting this exciting science. I am sure many readers would benefit from in depth treatments given by their published work.

The biographical notes, which focus on the human side of some key actors, are based on a subjective collection of information from several historiographic books on post-nineteenth century physics. Remarkable Physicists *by Ioan James [59] is an excellent introduction to many exciting lives and has served as the backbone of what is too briefly presented here.*

Some readers may not be satisfied with pure science and need to rely on possible concrete examples. They will find motivation to further endure more concepts and calculations in a one-sentence box at the beginning of each chapter. This gives a glimpse of some applications that will be treated, or briefly evoked.

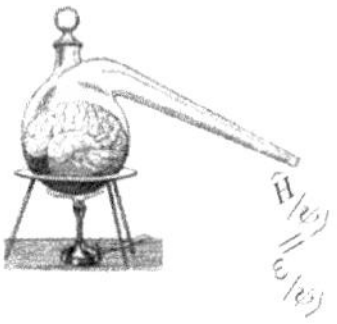

To safely wrap up the chapter, another box summarizes the most important results in terms of ideas, expressions or formula. These will be essential for understanding the remainder of the book, or are quite simply a "must have" for each future quantum-educated scientist.

Questions marked with ✍ include important results which are relevant for better understanding later developments given in the main text. It is thus advisable to spend the little additional necessary time to work them out.

Finally, I wish to express my deep gratitude to all the wonderful people and colleagues without whom this book would have never been anything but a teacher's fantasy.

Among them, I am profoundly indebted to thousands of students at Ecole Centrale Beijing (ECPKn) and CentraleSupélec (CS) Paris who had to study hard to keep up with our high expectations in basic sciences. Their enthusiasm for the Quantum and Statistical Physics course never ceased to grow over the last 10 years and was the true motivation for proposing a written version for a much broader and international readership. The original lecture notes benefited from students' numerous remarks, and additions suggested by colleagues and friends in the Physics department: Pierre Becker, Pietro Cortona, Bruno Palpant, Thomas Antoni and Pierre-Eymeric Janolin, to name but a few.

Devinderjit Sivia had the brilliant idea that I should spend some time as a visiting scholar at St John's College in Oxford where the project for this book really took shape. Claude Lecomte, Sébastien Candel and Julie McDonald thoughtfully helped me make it happen. In Oxford, Richard Compton had the generosity to believe in the project and introduced me to Laurent Chaminade at World Scientific. Over the years of writing, Jennifer Brough constantly provided help and encouragement.

Nevertheless, none of this would have been possible without four amazing human beings.

Julie McDonald at CS has already been mentioned and has now become a dear friend. But also, starting long before the Oxford period, she has been the most supportive English teacher I have ever met. No English version would exist without her encouragement and never-ending dedication to spot and correct my numerous spelling, grammar and stylistic deficiencies.

Guillaume Merle at ECPKn spent hours reading and correcting the manuscript with an incredible ability to detect incoherences, weaknesses in the explanations and mistakes. He is a true (and generous) master of the red pen.

Jean-Christophe Pain at CEA was my second reader. His enthusiastic endorsement, suggestions and ideas were a valuable source of inspiration to help make this book a real pedagogical tool.

Last, but far from least, my family has been very compassionate and understanding over the writing period. I am fully aware of how much they had to endure because of my mental and physical absence. In particular, it is essential to emphasize the key role played by my wife Nadine who kept me going with her loving patience and unstinting support.

Table 1. Table of notations used in this book. Note that vectors are written in bold face.

Notations	Descriptions	
$\widehat{A}$	operator	
$\{a_n, \|\phi_n\rangle\}$	eigenvalues and corresponding eigenkets of $\widehat{A}$	
$\|\psi\rangle$	state ket	
$\langle\psi\|$	state bra	
$\vec{\nabla}_r$	"nabla" operator acting on variable $\boldsymbol{r}$, as in:	
	$\vec{\nabla}_r f(\boldsymbol{r})$	gradient operator with respect to variable $\boldsymbol{r}$
	$\vec{\nabla}_r \cdot \boldsymbol{u}(\boldsymbol{r})$	divergence operator with respect to variable $\boldsymbol{r}$
	$\vec{\nabla}_r \times \boldsymbol{u}(\boldsymbol{r})$	curl operator with respect to variable $\boldsymbol{r}$
	$\nabla_r^2 f(\boldsymbol{r}) \equiv \vec{\nabla}_r.(\vec{\nabla}_r f(\boldsymbol{r}))$	Laplacian operator with respect to variable $\boldsymbol{r}$
	$\widetilde{\nabla}_r^2 \boldsymbol{u}(\boldsymbol{r}) \equiv (\vec{\nabla}_r.\vec{\nabla}_r^T)\boldsymbol{u}(\boldsymbol{r})$	Hessian matrix with respect to variable $\boldsymbol{r}$

Table 2. Table of constants used in this book.

Notations	Descriptions	Approximated values
c	speed of light in vacuum	$2.99792 \times 10^8\ \mathrm{m\,s^{-1}}$
ϵ_0	vacuum dielectric permittivity	$8.854 \times 10^{-12}\ \mathrm{A^2\,s^4\,kg^{-1}\,m^{-3}}$
μ_0	vacuum magnetic permittivity	$4\pi \times 10^{-7}\ \mathrm{kg\,m\,A^{-2}\,s^{-2}}$
e	elementary charge	$1.602 \times 10^{-19}\ \mathrm{C}$
h	Planck's constant	$6.62607 \times 10^{-34}\ \mathrm{J\,s}$
$\hbar$	Planck's constant divided by 2π	$1.054589 \times 10^{-34}\ \mathrm{J\,s}$
k_B	Boltzmann's constant	$1.38066 \times 10^{-23}\ \mathrm{J\,K^{-1}}$
$\mathcal{N}_A$	Avogadro's number	$6.02214 \times 10^{23}\ \mathrm{mol^{-1}}$
σ	Stefan–Boltzmann's constant	$5.67 \times 10^{-8}\ \mathrm{W\,m^{-2}\,K^{-4}}$
m_e	mass of electron	$9.1094 \times 10^{-31}\ \mathrm{kg}$
m_p	mass of proton	$1.6726 \times 10^{-27}\ \mathrm{kg}$
m_n	mass of neutron	$1.6749 \times 10^{-27}\ \mathrm{kg}$
a_0	Bohr's radius	$4\pi\epsilon_0\hbar^2/(me^2) \approx 0.5292 \times 10^{-10}\ \mathrm{m}$

About the Author

Jean-Michel Gillet is Professor of Quantum and Statistical Physics at CentraleSupélec (Paris-Saclay University), France. He heads the Physics Department and his research field is in quantum crystallography, which combines quantum physics and chemistry theoretical methods with X-rays, electron and neutron scattering techniques. The contents of this book reflects the course taught at Bachelor's level to engineering students with a strong background in physics and mathematics.

Contents

Part I
Experimental Puzzles and Birth of a New Constant in Physics

Fig. I.1. William Thomson, Lord Kelvin of Largs (1824–1907). He was amazingly productive both in basic sciences (in his paper, "On the dynamical theory of heat", he postulates a state of complete rest which he called "the absolute zero", hence the Kelvin temperature scale) and engineering sciences (such as the development of a telegraph line between Ireland and Newfoundland which brought recognition from the financial investors and a knighthood from Queen Victoria in 1866). Always interested in maritime affairs (he delivered six speeches on this topic to the House of Lords), he invented a magnetic compass and sold over 10,000. Multiple patents brought him wealth at a time when the scientific community considered it beneath them to build partnerships with industries. In 1892, he became Baron Kelvin (named after the stream flowing through the University of Glasgow where he was a Professor for 53 years). A note next to a caricature in Vanity Fair (1897) mentioned "He knows all there is to know about heat, all that is yet known about magnetism, and all he can find out about electricity". Photo credits: National Galleries of Scotland Commons.

William Thomson (Fig. I.1) was not known for his overwhelming modesty. He was one of the most brilliant students, then scientists, of his time. But his time turned out to be the junction between two eras and he most certainly was the product of the nineteenth century ideas. As his fame grew and he became Lord Kelvin, some of his strong opinions and definitive statements had profound impacts on the scientific community. Of course, before the British Association for the Advancement of Science in 1900, his statement "There is nothing new to be discovered in physics now. All that remains (to be done) is more and more precise measurements" (also attributed to Michelson) could have appeared to drop the curtain on the stage of research in physics. But Lord Kelvin was not alone in this belief. In 1871, James Clerk Maxwell had captured the general mood and reported it in his "Introductory Lecture on Experimental Physics" (Cambridge) declaring "...the opinion seems to have got abroad, that in a few years all the great physical constants will have been approximately estimated, and that the only occupation which will then be left to men of science will be to carry these measurements to another place of decimals".

At the end of the nineteenth century, physics (and other natural sciences) could pride itself on having massively contributed to the industrial revolution. Many applications came from a deeper understanding of mechanics, i.e., the motion and equilibrium of solid objects, thanks to the glorious heritage of Newton and further developments by Lagrange and Hamilton (see Chapter 6). Astronomy was probably the one field where the laws of mechanics could show their mighty power of explanation and prediction. All these impressive theories were supported by ever more refined observations thanks to the development of better instruments. It was obvious that one parameter was the key: the gravitational constant G. The role of mechanics and laws of conservation should be considered as of paramount importance in the rise of thermodynamics which, in turn, allowed for (and was motivated by) the invention of more efficient heat engines. After all, heat was another form of energy transfer and temperature nothing but a measure of microscopic particles jiggling speed. Heat was known to be conducted in several ways. One of these ways seemed to be very similar to radiation as it could be carried through an empty space. With the discovery of Maxwell's equations, such connections between sciences that could previously have been considered as separate fields of knowledge were indeed also to be found in the unification of optics and electromagnetism. This particular link was symbolized by a finite value of light speed: c. Of course some $3 \times 10^8\,\mathrm{m\,s^{-1}}$ could in many circumstances be replaced by infinity, and the concept of rays at the centre of geometrical optics was sufficient to design and manufacture reading or opera glasses and telescopes. But diffraction or, more generally, interference phenomena were the signs of more subtle behaviour. Dispersion of colours by gratings found its explanation and better devices could be developed. It would give rise to a new field: spectroscopy of objects such as remote stars or terrestrial flames from burning gases. Every piece of Mother Nature's giant puzzle seemed to find its perfect match and fall into place. Finally, thanks to all these joint theoretical and experimental works, Nature was making sense!

The circumstances thus naturally favoured the general optimism and faith in Science embodied by Lord Kelvin's assertive statement. However, some concerned scientists may also have considered the necessity of "more precise measurements" as a hint that the whole story had not yet been told.

This part aims at emphasising the role of experimental physics in questioning well-established theories. We will thus build upon Enrico Fermi's proposition "when an experiment reproduces the theory, you've made a measurement. When it doesn't, you've made a discovery". The next two chapters report a subjective journey at the very roots of that which triggered the birth of quantum physics. It is presented as a round-trip starting from electromagnetic radiation wave field to reach the necessity of particles of light and then from massive particles to their wave-like behaviour. This trip should be considered as a progressive description of the need for a paradigm shift. It is a successive accumulation of facts that, in the early stages, forces the scientific community to bend the very physical laws that brought all those scientific, technological and even industrial successes. When bending turns out to be insufficient, it will then be time to break those laws and build new ones. But this can wait until the next part, and let us first examine what kind of difficult experiments were still causing trouble to the nineteenth century physical models.

1

From Waves to Particles

The content of this chapter will help you understand ***the shape of the fossil cosmic background radiation, photoelectricity (in printers and on the Moon)****, and related topics.*

This chapter explains how, on the basis of a set of counter-intuitive experimental results, the scientific community came to invent the notion of *photon*. Primarily introduced as a mere mathematical trick, the photon's actual role in many aspects of electromagnetic interaction with matter had to become an ontological fact.

We will thus explain how the full black-body radiation spectrum could only be accounted for by lifting the unnecessary constraint of continuity on possible mean energies carried by an electromagnetic field. This will be shown to be fully compatible with the frequency dependence of electron emission by a metal upon UV light irradiation. Finally, the assimilation of an electromagnetic energy bunch to a particle, travelling in vacuum at light speed, will be firmly established when it appears obvious that it does indeed show every single property to satisfy conservation laws of classical collision theory.

1.1. Short Wavelength Issue in Black-Body Radiation

Observation teaches us that every physical object brought to a temperature T radiates an electromagnetic wave. To express that it has reached a particularly high temperature, we say that it has been heated to "red" or even to "white", for example. The connection between a temperature and the radiated spectrum colour is thus entrenched in common knowledge. The usual (and classical) model upon which one relies to explain such a phenomenon is as follows: the heat, which is supplied to the system, increases the motion of particles. Locally, accelerated charges act as many antennas that emit an electromagnetic field in an incoherent manner, owing to their random and non-harmonic motion. This willingly partial and very approximate description only serves to justify the fact that the observed spectrum of radiation is continuous, unlike the emission lines from an antenna powered by a periodic electrical current. In parallel to this emission process, the absorption of an electromagnetic wave can be regarded as an excitation of tiny oscillators. The absorption and emission processes then clearly appear to be totally symmetrical.

By convention, the idiom *black-body* is used to describe a total (perfect) absorber: it absorbs all incident electromagnetic radiation without reflecting any of it. Its total temperature is kept to a fixed value T as a result of a luminance balance. At thermal equilibrium, all absorbed energy is converted into electromagnetic radiation. By definition, an ideal black-body emits radiation, the spectrum of which solely depends on its equilibrium temperature.

The simplest example, which was carefully studied by Wien, is that of a completely sealed oven. One of its walls is punched with a small hole. Radiation power, generated by thermal motion of the atoms and molecules inside, escapes the enclosure through this tiny aperture. The key assumption of such an ideal black-body source is that the chemical nature of the materials coating the cavity's walls does not play any role in the radiated electromagnetic spectrum.

Following Kirchhoff's first experiments (1859), Stefan noticed in 1879 that the energy emitted per second, and per unit area, varies in proportion to temperature to the fourth power. He very simply formalizes this empirical law in the form:

$$P = \sigma T^4 \tag{1.1}$$

where σ is the Stefan–Boltzmann constant. This formula is known as Stefan's law and today's σ experimental value is set to $5.67 \times 10^{-8}\,\mathrm{W\,m^{-2}\,K^{-4}}$.

With his punched oven device, supplemented with a spectrometer, Wien (1894) could then observe that the wavelength at which the energy flux is maximal varies as the inverse of the black-body's equilibrium temperature. It is Wien's displacement law:

$$\lambda_{\mathrm{max}} T = \text{const.} \approx 2.9\ \mathrm{mm\,K} \tag{1.2}$$

Question 1.1: **Black-body orders of magnitude.**

1. What is the temperature order of magnitude for a "red hot" material?
2. What is the wavelength at which the spectrum emitted by a black-body at room temperature is expected to be maximum?

Answer:

1. Firstly, it is important to ensure that the object radiates as a black body. Then, according to Wien's displacement law, if red corresponds to a wavelength of the order of 600 nm, we find: $T \approx 2.9/(0.6 \times 10^{-3}) \approx 5000$ K.

   Curiously, we know that this is not the colour that one observes at such a temperature. An incandescent lamp filament is about 3000 K. It should be borne in mind that the spectrum is broad. Many other wavelengths are thus also detected and, in particular yellow and blue (even small proportions), come to significantly change the colour perceived by the eye. Therefore, to see the red, it is not the maximum of the spectrum that is to be at 600 nm but a relatively small part of the spectrum. The main part of the spectrum is hence in the infrared range (which we cannot see with the naked eye). A temperature between 600 K and 800 K is sufficient.
2. At 300 K, Wien's displacement law states $\lambda_{\mathrm{max}} \approx 10\,\mu$m. The maximum is then in the near/mid infrared.

These findings push Rayleigh to model the black-body radiation phenomenon on the basis of electromagnetic standing waves physics. His idea, following the concept of the thermodynamics equipartition theorem, is to attribute $(k_B T)/2$ for each degree of freedom of the problem, where k_B is the Boltzmann constant.[1] For him, it is only a problem of waves. Following

[1] The calculations for this section are conducted in more detail in the statistical physics section of Volume 2.

Helmholtz and Maxwell, and similar to a guitar string, he imagines that only a (transverse) standing waves system can permanently be set in the above-mentioned cavity. Each of these waves represents a degree of freedom. By counting permitted wave numbers in the cavity (see Question *1.2*), it is then possible to obtain the law of Rayleigh–Jeans for spectral energy (per unit volume) radiated from the black-body:

$$dU = 8\pi k_B T/\lambda^4 d\lambda \tag{1.3a}$$

or, with $\lambda = c/\nu$,

$$dU = 8\pi\nu^2 k_B T/c^3 d\nu \tag{1.3b}$$

Question 1.2: **Counting electromagnetic modes in a cavity.** Calculate the number of electromagnetic modes per unit volume of the cavity in a frequency interval $d\nu$. Assume that the walls are perfectly reflective. Deduce, from the equipartition energy theorem, an electromagnetic energy spectrum expression as a function of frequency and temperature.

Answer: Let us consider one dimension only to start with. Let L be the cavity size. In steady state, the wave must cancel at the perfectly reflecting walls. This imposes a constraint on the wavelengths, so that $n\lambda = 2L$. In one dimension, the possible wave vectors are $2\pi/\lambda = k_n = n\pi/L$ where n is an integer. In three dimensions, choosing a cubic cavity for simplicity, standing wave vectors are $\boldsymbol{k} = \pi(n_x, n_y, n_z)/L$. Recall that a standing wave is a superposition of two waves, with identical module wave vectors but pointing to opposite directions. This implies that the description of standing waves only makes use of positive values of n_x, n_y and n_z. Standing wave vectors therefore occupy only one-eighth of wave vectors space (the octant corresponding to natural integer triplets).

A stationary mode therefore occupies a volume (in the space of wave vectors): $(\pi/L)^3$. Thus, the number of modes that correspond to a module lower than (or equal to) a given value k is $N_k = \frac{1}{8}\frac{(4/3)\pi k^3}{(\pi/L)^3}$. The number of modes per unit volume is: $n_k = \frac{k^3}{6\pi^2}$. Therefore, using $k = 2\pi/\lambda = 2\pi\nu/c$, the number of modes with frequency lower than ν is: $n_\nu = \frac{4\pi}{3c^3}\nu^3$ and the number of modes per unit frequency range is: $dn_\nu = \frac{4\pi}{c^3}(\nu^2)d\nu$.

Each electromagnetic wave vector is associated with an electric field and a magnetic field. The polarization of the wave is expressed (for a transverse wave) on the basis of two perpendicular directions. Thus, for a given standing wave vector, there are four degrees of freedom. Based on the equipartition energy theorem, one sets $k_BT/2$ per degree of freedom and naturally gets to a spectral distribution of energy density: $dU(\nu, T) = \frac{8\pi}{c^3}(\nu^2)k_BT d\nu$.

Although this expression gives a fair account of observations at low frequencies, it does however seem to disagree with Wien's displacement law (1.2): the theoretical spectrum described here is monotonous! What is worse is the resulting form of total energy density: a summation over contributions from all wavelengths is doomed to diverge! Referring to this, Ehrenfest coined the term "ultraviolet catastrophe": spectral density increases as the square of the frequency, with no limit. Although it is a very fine colour for a catastrophe, such inconsistency was to question the most established physical models.

In 1900, convinced that this is purely a problem of thermodynamics, Max Planck (Fig. 1.4) takes on the task of solving what he rightly sees as one of the key points in physics of this time and especially the macroscopic manifestation of microscopic processes by which energy is exchanged between electromagnetic radiation and matter.

Barely distorting his thought,[2] his line of reasoning can be summarized as follows. If one accepts that, in the cavity, there is indeed a number of stationary electromagnetic waves that have been established at thermodynamic equilibrium from the particles oscillating on the walls, without knowing the details, it can be assumed that counting $dN = 8\pi\nu^2/c^3 d\nu$ modes in the frequency range of $d\nu$ is necessarily correct. The problem must therefore come from the calculation of an oscillator's average energy. We can be even more precise[3]:

$$\langle\epsilon\rangle_T = \frac{\int_0^\infty \epsilon\, e^{-\epsilon/(k_B T)} d\epsilon}{\int_0^\infty e^{-\epsilon/(k_B T)} d\epsilon} = k_B T \tag{1.4}$$

where use is made of the Boltzmann distribution of energies at temperature T. Somehow this expression needs to compensate for the fact that the number of modes diverges at high frequency. The trick is then to consider the integral in terms of a discrete sum, using the usual method of rectangles with a sampling of constant step $\Delta\epsilon$. Substituting $n\Delta\epsilon$ to ϵ, the calculation

[2]The actual story is a little more complex than what is traced here in many accounts. Planck was confident that thermodynamics, particularly the second principle, was the underlying reason for the black-body's characteristic behaviour. His purpose was then to find a law of entropy behaviour for the black-body that could account for what happens both at low and high frequencies. Difficulties in reaching a result in line with measures of Rubens and Kurlbaum (more accurate than Wien's and Stephan's) led him to propose a purely phenomenological interpolation which proved later to be the origin of the discovery of a quantum for electromagnetic field energy.

[3]This calculation is fully developed in Volume 2.

for a mean energy (1.4) can thus be reformulated as

$$\langle \epsilon \rangle_T = \frac{\sum_{n=0}^{+\infty} n\Delta\epsilon \, e^{-n\Delta\epsilon/(k_B T)} \Delta\epsilon}{\sum_{n=0}^{+\infty} e^{-n\Delta\epsilon/(k_B T)} \Delta\epsilon} \tag{1.5}$$

As shown in Question *1.3*, it is thus found that

$$\langle \epsilon \rangle_T = \Delta\epsilon \frac{e^{-\Delta\epsilon/(k_B T)}}{1 - e^{-\Delta\epsilon/(k_B T)}} = \frac{\Delta\epsilon}{e^{\Delta\epsilon/(k_B T)} - 1} \tag{1.6}$$

It clearly appears that when the step size $\Delta\epsilon$ tends to 0, it inevitably plunges us into the throes of the ultraviolet catastrophe as the mean energy becomes independent of the frequency again. It is indeed acceptable to make $\Delta\epsilon$ become arbitrarily small only if one considers a very low frequency range since it is precisely in this area that the classical model best reproduces the experimental observations. As the apocryphal story is now being told, the idea put forward by Max Planck (Fig. 1.4) is both very simple and particularly daring: if an infinitely small step is only acceptable in the low frequency range but is required to take an appreciable value as high frequencies are reached, why not choose a step that would be a function of the vibrational mode frequency? He attempts a proportional relationship $\Delta\epsilon \propto \nu$, which, everyone would agree, would naturally fulfil a simplicity criterion. The proportionality coefficient[4] h would be determined later from experiments if this route of exploration was relevant. Thus, the average energy carried by a given mode now depends on its frequency ν:

$$\langle \epsilon \rangle_T = \frac{h\nu}{e^{h\nu/(k_B T)} - 1} \tag{1.7}$$

Using this mean energy for each mode, Planck can then derive the black-body radiation energy distribution:

$$dU = \left[\frac{h\nu}{e^{h\nu/(k_B T)} - 1}\right] \left[8\pi\nu^2/c^3 d\nu\right] \tag{1.8}$$

[4]The choice of the letter h comes from the German term *hilfe*, very appropriately associated with the need for *help* in this difficult matter.

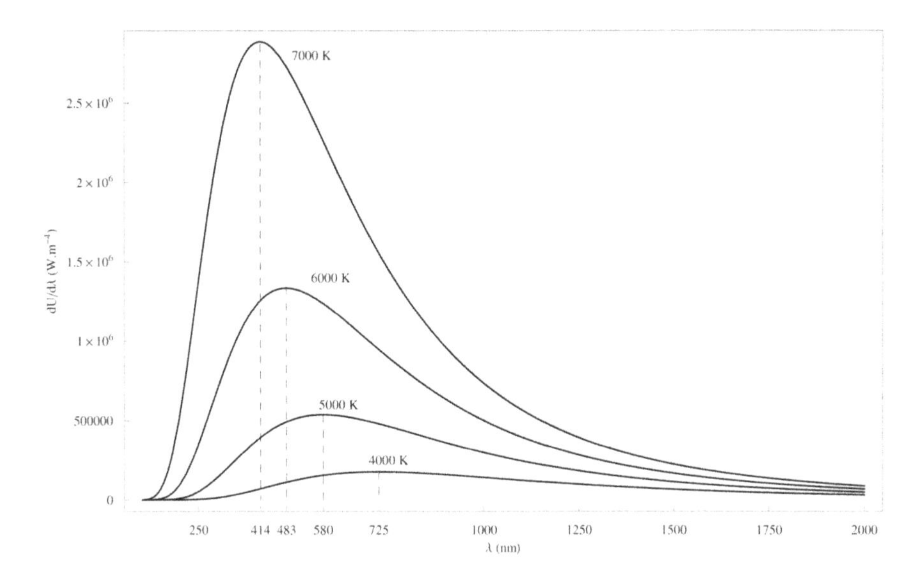

Fig. 1.1. Black-body spectrum for different temperatures.

The first term between brackets represents the energy carried by each mode. The second bracketed quantity is the number of modes (per unit volume) of radiation with frequency between ν and $\nu + d\nu$. The conversion into wavelength is readily obtained:

$$dU = \frac{8\pi hc}{\lambda^5} \frac{1}{e^{hc/(\lambda k_B T)} - 1} d\lambda \tag{1.9}$$

This black-body spectral energy density $\frac{dU}{d\lambda}$ is shown in Fig. 1.1 for several temperatures.

Question 1.3: **Computation of an electromagnetic mode mean energy.** Starting from (1.6), show (1.7) for the average energy carried by an electromagnetic wave of frequency ν. It is advisable to make use of $\sum_{n=0}^{+\infty} x^n = \frac{1}{1-x}$, providing that $e^{-\Delta\epsilon/(k_B T)} = x < 1$. In a similar line of thoughts, it will be useful to consider

$$\sum_{n=0}^{+\infty} nx^n = x \sum_{n=1}^{+\infty} nx^{n-1} = x \frac{\partial}{\partial x} \sum_{n=0}^{+\infty} x^n = \frac{x}{(1-x)^2}$$

Answer: According to the expression proposed by Planck, the mean energy is obtained by calculating

$$\langle \varepsilon \rangle_T = \frac{1}{\sum_n e^{-nh\nu/(k_B T)}} h\nu \sum_{n=0}^{\infty} n e^{-nh\nu/(k_B T)}$$

It is better to start with the calculation of the denominator which is nothing more than the sum of a series of ratio $e^{-h\nu/(k_B T)}$. The result is

$$\mathcal{N} = \sum_n e^{-nh\nu/(k_B T)} = \frac{1}{1 - e^{-h\nu/(k_B T)}}$$

We can turn to the calculation of the numerator. Let us set $x = h\nu/(k_B T)$, it appears that

$$\sum_{n=0}^{\infty} n e^{-nx} = \frac{-\partial}{\partial x} \sum_{n=0}^{\infty} e^{-nx}$$

This method, "differentiation under the integral sign", is obviously justified if the sum converges! We then obtain:

$$h\nu \sum_{n=0}^{\infty} n e^{-nh\nu/(k_B T)} = h\nu \frac{e^{-h\nu/(k_B T)}}{(1 - e^{-h\nu/(k_B T)})^2}$$

and it is then Planck's result:

$$\langle \epsilon \rangle_T = h\nu \frac{e^{-h\nu/(k_B T)}}{1 - e^{-h\nu/(k_B T)}} = \frac{h\nu}{e^{h\nu/(k_B T)} - 1}$$

Question 1.4: **The Rayleigh–Jeans limit.** Check that Planck's distribution for black-body radiation (1.8) reproduces Rayleigh–Jeans' result (Fig. 1.2) in the low frequency limit ($h\nu \ll k_B T$).

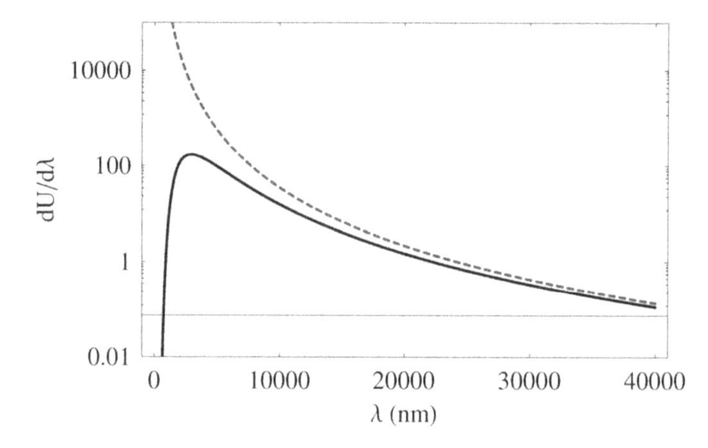

Fig. 1.2. Rayleigh–Jeans law (dotted line) is asymptotic to that of Planck (solid line) for large wavelengths. Here, $T = 1000$ K.

Answer: In the limit of low frequencies (i.e. large wavelengths) one can set $h\nu/k_B T \ll 1$. Thus, a Taylor series development of Planck's distribution carried to the zeroth order for the numerator and to the first order for the denominator yields

$$dU \approx \left[h\nu \frac{1}{(1 + h\nu/(k_B T)) - 1}\right] \left[8\pi\nu^2/c^3 d\nu\right]$$

or, after simplification,

$$dU \approx k_B T \left[8\pi\nu^2/c^3 d\nu\right]$$

which happens to be the law originally proposed by Rayleigh and Jeans (1.3).

Question 1.5: **Derivation of Wien's displacement law**. In the short wavelength limit (it would be advisable to use (1.9)), show that Planck's distribution has a maximum at a wavelength $\lambda_{\max}$ such that $\lambda_{\max} T = C$ where the value of C depends on Planck's constant h. Using the experimental value, $C = 2.9\,\mathrm{mm\,K}$, derive an approximate value for h.

Answer: In the short wavelength limit, Planck's distribution becomes

$$dU \approx \frac{8\pi hc}{\lambda^5} e^{-hc/(\lambda k_B T)} d\lambda$$

$dU/d\lambda$ cancels out for

$$\lambda_{\max} T \approx \frac{hc}{5k_B}$$

The product is indeed a constant that can be computed $C \approx h \times 0.434 \times 10^{31}\,\mathrm{m\,K}$. This result can then be compared to the experimental value $C \approx 2.9 \times 10^{-3}\,\mathrm{m\,K}$ to find an approximate value of $h \approx 6.68 \times 10^{-34}\,\mathrm{J\,s}$.

Planck's expression for the black-body radiation spectrum perfectly matches Rayleigh–Jeans large wavelength asymptotic behaviour but also accounts for Wien's displacement law (Question *1.5*) with only one parameter: the h constant which can be determined from experimental data. There is a progressive awareness that Planck's formula is certainly more than a mere interpolation function preventing the ultraviolet catastrophe.

We are witnessing the birth of a new constant, rightly called "Planck constant", the value of which is approximately $h \approx 6.68 \times 10^{-34}\,\mathrm{J\,s}$. Compared to those constants that we usually deal with in macroscopic physics, this value of h is incredibly tiny. However, one should bear in mind its essential role in Planck's interpolation model. Replacing an integral over a continuous function by a discrete sum clearly means that the oscillators, and therefore the electromagnetic modes in the cavity, can only have energies with values which are multiples of $h\nu$. Since the proportionality constant h cannot be replaced by 0 without causing ultraviolet catastrophe, it clearly means that for a given vibrational mode of frequency ν, all energies are not accessible. The possible energies of the modes are said to be "quantized" and cannot differ from $nh\nu$ where n is a natural integer.

Planck's hypothesis is presented here as a mathematical artifice. However, as the following will show, it actually marks the birth of an entity of utmost importance: a minuscule lump of energy carried by electromagnetic radiation, the photon.[5] Less than 200 years after the publication of *Opticks* by I. Newton and 35 years after J.-C. Maxwell's publication, *A Dynamical Theory of the Electromagnetic Field*, radiation seemed to exhibit particle-like behaviour, at least under certain circumstances.

1.1.1. *Applications of black-body radiation*

Black-body radiation is not only a significant component in the history of science but still a fundamental basis for a number of technological devices such as the infrared camera and more precise pyrometers that are contact-less thermometers. Since black-body radiation is so convenient and well documented, but all bodies do not exactly behave as black-bodies, a quantity called "emissivity" is introduced. It is defined as the ratio of the luminance[6] of the real body to that of a black-body. The "grey-body" hypothesis assumes that emissivity does not depend on the wavelength. Black-body spectrum is also used in astrophysics as an approximate model for estimating the temperature of distant stars (for an example, see Fig. 1.3) through the measurement of $\lambda_{\max}$ after it has been corrected for universe expansion Doppler shift. As spectra measurements become increasingly accurate, it clearly appears that very few objects do radiate as a black-body. However,

[5]The term was proposed in 1926 by G. Lewis.

[6]Luminance is defined as the power radiated per unit area (m^2) and solid angle (in steradian).

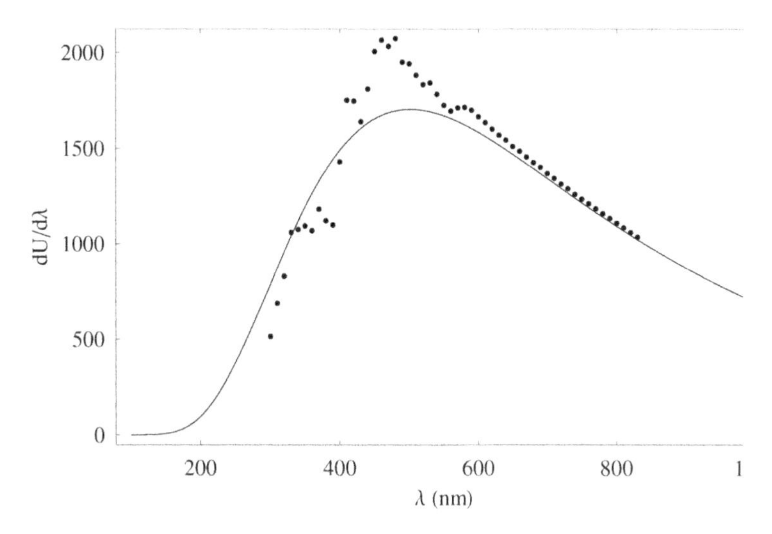

Fig. 1.3. Planck's law is consistent with the spectrum of the sun and the best match is obtained for $T = 5780\,\mathrm{K}$. While it turns out to be a fair estimate of the actual surface temperature, note differences with a perfect black body. Sun spectral data are taken from [12].

cosmic microwave background radiation (see Question *1.6*) and Hawking radiation from black holes seem to match it rather well.

> *Question 1.6:* **Fossil cosmic background radiation**. In the early 1930s, Jansky [60] detects a radio frequency radiation from the centre of our galaxy, the Milky Way. In 1964, other researchers at Bell Labs, Penzias and Wilson, after checking that they are not measuring some pigeons nesting activity in their antenna, report to have recorded isotropic microwave radiation. They are unable to determine its origin until they are informed of Dicke's calculations. Unaware of predictions previously made by Alpher and Gamow in 1948 [3], Dicke had predicted that the expansion of the universe observed by Hubble (see Question *2.1*) would result in a dilution of the radiation energy generated while matter was created. Numerical values were consistent, confirming Friedman's hypothesis [42]: there would have been a key moment of creation of matter, the so-called *Big Bang*. Observations [73] and theory [27] are published in two separate articles, but in the same issue of the same journal, following one another. However, only Penzias and Wilson will get the Nobel Prize in 1978. Figure 1.5 reports recent measurements collected by the far-infrared (FIRAS) spectrometer boarded on COsmic Background Explorer (COBE). Use the curve to propose an estimate for the cosmic microwave background temperature.

Fig. 1.4. Max Planck (1858–1947) was awarded the Nobel Prize in 1919 (but for year 1918). He was the offspring of a line of theologians (his grandfather and his great grandfather) and dedicated civil servants. He enjoyed a long friendship, playing the piano with Einstein who accompanied him on the violin. He even considered becoming a composer. While a faithful German patriot (who soon regretted the "Manifesto of the 93"), he tried to argue with Hitler in 1933 against racist laws which he presented as jeopardizing the nation's scientific progress. In 1938, he resigned as Secretary of Berlin Academy where he had contributed to creating a chair for Einstein to encourage his scientific production. He had four children by his first wife. Three died during the WWI period. The only survivor, Erwin, was executed by the Nazis following his (marginal) involvement in the July 1944 plot against Hitler. Max Planck was the PhD advisor to Max von Laue, who received the Nobel Prize in 1914 for measuring X-ray diffraction wavelength on a single crystal (see Volume 3). Photo credits: by courtesy of Bundesarchiv, Bild 183-R0116-504/Licence CC-BY-SA-3.0.

Answer: The best way is obviously to use a black-body expression with the temperature as a fitting parameter and adjust for an optimal matching with experimental data. An alternative is to make use of Wien's displacement law. The wave number at which the spectrum is maximum is about 7.3 cm^{-1}. It roughly corresponds to $\lambda_{\max} \approx 1370\ \mu$m.

$$T_{\mathrm{CMBR}} = \frac{hc}{5k_B\lambda_{\max}} \approx 2.1\ \mathrm{K}$$

Further analysis led the authors [38] to propose a temperature of 2.728 ± 0.004 K.

1.2. Frequency Dependence of Photoelectricity

In 1887, Heinrich Hertz reports the following experiment [53]. He builds an electromagnetic wave emitter by causing electric discharges in an induction coil with characteristic dimensions adapted to select a particular wavelength range. He then builds a receiver in the form of a copper wire extending from a copper ball and forming a single loop before returning to the immediate vicinity of the ball. According to Maxwell's theory, the existence of an

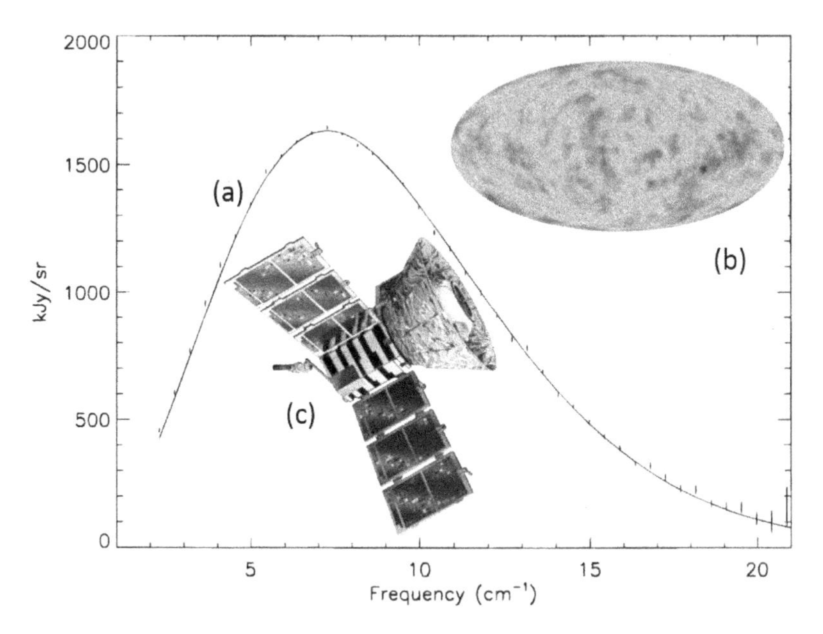

Fig. 1.5. (a) Spectrum of the microwave background radiation measured by FIRAS boarded on COBE (c). The authors superimposed a black-body spectrum to determine the temperature. Frequencies are expressed in wave number $\frac{1}{\lambda}$ and intensity is in kilo-Jansky per steradian (1 Jy $= 10^{-26}$ W Hz^{-1}). See also Question *1.6*. (b) All-sky image reconstructed from four years of data from COBE. Data from [38] and arXiv:astro-ph/9605054. Photo credits: ESA and the Planck Collaboration (b), and NASA (c).

electromagnetic wave excitation must generate an oscillation of charges and the possible subsequent occurrence of an electric arc between the lower end of the loop and the copper sphere.

The experiment is a success: Hertz now possesses an emitter and a receiver in working condition. However, he soon observes that the incident electromagnetic wave happens to be attenuated if certain materials are inserted on its way from the source to the receiver. In his report, almost anecdotally, he notices that ultraviolet radiation seems to create a more intense electric arc at the receiver.

The year following this communication, Wilhelm Hallwachs [48] observes that an electrically charged zinc plate, which is connected to a gold foil electroscope, discharges much more quickly when it is exposed to ultraviolet light from an arc lamp or an incandescent magnesium light bulb.

A decade passes before J.J. Thomson tackles the problem and proposes an explanation [88]. He claims that the phenomenon is not different from that observed for the cathode ray[7] in which electrons are emitted by a conductor subjected to a strong static electric field. Taking the description proposed by Maxwell, he develops a theory according to which the electric field carried by the incident wave induces a forced oscillation of charges with an amplitude proportional to the radiation field. As a consequence, the number of ejected electrons and their speed should increase as the intensity of the wave is augmented.

Three years later, in 1902, Lenard (Fig. 1.6) conducts the experiment that tumbles the old theory of light–matter interaction [66].

He generates a broad spectrum electromagnetic field from a particularly powerful carbon arc lamp. Therewith, he irradiates a metal plate. The electrons emitted by this cathode are detected by a second plate facing it: the collector. It is connected to the emitting plate by sensitive ammeter in order to measure the created current.

In order to estimate the speed of the emitted electrons, Lenard is able to impose a negative potential to a grid between the irradiated plate and the collector plate. A cancellation of the electrical current corresponds to a balance between the kinetic energy of electrons and the energy of the repulsive potential thus applied.

To his surprise, Lenard remarks that the equilibrium potential, and therefore the speed of the emitted electrons, is totally independent of the incident wave intensity. Of course, boosting the light intensity increases the number of electrons but with no detectable effect on their kinetic energy. However, favoured by the power of his lamp, Lenard can insert a grating which allows for the qualitative separation of incident long and short wavelength electromagnetic radiation. His measurements show without a doubt that ejected electrons gain speed as the frequency is increased.

[7] Cathode ray is the name originally given to an electron beam. In the twentieth century television technology, a cathode ray tube was at the origin of the moving image. The electron beam was deflected by a modulated electric field. The conversion of the electron image into a visible signal was then created by the beam impact onto a fluorescent screen.

Fig. 1.6. Philipp von Lenard (1862–1947) won the Nobel Prize in 1905 for his admirable experimental work with cathode rays. He was however still resentful towards the committee for not having recognized in 1901 that he was the "true mother of X-rays": "All Roentgen had to do was push a button, since all the groundwork had been prepared by me.... Without my help, the discovery of X-rays would not have been possible even today. Without me, the name of Roentgen would be unknown". In recognition for his actions promoting the "Aryan science" (with Stark), he treasured a letter from Hitler mentioning "With you, the National Socialists' thoughts have had a courageous supporter and brave fighter since the beginning, who effectively curtailed the Jewish influence on science and who always has been my faithful and appreciated colleague. This shall never be forgotten". One of his favourite targets, Einstein, the inventor of the "Jewish physics" theory which was "never intended to be true", was involved in a long lasting dispute with Lenard. He wrote to Laub: "Lenard must, however, in many things, be wound quite askew. His recent lecture on these fanciful ethers appears to me almost infantile. Further, the study he commanded of you (...) borders on the absurd. I am sorry that you must spend your time on such stupidity" (see [54]). Photo credits: by courtesy of Bundesarchiv, Bild 146-1978-069-26A/photographer: ohne Angabe/Licence CC-BY-SA 3.0.

Another 3 years had to pass before Einstein's "Annus Mirabilis" (1905). In one of his famous papers, Einstein (Fig. 1.8) expounds a model to explain the photo-emission behaviour that seemed to go against the basic principles of classical physics.

The wording of Einstein's proposal is actually quite simple. First, he abandons the idea of a force imposed on the electrons by the electric field carried by the wave and resolutely adopts an energy transfer point of view. Following Planck's model, he suggests that the electromagnetic wave energy,[8] which is transmitted to the electrons in the metal plate, is proportional to the frequency ν as $\varepsilon = h\nu$. Thus, the conservation

[8]Thus, Einstein abandons the concept of oscillators to explicitly focus on the quantification of the electromagnetic field itself.

law imposes that the kinetic energy of ejected electrons is the difference between the electromagnetic energy $h\nu$, which they have absorbed, and the energy they used to get away from the attractive potential of the nuclei in the cathode. This extraction energy is named *work function* and denoted by W (see Question *1.7*). The kinetic energy of an ejected electron is therefore

$$\varepsilon_k = h\nu - W \tag{1.10}$$

The proportionality constant h can be determined from Lenard's curve relating the potential of the collector necessary to stop the flow of electrons to the frequency of the exciting wave.

In his model, Einstein assigns a physical reality to the photon. It truly becomes the form by which the electromagnetic radiation exchanges energy with matter.

> *Question 1.7:* **The work function of zinc.** The work function, that is to say the minimum energy required to extract an electron from a solid, is 4.3 eV for zinc. Using the curve published by Lenard (Fig. 1.7), estimate the wavelength of the electromagnetic radiation that was used in this experiment.

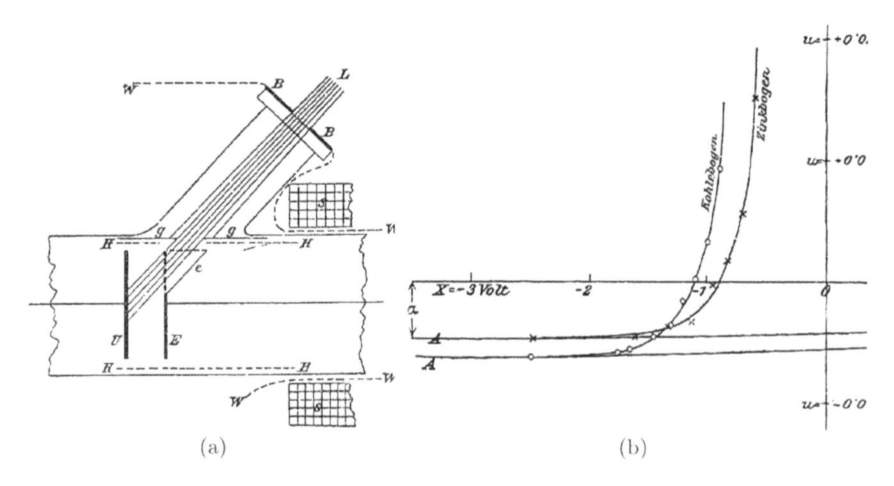

Fig. 1.7. (a) Lenard's experimental setup. The generated light comes through L and impinges on the cathode in U. An electric potential difference is applied between U and the collector E. An ammeter measures the transmitted current. (b) Curve measured by Lenard. It shows the current intensity as a function of decelerating potential. Reprinted with permission from [66]. Copyright © 1902 by John Wiley & Sons.

Fig. 1.8. Albert Einstein (1879–1955) at a radio exhibition. Some of Einstein's teachers described him as mentally slow, possibly because of his echolalic habits and that he could not speak fluently before reaching the age of seven. He had a great respect for Henrik A. Lorentz: "His genius was the torch which lighted the way from the teaching of Clerk Maxwell to the achievements of contemporary physics". Einstein's most productive period was the seven years spent as a patent examiner for electrotechnology in Bern. This left him enough time to carry out independent and original theoretical work that led to the three 1905 Annus Mirabilis papers: the special theory of relativity, Brownian motion and the laws of photoelectric effect. It is the latter which brought him the 1921 Nobel prize (received in 1922) while popular fame came by the first. It is believed that relativity was still considered too epistemological (according to H. Bergson, for example) and politically controversial (see Fig. 1.6) to match the Nobel committee's criteria. The prize money was used to settle a divorce from his first wife Mileva. His independence and a lack of taste for social and family life, caused him to loosen ties with his children and two successive wives. During his last 20 years in Princeton, he became even more isolated "...my standard of decent behaviour has risen as I grew older: I cannot be sociable with people whose fame has gone to their heads". He remained highly sceptical about the non-deterministic aspects of quantum physics prevailing interpretation and repeatedly declared that "God doesn't play dice". Photo credits: by courtesy of Bundesarchiv, Bild 102-10300/Licence CC-BY-SA 3.0.

Answer: Lenard's curve represents the current intensity measured in the micro-ammeter according to the decelerating potential of the gate. It can be seen that the potential difference when the current ceases to be zero, i.e. the one for which electrons begin to be ejected is about 1.5–2 V. This means that 1.5–2 eV need to be spent to compensate for their ejection kinetic energy. The $h\nu$ energy absorbed by an electron is partly (4.3 eV) used for extraction and the remaining is converted into kinetic energy (1.5–2 eV). The total energy absorbed by an electron in zinc is at least 5.8 eV. This suggests that $\lambda = c/\nu = hc/(5.8 \times 1.6 \times 10^{-19}) = 0.32 \times 10^{-7}$ m. This is well within the range of ultraviolet radiation.

As electromagnetic energy comes in discrete $h\nu$ lump, it should be emphasized that there is consequently a minimum frequency below which electrons cannot absorb enough energy to achieve their extraction from the plate: $\nu_{\text{min}} = W/h$. This explains why, for any light intensity, no ejected electrons could be observed at low frequency.

1.2.1. *Applications of the photoelectric effect*

The explanation above could give the false idea that electrons are initially motionless. As it will become apparent in Volume 2, it is an oversimplification. The electron kinetic energy in a solid is an important component that drives many of the macroscopic properties. Lenard's experiment provides a way to gain access to such information. This is now called "photo-emission" spectroscopy. It is a technique for the study of the electrons' state (mainly in solids) by careful analysis of their ejection speed vector under UV or X-ray monochromatic light. One can thereby deduce valuable atomic scale information explaining macroscopic properties such as cohesion, adhesion, conductivity and oxidation states of the elements. The famous photoelectric solar cell is nothing but an everyday transposition of Lenard's experiment. The kinetic energy of photo-ejected electrons is then converted back to potential energy stored in batteries or directly used to power electrical devices.

A popular application that sometimes tends to be overlooked is the photoelectric effect as a key element of copy machines and laser printers (Fig. 1.9). The component that concerns us the most is the drum. It is generally made of aluminium covered with a photoconductive material and, at first, uniformly charged. As the drum rotates around its axis, an impinging laser beam photo-ejects charges on each line, thereby reproducing on the drum what has been scanned from the original document. The drum is then put into contact with negatively charged ink powder which will only stick to discharged regions and will be expelled from other negative areas. Transfer to a positively charged sheet of paper and securing the bonding with fusion finalizes the printing.

A picturesque consequence of the photoelectric effect is its influence on "Moon dust". Under the bright light of the sun (its radiation is not filtered

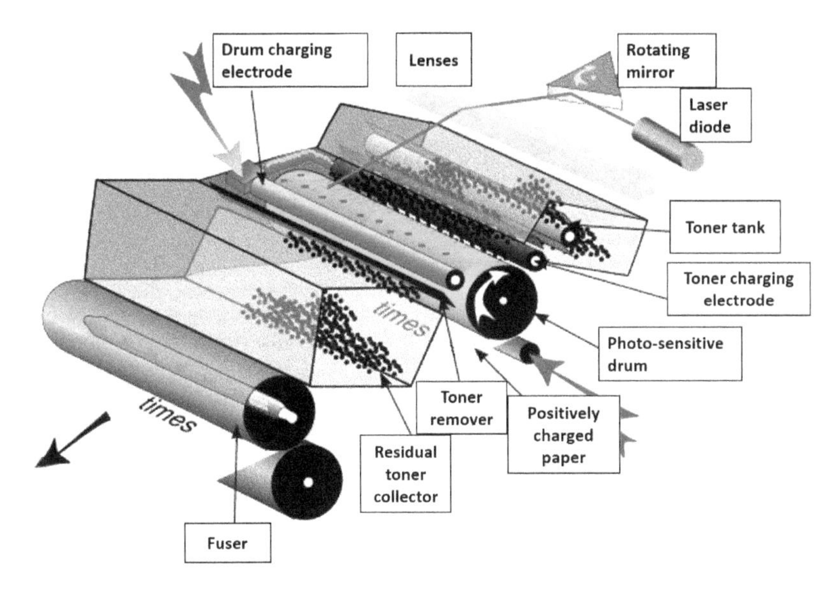

Fig. 1.9. Schematic general view of a laser printer with its main components. with details of the paper sheet path, laser-induced photo-emission from the drum. After Creative Commons Licence (CC-BY-SA 3.0).

by any atmosphere), the dust on the moon surface becomes charged by photoelectric effect. These grains are subject to a rather low gravitational field and thus tend to repel each other. This repelling effect creates a levitating dust layer above the surface of the Moon. This striking observation (Fig. 1.10) was first made by Apollo 17 and Surveyor spacecrafts [77] in the 1970s.

1.3. Compton, Checking on Electrons' Speed

In 1905, when Einstein proposed the photon concept, i.e. an elementary electromagnetic radiation energy lump $\varepsilon = h\nu$, he was then able to explain that the photoelectric effect was generated by the minimal absorption of a photon. Its energy had to be greater than the work function of an electron in a metal. But he considered that the discrete Planck sum was, for each frequency, over an average number of photons in the cavity. He then returned to a "particle-like" view of light similar to that which, centuries before, Newton had asserted.

Fig. 1.10. A reconstruction of the western horizon of the Moon from 5 photographs of Surveyor-7 shortly after sunset. The circle S.D. indicates the position of the solar disk. The arrow points at Moon dust layer levitating about 30 cm above the ground. Reprinted with permission from [77]. Copyright © 1974 by D. Reidel Publishing Company and Reprinted with permission from [81]. Copyright © 2000 by the American Physical Society.

Extending his idea, in 1917, Einstein postulated that if the photon had to be considered a particle,[9] it ought to have a momentum $\boldsymbol{p}$ so that its kinetic energy would be $\varepsilon = pc$. But Planck's black-body theory and the interpretation of the photoelectric effect required an $\varepsilon = h\nu$ energy for the photon associated with an electromagnetic wave with frequency $\nu = c/\lambda = ck/(2\pi)$, where k is the wave vector. It is therefore clear that some kind of relationship must exist between the photon momentum $\boldsymbol{p}$ and the electromagnetic wave vector $\boldsymbol{k}$:

$$\boldsymbol{p} = \hbar \boldsymbol{k} \tag{1.11}$$

where, for further convenience, we have introduced a Planck reduced constant $\hbar = h/2\pi$. This notation will be used intensively throughout the rest of this book.

Because they involve the mechanistic laws of conservation, photon–electron collision experiments should provide confirmation of the corpuscular nature of a photon. A. H. Compton (Fig. 1.11) had repeatedly observed an increase in the wavelength of an X-ray beam scattered from a graphite target. In 1923, he suggests [20] attributing this phenomenon to an inelastic

[9]Einstein, in his theory of relativity [32], had deduced that the energy of any particle moving momentum p was to take the form $\varepsilon = \sqrt{p^2c^2 + m_0^2c^4}$ where m_0 is the rest mass of the particle. According to the principle of constancy of the velocity of light, the photon cannot be at rest: its mass is then zero and $\varepsilon = pc$.

Fig. 1.11. Arthur Holly Compton (1892–1962) was awarded the Nobel Prize in 1927 (shared with Charles T. R. Wilson). This photograph is from his Manhattan Project badge which mentions a fake name for security reasons. As a son of an ordained Presbyterian minister with a "scientia et religio ex uno" motto, he considered a possible career in religion after his PhD. For two years, he was a Research Engineer of the Westinghouse Lamp Company working on sodium lamps and at Signal Corps developing aircraft instruments. In the 1930s working as a consultant for General Electric, he initiated a project which will result in the development of fluorescent lamp industry. While in Cambridge at Ernest Rutherford's lab, he observed that scattered radiation was more easily absorbed than the primary beam. This was the first hint of an energy loss radiation–matter collision which led him to his groundbreaking experiment. Starting in 1941 (just before Pearl Harbor), he became involved full time into the war effort to obtain a chain reaction. As head of the Manhattan Project's Metallurgical Laboratory, he was responsible for the conversion of uranium into plutonium and recruited Robert Oppenheimer to be in charge of the atomic bomb design in Chicago. Shielding against radiation was not his only concern and he hired N. Hilberry to receive "all kicks intended for the project leader, and carry through all the unpleasant tasks from which the project leader wants to escape". After the war, he gradually became more interested in philosophical and religious matters. A Moon crater was named after two of the Compton brothers (Karl who also served as President of the MIT). From [2]. Photo credits: public domain, by courtesy of Los Alamos National Laboratory.

collision, i.e. with an energy transfer, of photons with electrons in the target.[10] Following Einstein's model, the energy and momentum conservations

[10]The light as particles interpretation was published almost simultaneously by Debye [25].

are written as

$$h\nu_1 + \varepsilon_i = h\nu_2 + \varepsilon_f \tag{1.12a}$$

$$\boldsymbol{p}_1 + \boldsymbol{p}_i = \boldsymbol{p}_2 + \boldsymbol{p}_f \tag{1.12b}$$

where the indices i and f respectively refer to initial and final energies and momenta of electrons in the target. The indices 1 and 2 respectively relate to incident and scattered photons.

For the sake of simplification, it is assumed that the electron is free and initially at rest in the laboratory reference frame (Fig. 1.12). Its energy is thus given by the rest mass $\varepsilon_i = m_e c^2$ and its momentum is set to $p_i = 0$. After interaction with the electromagnetic field, the total energy becomes $\varepsilon_f = \sqrt{p_f^2 c^2 + m_e^2 c^4}$. On the photon side, $p = h\nu/c$ and $\varepsilon = h\nu$. Setting to 2θ the angle between $\boldsymbol{p}_1$ and $\boldsymbol{p}_2$, we get the energy conservation condition:

$$\left[\frac{h}{c}(\nu_1 - \nu_2) + m_e c\right]^2 = \frac{h^2}{c^2}\left(\nu_1^2 + \nu_2^2 - 2\nu_1\nu_2 \cos 2\theta\right) + m_e^2 c^2 \tag{1.13}$$

which simplifies to

$$\nu_1 - \nu_2 = 2\frac{h\nu_1\nu_2}{m_e c^2}\sin^2\theta \tag{1.14}$$

As the quantity on the right side of the equality is always positive, it clearly shows that the scattered wave frequency is bound to be smaller than the incident one. This is exactly what Compton observes in the graphite experiment:

$$\lambda_2 - \lambda_1 = 2\frac{hc}{m_e c^2}\sin^2\theta \tag{1.15}$$

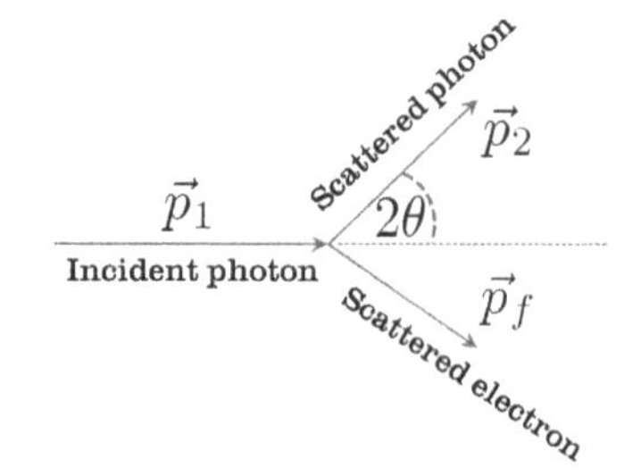

Fig. 1.12. Geometry and notations for modelling Compton's experiment. The electron is assumed to be initially at rest.

The maximum wavelength shift is obtained for a *backscattered photon*, i.e. $2\theta = \pi$. The *Compton wavelength* $\lambda_c = hc/(m_e c^2) \approx 2.42 \times 10^{-12}$ m is the wavelength shift that corresponds to a photon observed at a 90° scattering angle.

Note that this reasoning is entirely based on the hypothesis that the electromagnetic wave can be reduced to a collection of photons with momentum $\boldsymbol{p} = \hbar \boldsymbol{k}$. It then expresses the reality of the particle nature of light under such circumstance. The agreement of this collision model with Compton's experimental observations is excellent (Figs. 1.13 and 1.14). The assumption of electromagnetic radiation dual behaviour is then validated and paves the way for the idea of a possible "wave-particle duality".

1.3.1. *Applications and illustrations of Compton scattering*

Having long been considered as an unwanted noise for high-energy scattering and X-ray diffraction experiments, Compton scattering has now acquired better recognition, especially since the advent of intense X-ray

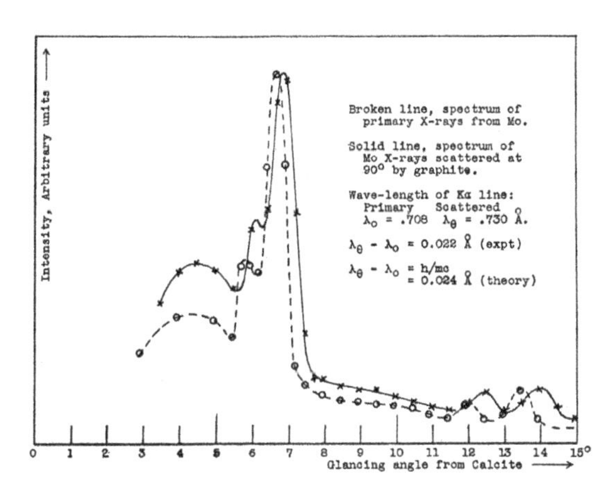

Fig. 1.13. The fluorescence emission spectrum of the molybdenum is used as incident radiation (see Volume 2). There is a shift between the incident spectrum and the spectrum scattered from the graphite target. The abscissa is expressed in degrees of inclination of the calcite crystal used as an energy analyzer of the detected photons (correspondence between orientation angle and wavelength is given by Bragg's formula $\lambda = 2d \sin \theta_B$). Reprinted figures with permission from [20]. Copyright © 1923 by the American Physical Society.

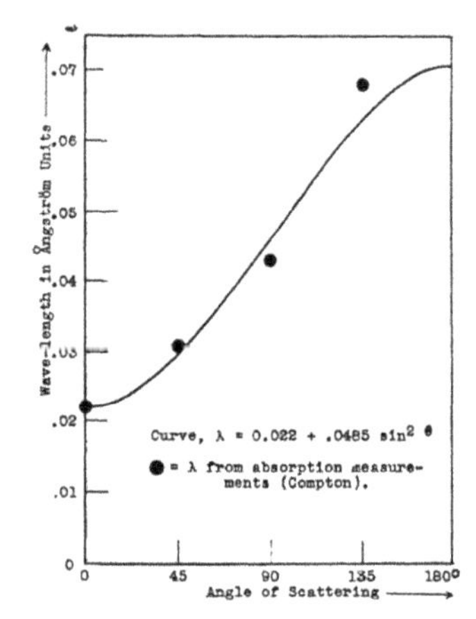

Fig. 1.14. The wavelength shift indeed follows the $\sin^2\theta$ law. Reprinted with permission from [20]. Copyright © 1923 by the American Physical Society.

sources (such as synchrotron radiation produced in large storage rings). It is known to provide very useful information on the electron speed distribution in solids (see Question *1.8*). At much higher energies, as opposed to the aforementioned Compton scattering processes, there is a possibility that an ultra-relativistic electron loses energy and transfers it to an incoming photon. This relativistic effect [8] is often called *inverse Compton scattering*. This is sometimes observed in astrophysical objects, where photons of energy $h\nu$ are backscattered to some $\gamma^2 h\nu$ with $\gamma = \varepsilon/(m_0c^2)$, ε being the electron initial high kinetic energy and m_0 its rest mass. This is believed to be the case for loud radioactive galaxies such as Centaurus A or Cygnus A quasars. Powered by a massive black hole at their centre, matter from the accretion region is ejected in two opposite directions at very high speed, forming the so-called *jets*. Electrons at such relativistic speed, when subject to a magnetic field, create synchrotron radiation. These jets are thus the locus of a radio wave emission from which quasars[11] were first detected. But the electrons have such a high kinetic energy, with γ close to a thousand, that the mere Cosmic Background 3 K Radiation is suspected [49] to be "Compton backscattered" and at the origin of the observed X-ray emission (Fig. 1.15).

[11] "Quasar" comes from the early 1970s contraction of "quasi-stellar radio source".

Fig. 1.15. An X-ray image of a Centaurus A quasar jet. The picture is made from photons collected in the 0.4–2.5 keV window by Chandra X-ray space observatory. Reprinted with permission from [49]. Copyright © 2005 by the Royal Society.

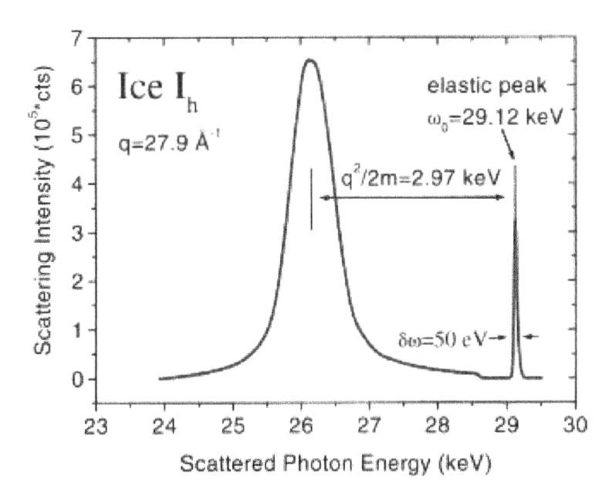

Fig. 1.16. Compton scattering spectrum for ice measured at European Synchrotron Radiation Facility. As a reference, the narrow elastic scattering peak is shown and, at lower energies, the broader Compton peak is a function of electron momentum distribution. See Question *1.8*. Reprinted with permission from [57]. Copyright © 1999 by the American Physical Society.

Question 1.8: **Compton spectroscopy**. The calculation presented above assumes, wrongly, that the electron is initially at rest. As it will be shown later in this volume, it turns out that this is never the case. It is proposed here to show that Compton scattering is also a technique for measuring the electrons' velocity in a target. We assume that the electron has an initial momentum $\boldsymbol{p}_i$ that is changed to $\boldsymbol{p}_f$ upon interaction. Neglecting relativistic effects, we here set the respective initial and final energies to: $\varepsilon_i = p_i^2/(2m)$ and $\varepsilon_f = p_f^2/(2m)$. Show that if one assumes a negligible change in the electron momentum, $p_1 p_2 \approx p_1^2$, i.e. a weak energy transfer, then

$$\lambda_2 - \lambda_1 \approx 2\lambda_c \sin^2\theta + 2\frac{\lambda_1 p_{iz}}{m_e c}\sin\theta$$

where p_{iz} is the electron initial momentum component along the so-called *scattering vector* $\boldsymbol{K} = \boldsymbol{k}_1 - \boldsymbol{k}_2$. Comment on Fig. 1.16.

Answer: The energy and momentum conservation for photon+electron systems is

$$h\nu_1 + \varepsilon_i = h\nu_2 + \varepsilon_f$$

and

$$\boldsymbol{p}_1 + \boldsymbol{p}_i = \boldsymbol{p}_2 + \boldsymbol{p}_f$$

We are not interested in the electron final momentum, so

$$\boldsymbol{p}_f = \boldsymbol{p}_i + \boldsymbol{p}_1 - \boldsymbol{p}_2$$

The electron final kinetic energy thus becomes

$$\frac{p_f^2}{2m} = \frac{1}{2m}(p_i^2 + |\boldsymbol{p}_1 - \boldsymbol{p}_2|^2 + 2\boldsymbol{p}_i.(\boldsymbol{p}_1 - \boldsymbol{p}_2))$$

By convention, vector $\boldsymbol{p}_1 - \boldsymbol{p}_2$ is chosen to point towards the $\boldsymbol{e}_z$-axis. As in the previous question, we denote the scattering angle (between $\boldsymbol{p}_1$ and $\boldsymbol{p}_2$) by 2θ, so that

$$\frac{p_f^2}{2m} = \frac{p_i^2}{2m} + \frac{1}{2m}(p_1^2 + p_2^2 - 2p_1p_2\cos 2\theta + 2p_{i,z}\sqrt{p_1^2 + p_2^2 - 2p_1p_2\cos 2\theta}))$$

where $\boldsymbol{p}_i.\boldsymbol{e}_z = p_{i,z}$.

If we assume a negligible variation for the photon energy, the electron kinetic energy after the collision can be simplified. We thus set $p_1 \approx p_2$ (for the module, only) and

$$\frac{p_f^2}{2m} \approx \frac{p_i^2}{2m} + 2\frac{p_1^2}{2m}(1 - \cos 2\theta) + \frac{p_1}{m}p_{i,z}\sqrt{2}\sqrt{1 - \cos 2\theta}$$

We then convert to sine:

$$\frac{p_f^2}{2m} \approx \frac{p_i^2}{2m} + 4\frac{p_1^2}{2m}\sin^2\theta + 2\frac{p_1}{m}p_{i,z}|\sin\theta|$$

The photon energy change is

$$h(\nu_1 - \nu_2) \approx 4\frac{p_1^2}{2m}\sin^2\theta + 2\frac{p_1}{m}p_{i,z}|\sin\theta|$$

We then make use of $\nu = c/\lambda$ and $p = \hbar k = h/\lambda$ and find the shift in wavelength:

$$\frac{1}{\lambda_1} - \frac{1}{\lambda_2} \approx 4\frac{h}{2mc\lambda_1^2}\sin^2\theta + 2\frac{1}{mc\lambda_1}p_{i,z}|\sin\theta|$$

which can be reduced to

$$\lambda_2 - \lambda_1 \approx 2\frac{h}{mc}\sin^2\theta + \frac{2}{mc}\lambda_1 p_{i,z}|\sin\theta|$$

This expression indicates that, if the electron has an initial non-zero momentum, the radiation wavelength shift provides a way to measure the component of $\boldsymbol{p}_i$ along $\boldsymbol{e}_z$. It is therefore nothing but a Doppler effect with a significant difference: even if the electron is at rest, a change in wavelength can still exist.

Figure 1.16 shows that the variation in photon energy (for a given angle 2θ) is actually not fixed. It is noted that, around the 3 keV Compton shift, there is a distribution of photon energies (of approximately Gaussian form). This distribution is directly connected to the electron momentum distribution in the ice sample.

Thus, for a monochromatic incident radiation, instead of a very sharp peak centred at $2\lambda_c\sin^2\theta$, we observe an intensity distribution directly related to the electron speed distribution. This method provides fine details on the microscopic mechanisms responsible for conduction properties in solids or helps characterize the nature of chemical bonds.

Chapter 1: Nuts & Bolts

- *Light can be described in terms of a flow of energy lump, called "**photons**".*
- *For a given **frequency** ν, the **energy** of each photon is $h\nu = \hbar\omega$ with the reduced Planck constant $\hbar = \frac{h}{2\pi} \approx 10^{-34}\,Js$.*
- *In a **photoelectric effect**, the ejected electron thus has the kinetic energy $\varepsilon = h\nu - W$ where W is the metal work function.*
- *If the electromagnetic wave propagates with a **wave vector** $\mathbf{k}$, each photon has a **momentum** $\mathbf{p} = \hbar\mathbf{k}$.*

2

From Particles to Wave Fields

The content of this chapter will help you understand ***fluorescence in art history or for X-ray production, the speed of galaxies, single-molecule imaging****, and related topics.*

We have reviewed experimental evidence in support of the description of a monochromatic electromagnetic wave (of frequency ν and wave vector $\boldsymbol{k}$) as a flow of particles, indivisible motes of energy $\varepsilon = h\nu$, moving at the speed of light with a momentum $\boldsymbol{p} = \hbar\boldsymbol{k}$.

In this chapter, we will see that if we agree that it is necessary to describe light as an assembly of particles when it interacts with matter, we are forced to admit that, in return, massive objects sometimes behave as waves. The first hint will come from the unusual spectrum radiated by an excited gas calling for drastic change in well-established laws of physics. It will become obvious that we cannot circumvent the wave nature of matter when it is observed that interference patterns can be created from a beam of massive particles.

2.1. Bohr Orbits Ground-Breaking Model

When subjected to a 5000-volt excitation, hydrogen gas becomes highly excited. The de-excitation leads to the emission of a well-known light spectrum.[1] However, details about the light emission mechanism were not fully understood until 1885 when Balmer (a mathematics professor) proposed an empirical formula, later generalized by Ritz (1908), to monitor the narrow emission line wavelength values λ from the hydrogen atom (in the weakly excited case):

$$\frac{1}{\lambda} = R_\infty \left(\frac{1}{2^2} - \frac{1}{n^2} \right) \tag{2.1}$$

where n is the radiation line number and $R_\infty \approx 1.097 \times 10^7\ \mathrm{m}^{-1}$ is a constant. This is determined by observing that the first visible emission lines are respectively red (0.656 μm), blue (0.486 μm), indigo (0.434 μm) and purple (0.410 μm). Additional transitions following a similar sequence law were also observed for alkali metals or ionized helium, which will be later described as "monovalent atoms".

The existence of a discrete spectrum of this kind was something of a puzzle because nothing in the nineteenth century physics could provide a satisfactory explanation. In 1913, Niels Bohr (Fig. 2.1) crafted the elements of a theory, focusing on the excitation and de-excitation process in the hydrogen atom, to shed some light onto this awkward line spectrum sequence. To achieve his goal, however, he had to undermine some well-established physical principles: Starting from an atomic structure demonstrated by Rutherford (a positive nucleus at the core, and negative electrons on the outskirts), Bohr interpreted it as if the electron were on a stationary orbit, circling the nucleus like a satellite around its planet. Moreover, he added a daring assumption: "not all orbits are possible". Thus, going against the widely accepted (and observed) law according to which an accelerated charged particle should radiate, Bohr's proposition here boils down

[1]The spectrum was in fact exploited by Anders Angström to confirm the presence of hydrogen in the sun.

Fig. 2.1. Niels Bohr (1885–1962) was awarded the Nobel Prize in 1922. He and his brother were very enthusiastic promoters of football in Denmark (Harald, who was to become Director of the Copenhagen Mathematical Institute, was captain of the National Team at the 1908 Olympic games). In 1911, Niels started a close relationship with Rutherford, whose atom planetary model was the starting point for Bohr's construction of quantum mechanics. While very passionate in a one-to-one scientific discussion, he was described as a mumbler, whisperer and hard-to-understand lecturer. In addition to his own strong ideas (and scientific duel with Einstein and Schrödinger), he was eager to establish his home, then the institute, as a scientific haven to many colleagues in those troubled times. The "Great Dane" also had a major influence on the understanding of nuclear fission [10] and his connection to Kapitza did not match Churchill's 1944 tastes: "the President (Roosevelt) and I are much worried about Professor Bohr. How did he come into this business? (...). He says that he is in close correspondence with a Russian professor, an old friend of his in Russia, to whom he had written about the matter and may be writing still.(...) What is all this about? It seems to me that Bohr ought to be confined or at any rate made to see that he is very near the edge of mortal crimes... I do not like it at all." Niels' son, Aage, also won the Nobel Prize for physics in 1975. Photo credits: Library of Congress via Wikimedia Commons.

to the statement that there must be something special about electrons: they do not radiate when they sit on these stationary orbits and radiation only occurs when they jump from one particular permitted orbit to another. Let us consider the system in the centre-of-mass reference frame. This approximately[2] coincides with the position of the nucleus. To the same accuracy, the reduced mass m is that of the electron m_e. Bearing in mind Kepler's first law of elliptical orbital motions, we will restrict ourselves

[2]They differ by less than 0.1%, the relative difference between the nucleus and the atom's masses. The same approximation will be used for a thorough quantum treatment of atomic systems.

to the circular case. Consider the conserved quantities of this rotating system:

- its angular momentum: $\boldsymbol{L} = \boldsymbol{r} \times \boldsymbol{p}$;
- its total energy ε is the sum of the kinetic rotation energy $T = \frac{L^2}{2mr^2}$ and the Coulomb electrostatic interaction potential $V = -\frac{e^2}{4\pi\epsilon_0 r}$.

A stationary orbit is such that the radius corresponds to a constant energy $\partial\varepsilon/\partial r = 0$, which is equivalent to a balance between the centrifugal force and the Coulomb attraction:

$$\frac{L^2}{mr^3} = \frac{e^2}{4\pi\epsilon_0 r^2}$$

The orbit radius is thus obtained:

$$r = \frac{L^2 4\pi\epsilon_0}{me^2}$$

and, as a sole function of the angular momentum, the energy is

$$\varepsilon = -\frac{m}{2L^2}\left(\frac{e^2}{4\pi\epsilon_0}\right)^2$$

Thus, according to Bohr's model of stationary orbits, an orbit modification corresponds to a variation of angular momentum and a consequent energy change:

$$\varepsilon_1 - \varepsilon_2 = -\frac{m}{2}\left(\frac{e^2}{4\pi\epsilon_0}\right)^2\left(\frac{1}{L_1^2} - \frac{1}{L_2^2}\right)$$

When we compare this expression with that of Balmer (2.1), we are led to assume that the orbiting electron must change its angular momentum by some discrete values. Numerical estimates of R_∞ show that these values are multiples of an increasingly well-known constant: $\hbar$ with $\hbar = h/(2\pi)$. Obviously, this cannot be a coincidence. Planck's constant action range is not limited to radiation. It now contaminates matter! The constant R_∞, introduced by Balmer, is now known as the *Rydberg constant* and can be expressed as $R_\infty = \frac{m}{4\pi c\hbar^3}(\frac{e^2}{4\pi\epsilon_0})^2$. This astonishing result is a direct consequence of Bohr's "discrete orbit" hypothesis and, inevitably, raises questions about its compatibility with the mechanistic approach in physics.

The fact remains that, after having forced the idea of energy quantization of electromagnetic radiation to avoid the ultraviolet catastrophe, another mystery, nothing less than the stability of the atom and its emission spectrum, seemed to be (partially) solved by introducing a quantization of its angular momentum. The constants of motion appeared to be the prime target of this emerging physics where $\hbar$ plays the leading role. Momentum's turn was soon to come.

2.1.1. *Applications of atomic radiation spectra*

Understanding spectral lines from atomic radiation emissions has contributed to the development of many areas of physics and chemistry. Each atom has a well-defined characteristic spectrum. It is therefore possible by using this property to detect the presence of a particular element in a complex object. For example, all X-ray fluorescence techniques are based on this principle.[3] First, the sample is excited with a high-energy radiation such as hard X-rays. The fluorescence photon emission accompanying the subsequent de-excitation is then detected. Each radiation line of this electromagnetic spectrum can be attributed to a specific atomic element (and only one, if the wavelength resolution is sufficient). This allows a non-destructive contact-free chemical analysis of the sample. It is possible to go even further and extend this technique to imaging: choosing to detect only one particular wavelength, known to be specific to a given element fluorescence spectrum, makes us blind to any other contribution in the object but this particular chemical element. This is named "element specific imaging" with a wide range of applications in many domains outside physics, including art history.

As an example, let us consider one among numerous cases of Goya's recycled paintings [61]. In 1823, Goya paints a portrait of a senior magistrate at the Sala de Alcaldes de la Casa y Corte at Madrid territorial court. This powerful gentleman is Ramón Satué, as written in Goya's own hand in the lower left corner of the picture. This painting is usually considered to be one of the finest examples of Goya's mastery of colour balance with a vivid red waistcoat contrasting with the austere black and white which dominates

[3]Details about the fluorescence process are given in Volumes 2 and 3.

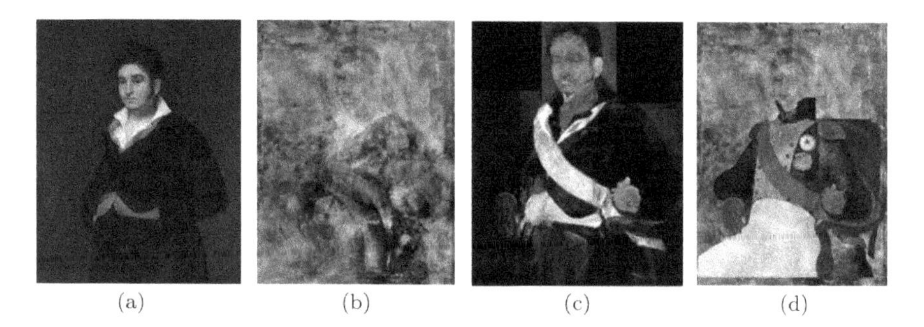

Fig. 2.2. Progressive recovery of a hidden portrait. (a) 1823 portrait of Don Ramon Satué, Spanish high court judge, by Goya. (b) Conventional X-ray radiograph showing the underpaint portrait of a sitter. (c) Mercury fluorescence image emphasizing the red parts of the two portraits. (d) Possible reconstruction of the sitter's portrait on the basis of mercury, antimony and iron fluorescence images. Reproduced with permission from [61]. Copyright © 2013 by Annual Review.

(Fig. 2.2(a)). The casual appearance of Satué is probably designed to show Goya's proximity to the man, soon to be appointed as a minister. However, X-ray radiographs (Fig. 2.2(b)) show evidence of an underpainting covered by this portrait and two questions then naturally arise: who was the author of the painting underneath and why was it covered? The conventional X-ray radiograph, as used in dentistry, is mostly sensitive to high-Z elements (such as lead) and lacks resolution to distinguish the two paintings. One can just see the portrait of a sitter, in a more formal posture, probably wearing a uniform. Taking advantage of the lack of colour (except the famous red regions) in Satué's picture, the procedure involves looking for the X-ray fluorescence of elements known to be employed in typical pigments such as mercury (for vermilion), antimony (for Naples yellow) or iron (in darker regions). For example, a mapping of mercury distribution, in the vicinity of 10 keV photon energy (Fig. 2.2(c)), reveals that the sitter is wearing a "vermilion sash from right shoulder to left hip, with a pendant star on the sash at the waist, plus another on a rayonnant silver plaque on the left breast."[4] Additionally, the antimony and iron fluorescence signals

[4]This citation is taken from a fascinating and very readable article by the investigators and promoters of X-ray fluorescence techniques for deciphering hidden art works [15]. Other examples can be found in [61].

reveal that the gold epaulettes and a row of gold buttons on a plastron are associated with a parade uniform of Napoleon's army (Fig. 2.2(d)). The decorations, the five-pointed star (Orden Real) and the rayonnant plaque, instituted by Napoleon's brother, Joseph Bonaparte, and worn at his court while he was monarch of Spain, cannot clearly identify the overpainted sitter. It might even well be Joseph Bonaparte himself. Whatever the truth may be, it is further evidence of Goya's acquaintances with the highest ranked officers of the First Empire. Keeping such a portrait after Joseph's defeat in 1813 would have not been a wise decision and it could have even been considered as a provocation to Ferdinand when he was restored to the throne in 1823. Brush strokes and style are considered to be evidences in support of the attribution of the hidden portrait to Francisco Goya.

The intensity of fluorescence radiation is proportional to the number of excited atoms. This emission process is also very useful because it allows for almost monochromatic X-ray sources. It is on this principle that the X-ray tubes used in medicine or radio-crystallography operate. A schematic representation of such a device is given in Fig. 2.3. Electrons are emitted from a hot tungsten filament. Once they are free in a vacuum, they are accelerated by a high voltage V to a so-called *anti-cathode* made of a metallic target in contact with a large copper block. As the electrons, with a speed $\sqrt{2Ve/m}$,

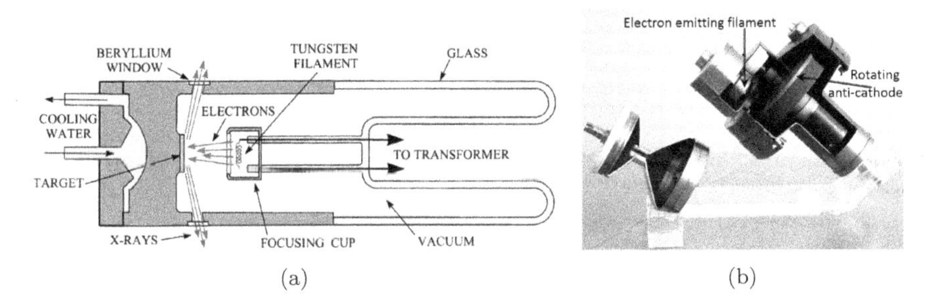

(a) (b)

Fig. 2.3. Portable X-ray source as used in medical imaging. (a) Schematic representation showing the filament from which the electrons are emitted and the target (anticathode). The element from which the latter is made conditions the energy of the X-ray fluorescence photons that are emitted through the windows. Photo credits: U.S. Geological Survey Department of the Interior. (b) A Straton rotating X-ray tube used for computed tomography scanning. Photons are emitted from a tungsten coated target (the "rotating anti-cathode"). Copyright © by Siemens AG.

enter the target, they transfer their kinetic energy to the atoms in a number of ways. Most of this energy (more than 90%) is wasted as it is merely converted to heat. This is why the target needs to keep an intimate contact with the underneath water-cooled copper conductor. However, some of the energy brought by the incident electrons is used to excite and, in the large majority of cases, ionize the atoms by ejecting some of the electrons from inner Bohr orbits, i.e., those which would have the smallest radii in the planetary picture. Atoms which are thus left in highly excited states will progressively recover their equilibrium by undertaking a series of electron transitions to fill the lowest orbits. According to Bohr's model, such changes in atomic states are accompanied by the emission of electromagnetic waves, or photons, the frequency of which is the difference between the two energy states connected by the transition. The characteristic spectrum of an X-ray tube can then be observed (Fig. 2.4): a broad distribution of frequencies, which is basically independent of the target material, is generated by the deceleration of electrons (Bremsstrahlung radiation) and fluorescence lines, at specific wavelengths, which differ from one metallic target to another.

Finally, in astrophysics again, the Hubble law provides a precious relation between the distance to a star and its relative speed to the observer. As detailed in Question *2.1*, an accurate determination of a particular radiation spectrum allows for an estimate of interstellar distances. In the same line of thought, Question *2.2* gives the premises of Mössbauer spectroscopy which also capitalizes on the Doppler shift of radiation spectra, but in the laboratory frame of reference.

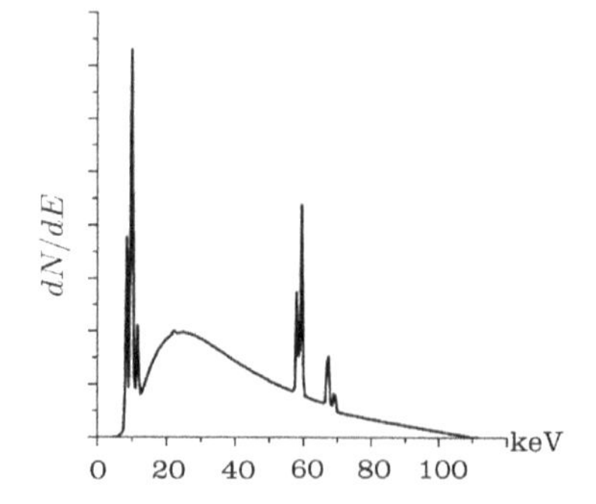

Fig. 2.4. X-ray spectrum emitted from a source similar to that of Figure 2.3. The target is made of tungsten which, after excitation by incident electrons, has fluorescent X-ray emissions at 59.3 keV, 57.98 keV and 67.24 keV. Lower energy photons are also emitted (but seldom used) in the 8–11 keV range. Photo credits: with kind permission of U. Ankerhold and Physikalisch-Technische Bundesanstalt.

Question 2.1: **Expansion of the universe: Doppler sees red.** In 1929, Hubble published an article [56] exposing the almost linear relationship between the distance of a star from the point of observation and its relative speed. From this discovery, the experimental value for the proportionality coefficient H_0 will be constantly refined:

$$v = H_0 d$$

where v is the distancing velocity of the star and d its distance from the observer. Distances are usually expressed in Mpc (mega-parsec), with 1 pc=648000 astronomical units $= 3.085 \times 10^{16}$ m. The velocity is in km/s. Today, the Hubble constant is set to $H_0 \approx 70\,\text{km/(s.\,Mpc)}$.

This law is considered of paramount importance in astronomy because it provides a method for determining the distance to a stellar object as soon as we know its separation speed. While measurements of rather short distances from the surface of the Earth can be done quite easily (e.g. laser ranging or triangulation methods), it is much more challenging for remote stars. However, the expansion of the universe implies that all galaxies are moving away from each other. The more distant they are, the faster they separate. This is precisely what Hubble's formula states.

It thus becomes essential to measure separation speeds. Since the stars are usually emitters of electromagnetic radiation from the de-excitation of their constituent atoms, it is worth considering the Doppler shift of the spectrum emitted by these sources in motion.

1. Let us first recall the non-relativistic Doppler effect. Consider a source that periodically emits a series of pulses separated by T_E. This source moves away from an observer (at rest) with a velocity vector $\boldsymbol{v}$ collinear with the line of sight. Let c be the propagation speed of the pulse in the observer's frame of reference. What is the distance between two consecutive pulses?
2. What is the period T_O of the series of pulses measured by the motionless observer? Deduce the Doppler shift between ν_e and ν_O, the emitted and the observation frequencies, respectively.
3. The expression that we have found applies to sound waves and is widely exploited in medical ultrasound Doppler imaging for example. If we are not dealing with pulses but rather an electromagnetic wave, spectral shifts become observable only for significant speeds, and non-relativistic classical mechanics is no longer suitable. When two objects are in relative motion, special relativity[a] establishes that the two local times (one of the source and that of the observer) are different. Calling Δt_E a time interval in the moving source local reference frame, it becomes $\Delta t_O = \gamma \Delta t_E$

[a]See, for example, *Feynman Lectures on Physics* now available at http://www.feynmanlectures.caltech.edu.

Question 2.1: (continued)

for the motionless observer, with $\gamma = 1/\sqrt{1 - v^2/c^2}$. What is the relativistic Doppler shift relationship between the emitted wavelength λ_E and the one detected by the observer λ_O?

4. Check that, in the limit of slow moving sources, we find the classical result. Show that, in all cases, a source moving away from the observer is seen with a larger wavelength than the one that was emitted. This effect is called "red shift".
5. A "red shift" is expressed by the quantity:

$$z = \frac{\lambda_O - \lambda_E}{\lambda_E}$$

Show that the knowledge of z can help estimate the distance to the source.

6. The Cygnus A quasar has a red shift $z = 0.056052$. Estimate the distance between the Earth and Cygnus A.
7. Hydrogen-excited radiation from the Andromeda galaxy is measured at $\lambda_O = 6556$ Å. This corresponds to $n = 3$ (line Hα) in Balmer's formula (2.1). Deduce the direction and the velocity magnitude of this galaxy relative to the Earth.

Answer:

1. Suppose that the first pulse is emitted toward the observer at time $t = 0$. After T_E, i.e. at the time of emission of the next pulse, the first pulse has travelled a distance $d_1 = cT_E$. But since the source moves at velocity v, the second pulse is emitted at $t = T_E$ from $d_2 = -vT_E$. Thus the distance between the two pulses is $\Delta d = (c + v)T_E$. If the source was approaching, we would of course find: $\Delta d = (c - v)T_E$.
2. Once issued, the pulses propagate at speed c. The time interval between their arrivals at the detector is therefore an apparent period seen by the observer: $T_O = \Delta d/c = (1 + v/c)T_E$ and the associated frequency ($T = 1/\nu$) is: $\nu_O = \nu_E c/(v + c)$.
3. Because clocks beat different measures, it is necessary to reconsider the former reasoning from a Lorentz perspective. Thus, if the source emits a wave with period T_E in its own reference frame, these time intervals will turn into γT_E in the fixed observer reference frame. It follows that the period measured by the observer will become $T_O = \gamma(1 + v/c)T_E$ and the wavelength

$$\lambda_O = c\gamma\left(1 + \frac{v}{c}\right)T_E = \sqrt{\frac{c + v}{c - v}}\lambda_E$$

4. With the adopted conventions, $v > 0$ for a source that moves away, and we see indeed that $\lambda_O > \lambda_E$.

   If $v \ll c$, the denominator can be approximated by a Taylor series expansion in powers of v/c, so that we get the previous non-relativistic result $\lambda_O \approx (1 + v/c)\lambda_E$.
5. The redshift quantifies the Doppler effect on radiation, usually the de-excitation spectrum of an atomic species and often that of hydrogen — which is the most abundant. Assuming a low redshift (i.e. $v \ll c$):

$$\frac{\lambda_O}{\lambda_E} = z + 1 = \sqrt{\frac{c + v}{c - v}} \approx 1 + \frac{v}{c}$$

We end up with the approximate expression $v \approx z \times c$, thus $d \approx z \times c/H_0$.

Answer: (continued)

6. For $z \approx 0.056162$, by directly applying the previously found relationship, we obtain a distance to Cygnus A $d \approx z \times c/H_0 \approx 231\,\text{Mpc}$.
7. The line matching $n = 3$ in Balmer's formula (2.1) corresponds to an emission wavelength $\lambda_E = 6563$ Å$> \lambda_O$. This value is greater than that which is measured by the observer. The Doppler shift then corresponds to a source which comes closer to the Earth. Furthermore, we can quantify the approaching velocity as $v = c \times z \approx -320\,\text{km/s}$. In this particular case, this effect is not related to the expansion of universe but to an expected merging of Andromeda galaxy with our Milky Way, in about four billion years.

Question 2.2: **Transmission and absorption, with a step back.**
In 1904, Robert Wood [93] discovered the electromagnetic resonant absorption phenomenon in sodium. The explanation had to wait until Bohr's theory of transitions between atomic levels appeared. However, Wood reported already that he was able to observe that the yellow light (589 nm line) emitted from the de-excitation of a sodium vapour — so typical of certain street lighting — could itself be reabsorbed by another sodium vapour placed at room temperature. This resonant absorption phenomenon can be reproduced by means of table salt and a spectral lamp (or by following Wood's indications in [94]).

This demonstration of resonant transitions between energy levels of two identical atoms was not easy to reproduce in the case of γ rays that are emitted (and subsequently absorbed) by nuclei. Werner Kuhn [65] remarked in 1929 that this was probably because a photon emission is necessarily accompanied by a recoil from the emitter. We denote this recoil energy by $\Delta\varepsilon$. One should expect a similar effect on the absorber side as it will also need to satisfy conservation laws and therefore take a portion of the energy from the incoming photon. Even if they are identical, a $2\Delta\varepsilon$ energy shift should be expected between a free emitter and a free absorber. While studying this phenomenon, Rudolph Mössbauer discovered one of the finest local spectroscopy methods (see Volume 2).

1. The $^{191}_{77}$Ir nucleus emits 129 keV γ photons. Nowadays, ^{57}Fe is overwhelmingly dominant among the elements that are studied by means of Mössbauer spectroscopy. It emits 14.4 keV photons. What are the respective frequencies of both electromagnetic radiations?

2. Start from momentum and energy conservations (assuming initially motionless nuclei) and compute how much the recoil modifies the emitted radiation frequency. What is the photon energy loss (in eV)? Compare with sodium's recoil energy (^{23}Na) associated with the emission of its yellow light.

3. It is instructive to consider low energy phenomena from a temperature perspective. To do this, we define the following conversion: $T = \varepsilon/k_B$. What are the temperature regimes which correspond to energy losses for photon emissions from $^{191}_{77}$Ir, ^{57}Fe and ^{23}Na?

4. Recall that for a classical monoatomic ideal gas, the average kinetic energy is $3k_BT/2$. Explain why it is easier to measure a resonant absorption of radiation with the yellow sodium.

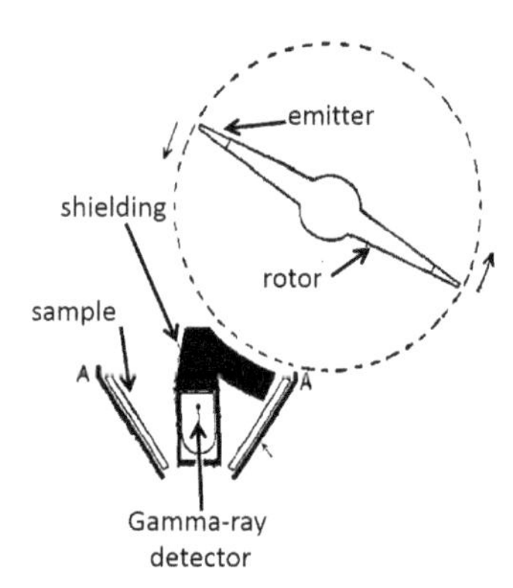

Fig. 2.5. Experimental device designed by Moon. The emitter is positioned at the tip of the rotor. The absorbing sample is positioned in front and the re-emitted (or resonantly scattered) photons are detected by the gamma counter. The counter's position is such that the photons associated with a Compton process can be readily removed. Reproduced with permission from [72]. Copyright © by IOP Publishing.

Question 2.2: (continued)

5. While some of his colleagues decided to heat the samples to make them acquire the speed necessary to compensate for the recoil, Moon [72] decided to explore a mechanical path and built the device pictured in Fig. 2.5. This time an ^{198}Au sample is considered. Gold shows a nuclear transition at 411 keV. While remaining at room temperature, Moon hopes to provide sufficient Doppler correction to the emitter. Estimate the tangential speed at which the source should move with the depicted rotational system for detecting a resonant absorption.

Answer:

1. The frequencies can be computed from $\nu = \varepsilon/h$ (energy has to be converted to joules). To make conversion faster, we can first compute $e/h = 2.418 \times 10^{14}$ Hz/eV. The frequency then reads $\nu_{\text{Ir-191}} = 3.12 \times 10^{19}$ Hz and $\nu_{\text{Fe-57}} = 3.48 \times 10^{18}$ Hz.
2. The conservation of momentum implies that the nucleus of mass M gains $MV = h\nu/c$ and the kinetic recoil energy carried by the nucleus is then: $\varepsilon = \frac{1}{2}MV^2 = \frac{1}{2Mc^2}(h\nu)^2$. This energy is taken from the emitted photon and it follows that the frequency of the γ radiation is reduced by an amount of:

$$\Delta\nu = \frac{1}{2Mc^2}h\nu^2$$

We thus find $\Delta\nu_{\text{Ir-191}} = 1.13 \times 10^{13}$ Hz and $\Delta\nu_{\text{Fe-57}} = 4.72 \times 10^{11}$ Hz.
The emitted photon loses respectively $h\Delta\nu_{\text{I-191}} = 0.0468$ eV and $h\Delta\nu_{\text{Fe-57}} = 0.00195$ eV. A similar calculation for the sodium yellow line yields $h\Delta\nu = 10^{-10}$ eV or a 25 kHz frequency shift.
3. The energy lost by the photon in the recoil process can be compared to the heat provided by a thermostat reservoir at temperature T. We find $T_{\text{Fe-57}} = h\nu_{\text{Fe-57}}/k_B = 22.6$ K, $T_{\text{Ir-191}} = 543$ K and $T_{\text{Na}} = 1.2 \times 10^{-6}$ K.
4. With an ambient temperature of 300 K, the distribution of sodium atom speeds is more than enough to compensate for its emission or absorption shifts.
5. The source is placed on a trolley and one measures the γ photon absorption by the target (usually fixed) as a function of the relative velocity. To achieve this, the speed of the source, as a whole, must be such that the Doppler effect is sufficient to compensate for both the emitter and the absorber recoil speeds. We then find

$$V_{\text{Au-198}}^{\text{trolley}} = 2c\frac{\Delta\nu_{\text{Au-198}}}{\nu_{\text{Au-198}}} = 668\,\text{m/s}$$

Answer: (continued)

the factor 2 comes from the global recoil energy of $2\Delta\varepsilon$. The challenge is thus to accelerate the emitter and reach almost twice the speed of sound. Moon's device was successful enough to significantly increase the resonant absorption effect (and a subsequent re-emission). It is on this relative motion principle that Mössbauer built, almost 10 years later, the apparatus which earned him the Nobel Prize. However, it will turn out that the decisive trick was to limit the absorber's recoil by nailing it down to an embedding crystal lattice (see Volume 2).

2.2. Louis de Broglie Introduces Particle Waves

The dual wave-particle nature of electromagnetic radiation began to take shape. It seemed quite clear that light was still to be considered as a wave, for example, to explain interference phenomena, but a proper account of its interaction with matter would call for a description in terms of particles. Absorption, especially in small amounts, emission, mainly of high frequencies and inelastic scattering, especially in backscattering geometry, strongly support a corpuscular picture of light particles with energy $\hbar\omega = h\nu$ and momentum $\hbar\boldsymbol{k}$.

The question may then arise as to the relevance of a reverse reasoning: would particles with a mass, like the electron, the neutron, the proton, the helium nucleus (alpha particle) and even heavier objects, present wave-like properties? This is what Louis de Broglie (Fig. 2.6) postulated in 1925 in his doctoral thesis [24]. He hypothesized that any particle has a wave nature and if its energy is ε and its momentum $\boldsymbol{p}$, it should be associated with a wave defined by (at least) an angular frequency $\omega = \varepsilon/\hbar$ and a wave vector $\boldsymbol{k} = \boldsymbol{p}/\hbar$ or a *de Broglie wavelength* $\lambda_{\mathrm{dB}} = h/p$. Thus, for L. de Broglie and soon the rest of the scientific community, the relationships valid for light should bear some universality.

Note that de Broglie's hypothesis was in line with Bohr's planetary model. Indeed, if a particle is constrained to a particular stationary orbit of radius r, then the associated wave must also exist on that orbit. This wave should then bear a 2π rotational invariance property. This is a condition for the wave not to interfere destructively with itself on the orbit (Fig. 2.7). As an example, suppose that the movement is circular in the (x, y) plane. The circumference of the circle, which is described by this trajectory, should correspond to an integer number of wavelengths: $n\lambda = 2\pi r$.

Fig. 2.6. Louis de Broglie (1892–1987) was awarded the Nobel Prize in 1929. He was perceived to be a lonely person who hardly travelled or interacted with the scientific community (he was not at ease with any other language but French). He was particularly inspired by Poincaré's writings (1910) and Langevin's lectures (1919). According to Ewald, Louis's brother (duke Maurice) had his own lab installed "in his private hôtel in the rue Lord Byron (Paris) where cables for the current came in by holes cut in the Gobelins adorning the walls". There, he was working on the photoelectric effect and X-ray emission. He reported that during WWI Louis was "able to serve his country while working as an electrician, taking care of machines and wireless transmissions [at the Eiffel Tower] and perfecting heterodyne amplifiers then in their infancy". Louis admitted that this gave him a useful introduction to practical aspects of wave science. Prince, then Duke when his brother died, Louis lectured for thirty-three years at the Sorbonne and was in his mid-life described as "wearing a dark blue suit which even then seemed slightly old-fashioned, with a wing collar and a pearl in his neck-tie" and a curious high-pitched voice. Photo credits: public domain from Mac Tutor History of Mathematics Archives.

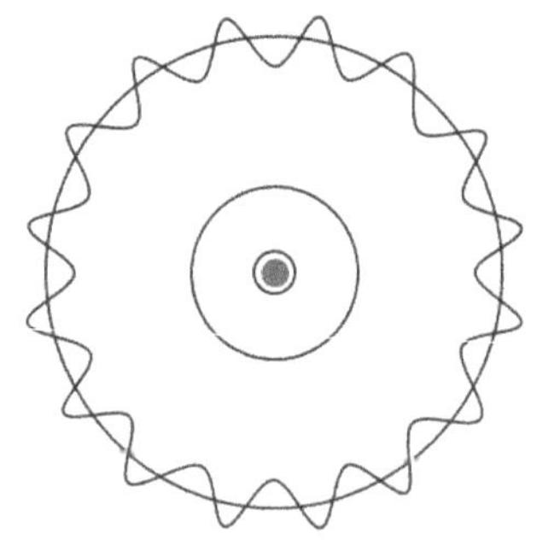

Fig. 2.7. The wave associated with an electron on an orbit must have a wavelength such that it cannot interfere destructively with itself. It is thus necessary that $\lambda_{\text{dB}} = 2\pi \times r/n$ where r is the orbit radius and n a non-null natural integer. Warning: the wiggling line does not represent the actual path of the electron around its orbit. At this stage, it is only assumed that the electron remains on the orbit. The mere purpose is to sketch wave phase modulations with no particular unit.

With the use of $\lambda = 2\pi/k = 2\pi\hbar/p$, we immediately find $pr = n\hbar$. Since the motion is circular, $\boldsymbol{p}$ must be perpendicular to $\boldsymbol{r}$ and we are led to write the quantization condition for the z component[5] of the orbital angular

[5]In this case, n is a positive or negative integer because it depends on the relative orientation of the rotation speed (momentum) and the radius r. The z component of angular momentum can thus be negative.

momentum $L_z = n\hbar$, with $n \in \mathbb{Z}$. This argument of angular momentum quantization will be developed in a much more formal and rigorous manner in Volume 2.

2.3. The Franck and Hertz Energy Loss Experiment

Bohr's atomic theory was aiming at providing a sound explanation for the light emission spectra observed from excited atoms. It was a phenomenological description and the reality of discrete stationary states had to be supported by other independent experimental evidence.

Shortly before the outbreak of World War I, James Franck and Gustav Hertz (Fig. 2.8) started to investigate the interaction of electrons with an atomic gas. The purpose was to challenge Townsend's hypothesis

Fig. 2.8. Gustav Hertz (left) and James Franck (right) received the Nobel Prize in 1925. Lise Meitner (ignored by the Nobel committee despite her seminal work leading to the discovery of nuclear fission) is here at the centre of the photograph. Hertz came back severely wounded from WWI. In the early 1920s, he worked at Philips Incandescent Lamp Factory's physics lab at Eindhoven and after a short period as an academic, he became head of research at Siemens and after WWII he went back to fundamental research in Soviet Union (where he partly worked on isotope separation) and finally in Germany. He was the nephew of Heinrich Hertz who proved the reality of Maxwell's electromagnetic waves.

When the Nazis started to rule Germany, Franck courageously demonstrated against the racial laws and later in protest resigned from the University of Gottingen. He then moved to the USA where he served as the director of Chemistry division in the famous Metallurgical lab in Chicago, under A.H. Compton, as part of the Manhattan project. He tried to oppose the first use of the nuclear weapon on Japanese civilians by encouraging an open demonstration in a desert area. After the war, he shifted to become the head of a photosynthesis research group. Photo credits: reprinted with kind permission from Oliver Hertz.

that, whatever the incoming electrons' energy, there should always be the possibility for it to exchange energy with the atoms. At this time, Bohr's model was not well known throughout the scientific community and atoms were generally considered as the locus of electron oscillations upon outside excitation, following Thomson's mechanism. The Franck and Hertz experimental setup was very similar to Lenard's except that the electrons were not generated by photoelectric effect but by thermoemission from a heated filament. The electrons were accelerated by an adjustable electric potential between the filament and a grid. Therefore, kinetic energy of the electrons could be varied at will and its influence on the current transmitted through the atomic gas was recorded. Figure 2.9 shows a typical result obtained for a vapour of mercury. It first shows that there is a minimum value of the electron kinetic energy required to induce an energy transfer to the atom. The most important aspect is certainly the periodicity in current drop as the accelerating voltage is increased. In the case of mercury, the voltage between two consecutive drops is about 4.9 V. Franck and Hertz interpreted this result as follows: if the electron had insufficient kinetic energy, no energy transfer was possible. Its speed would remain unchanged and the electron would either be transmitted or elastically reflected. However, above a characteristic threshold, the electron is fast enough to provide the atoms with the minimal energy that is required to obtain ionization. As proven later by Tate, it actually takes 10 V, around the value "expected on the basis of Bohr's theory of the atom" [86, 90], to actually trigger ionization. So what is this 4.9 eV energy transfer magic value that Franck and Hertz

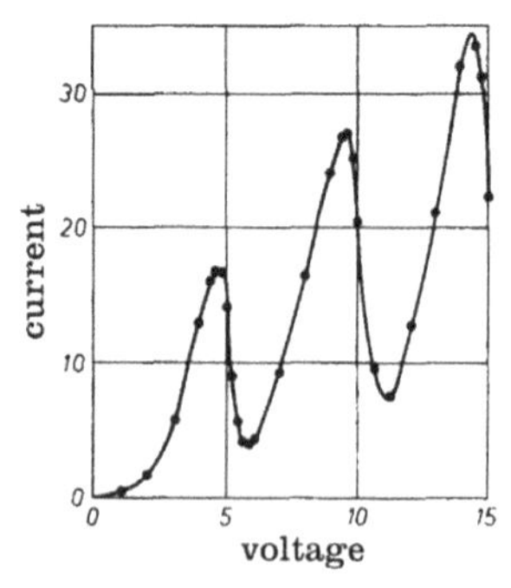

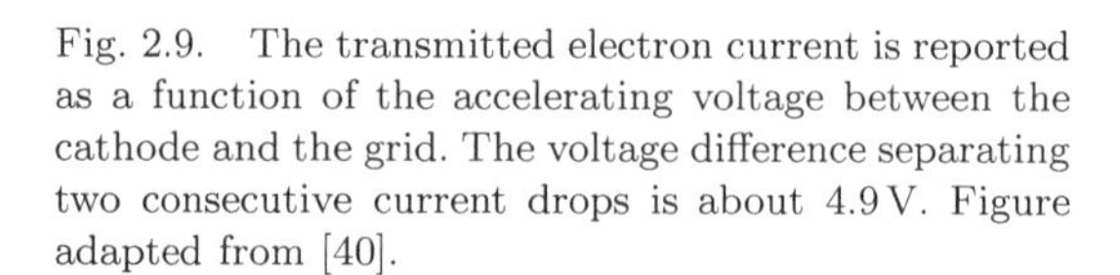
Fig. 2.9. The transmitted electron current is reported as a function of the accelerating voltage between the cathode and the grid. The voltage difference separating two consecutive current drops is about 4.9 V. Figure adapted from [40].

were measuring? It obviously corresponds to a process involving a change in the atoms. A clue comes from the fact that such an energy is associated with a 2.535 Å wavelength and matches one of the electromagnetic emission lines of mercury. As it turned out, this radiation was indeed emitted with augmenting intensity as an increasing number of current drops were observed. It thus became clear that the electrons were transferring their energy to the atoms which would release it by emitting light. As Bohr's model stated it, photons could only be emitted if electrons were changing states from one permitted stationary orbit to another. The conclusion was straightforward[6]: it was clear-cut evidence of an electron collision-induced atomic transition between two orbits. A set of carefully designed and conducted further experiments revealed different transitions and their corresponding emissions of electromagnetic radiation with matching wavelength.

2.4. Davisson and Germer Diffract Matter Particles

Since their discovery in 1895 by Roentgen, X-rays were thought to be the shortest wavelength electromagnetic radiation known to scientists. However, the only confirmation of their wave nature could come from producing interferences when X-rays encounter a periodic grating. After Laue's suggestion,[7] Friedrich and Knipping went onto check that what Haüy had called the "molécules intégrantes" in a crystal were periodically separated by short enough distances to allow for an interference experiment. They thereby performed the first successful X-ray diffraction from a three-dimensional (3D) crystal lattice. The experiment, that earned Laue the Nobel Prize in 1920, was a double proof: it confirmed both Haüy's intuition of a 3D periodic arrangement of atoms in crystals and the wave nature of X-rays. As a by-product, it provided a powerful method for next generation physicists,

[6]But it is unclear why it took Franck and Hertz the duration of the war to acknowledge a connection to Bohr's model.
[7]An interesting account of Laue's intuition is given by Paul Ewald in [29].

Fig. 2.10. Clinton Davisson (1881–1958) and Lester Germer (1896–1971) were awarded the Nobel Prize in 1937. Their work was motivated by a patent suit between their employer, Western Electric (to become Bell labs), and General Electric. Davisson and his assistant Germer were asked to test the effect of shooting positive ions at a hot metal surface and their subsequent emission of electrons. After the settlement (won by General Electric) they decided to use electrons for bombardment and reproduce what Rutherford had observed with alpha particles. It is thus by accident that they confirmed the wave-like behaviour of electrons (see text). Photograph reused with permission of Nokia Corporation.

chemists, biologists and engineers to investigate crystallized materials at the atomic scale with subatomic resolution.[8]

Clinton Davisson and Lester Germer (Fig. 2.10) published an article in 1927 that shook every physicist's view of matter. It is probably best to refer to their own account of this work [21]: "The investigation reported in this paper was begun as the result of an accident which occurred in this laboratory in April 1925. At that time, we were continuing an investigation (...) of the distribution-in-angle of electrons scattered by a target of ordinary (...) nickel. During the course of this work, a liquid–air bottle exploded at a time when the target was at high temperature; the experimental tube was broken, and the target heavily oxidized by the inrushing air". They go on to explain: "When the experiments were continued it was found that the distribution-in-angle of the scattered electrons had been completely changed". Finally, they observe: "This marked alteration in the scattering pattern was traced to a re-crystallization of the target. (...) Before the accident we had been bombarding many small crystals but

[8]See Volume 3 for more applications of X-ray diffraction.

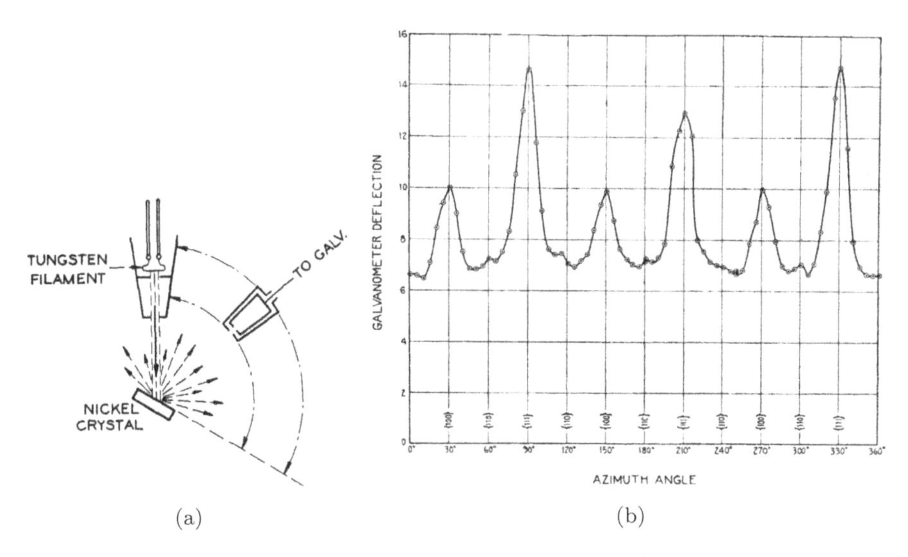

Fig. 2.11. Electron diffraction by a nickel single crystal. (a) Experimental setup. Reprinted with permission from [22]. Copyright © 1928 by Elsevier. (b) Detected intensity of electron flux as a function of the crystal orientation angle with respect to the incident beam. Reprinted with permission from [21]. Copyright © 1927 by the American Physical Society.

in tests subsequent to the accident we were bombarding only a few (order of 10) large ones."

While they could have decided to change the sample for a "fresh" one to proceed with their usual work,[9] their curiosity pushed them to investigate this unexpected "distribution-in-angle" of scattered electrons (see Fig. 2.11). And soon enough, they noticed that "the most striking characteristic of these beams is one-to-one correspondence which (...they) bear to the Laue beams that would be found issuing from the same crystal if the incident beam were a beam of X-rays". A year later, in a conference appropriately titled "Are Electrons Waves?", Davisson rephrased more clearly the pending question: "We have been accustomed to think of the atom as

[9]They were exploring the electronic shell structure of several metallic atoms by applying Bohr's model to the measurements of electron scattering angle dependence.

rather like the solar system — a massive nuclear sun surrounded by planetary electrons moving in closed orbits. On this view the electron which strikes into a metal surface is like a comet plunging into a region rather densely packed with solar systems. (...) The direction taken by such an electron as it leaves the metal should be a matter of private treaty between the electron and the individual atom. One does not see how the neighbouring atoms could have any voice in the matter. (...) Of course, if electrons were waves there would be no difficulty. (...) Is it possible that we are mistaken about electrons? Is it possible that we have been wrong all this time in supposing that they are particles and that actually they are waves?" [22].

As many experiments have shown that electrons have a charge and a mass, which are definitely particle properties, a very legitimate question would then be: "what kind of wave are we talking about?". This will be the starting point of the next chapter.

Since this first electron diffraction by a nickel crystal, numerous other experiments of a similar kind have corroborated the observations of Davisson and Germer for particles of different natures. In particular, interference experiments by a slotted system (Young's experiment) were made with electrons, neutrons, helium nuclei and even C_{60} and C_{70} molecules.

Question 2.3: **Helium atoms seen through Young's slits.** The ability of particles with mass to form interference patterns is not demonstrated only by the diffraction of electrons. Much heavier objects can exhibit behaviour previously thought to belong only to light or sound waves. Let us consider what happens to helium atoms. Helium is an inert gas: it will interact only weakly with the environment and its size prevents it from significantly penetrating material.

1. Figure 2.12 gives a description of the experimental setup as published by Carnal and Mlynek [17]. Helium atoms (an atomic mass number 4) exit a reservoir at temperature T. A first device aims at extracting the beam portion that is transversally coherent. If the mean de Broglie wavelength for these helium atoms is λ_{dB} =0.56 Å, give an estimate of the oven temperature. What is the temperature for $\lambda_{\mathrm{dB}} = 1.03$ Å?
2. What is the angular value for two successive fringes in Fig. 2.13?
3. Is the result on Fig. 2.13 compatible with an interference phenomenon?

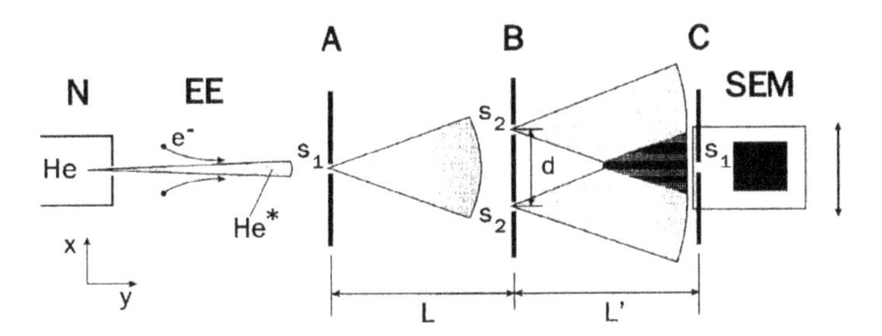

Fig. 2.12. Helium atoms exit from an oven N at temperature T. An electron beam (EE) prepares them in a given excited state allowing for easier detection in the end. A primary 2 μm slit in a gold foil mask (A) extracts the coherent part of the He beam. After a $L = 64$ cm path, the spreading (see also Fig. 3.6) allows for a uniform bombardment of two secondary slits (B) separated by $d = 8\,\mu$m. The detection system (SEM+C), located at $L' = 64$ cm, counts the number of collected helium atoms as a function of the transverse position by 1.88 μm steps. Reprinted with permission from [17]. Copyright © 1991 by the American Physical Society.

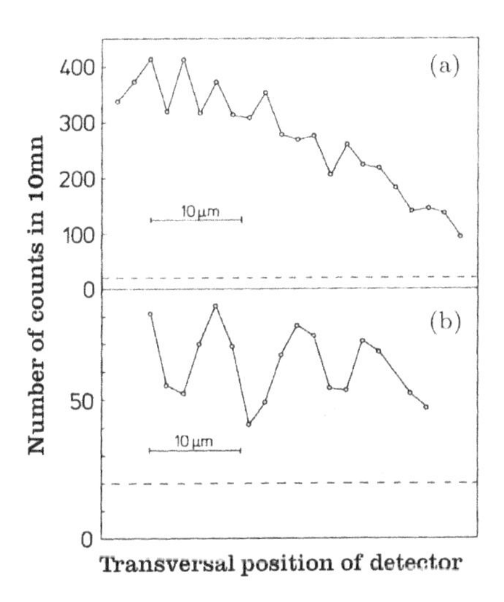

Fig. 2.13. Interference pattern formed by helium atoms after Young's slits interferometer (Fig. 2.4). The two curves correspond to different wavelengths (in (a) $\lambda_{dB} = 0.56$ Å and in (b) $\lambda_{dB} = 1.03$ Å), respective interference fringes are separated by (a) $\Delta x \approx 4.5\,\mu$m and (b) $\Delta x \approx 8.5\,\mu$m. Reprinted with permission from [17]. Copyright © 1991 by the American Physical Society.

Answer:

1. At temperature T, assuming no interaction between particles, the mean kinetic energy is $\frac{p^2}{2m} = \frac{3}{2}k_B T$. Momentum p is derived from wave vector k with $p = \hbar k$ and $k = \frac{2\pi}{\lambda_{\rm dB}}$. We thus obtain

$$T = \frac{(2\pi\hbar)^2}{3mk_B\lambda_{\rm dB}^2}$$

From the Avogadro number and the mass number A, the weight is given by $A \times 10^{-3}/\mathcal{N}_A \approx 6.642 \times 10^{-27}$ kg, we find $T \approx 500$ K for $\lambda_{\rm dB} = 0.56$ Å and $T \approx 150$ K

Answer: (continued)

for $\lambda_{\mathrm{dB}} = 1.03$ Å. The authors of the paper give different values, 295 K and 83 K, respectively. It is thus likely that the atoms selected for the experiment do not move at the mean speed but at one corresponding to a higher value of the Boltzmann probability distribution.

2. The detector is located 64 cm away from the slits. For the different wavelengths, fringes are respectively separated by $\Delta x_{0.56} \approx 4.5\,\mu$m and $\Delta x_{1.03} \approx 8.5\,\mu$m thus an angular separation of $\Delta\theta_{0.56} \approx 7 \times 10^{-6}$ rad and $\Delta\theta_{1.03} \approx 1.33 \times 10^{-5}$ rad.
3. Two interfering waves originating from slits that are separated by d will produce peaks of intensity in directions separated by $d\Delta\theta \approx n\lambda$. We thus respectively obtain $\lambda_{\mathrm{dB}} = 0.56$ Å and $\lambda_{\mathrm{dB}} = 1.03$ Å, $\lambda_{\mathrm{dB}}/\Delta\theta \approx 8\,\mu$m and $\lambda_{\mathrm{dB}}/\Delta\theta \approx 7.7\,\mu$m. The concordance of these results validates that we actually witness an interference effect that is created by the possibility of passing by either slit separated by $8\,\mu$m. The common use of an interference or a diffraction experiment, whatever the particles employed, is to perform a measurement of relative distances between scatterers. Here the scatterers are the two slits in the gold foil.

2.4.1. *Applications of massive particles diffraction*

Electron diffraction is one possible mode of electron microscopy. In contrast with X-rays, it is not difficult to shape very thin intense beams and, when coupled with regular microscopy, it allows for observation of highly localized microscopic structure. The low penetrating power of electrons, however, limits their routine use to thin samples or to surfaces.

In the following example, we will see that today's state-of-the-art experiments rely on a number of effects that were reported in the present chapter. We will later explain in more detail (Volume 3) how the interaction of short wavelength radiation beams (including X-rays, neutrons and electrons) with crystalline solids can be employed to obtain a highly accurate description of atomic positions in materials for electronic devices or in molecules such as those used in drug pills, and even of their electron distributions. However, not all molecules can be crystallized and there is intense ongoing research aiming at a determination of atomic positions in an individual isolated molecule. Few microscopy techniques[10] are currently able to provide such information because the relative distances between the nuclei are in the order of 2 Å. The only reasonable hope comes from the imaging of an interference pattern created by the scattering from the atoms constituting the molecule. Hensley, Yang and Centurion [52] used the electric field

[10]For a description of Scanning Tunnelling Microscope see section 4.9.2. For Atomic Force Microscope see Volume 2.

carried by a 300×10^{-15} s pulse of an intense (2.2×10^{13} W/cm^2) laser beam, with a linear polarization, to orient a CF_3I molecule within a jet. In such a gas phase, the molecules are clearly separated and can be considered totally independent of each other. Prior to its interaction with the molecule, a small portion of the laser beam is extracted to be converted to a higher frequency electromagnetic wave.[11] The setup is designed such that the frequency is tripled. Each photon of this new beam thus carries three times the energy of those in the primary beam: $\varepsilon = 3\hbar\omega$. This is enough to extract electrons by photoelectric effect from a piece of metal. Since the original laser pulse is very short, it is also true for the electron beam (about 2000 electrons are ejected during 500 femtosecond). The electron pulse is then accelerated by an electric field so that the electron kinetic energy reaches 25 keV. The associated de Broglie wavelength is thus $\lambda = \sqrt{h^2/(2m\varepsilon)} \approx 7.8 \times 10^{-2}$ Å and can be scattered from the oriented individual molecules. As in the case of Davisson and Germer's experiment, electrons are diffracted and their resulting angular distribution can be visualized on a fluorescent screen. The image is then recorded by a CCD camera (Fig. 2.14).

Several interference patterns (Fig. 2.15(b)) are recorded, each corresponding to a different molecule orientation (i.e. orientation of the laser polarization) relative to the electron beam. As explained in Volume 3, there

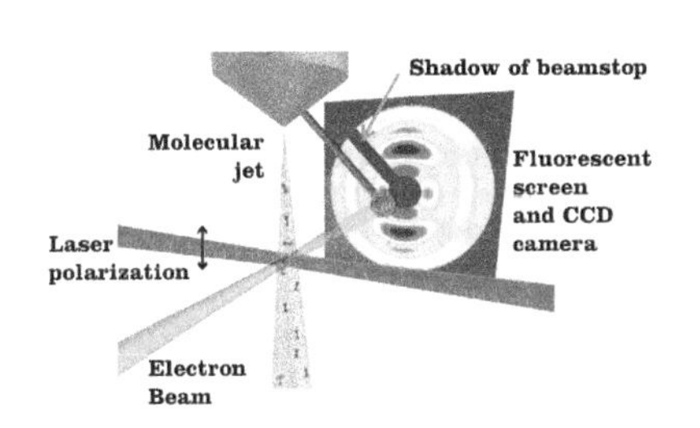

Fig. 2.14. Molecules are ejected from the nozzle at the top. At the crossing region of the photons' (in red) and electrons' (in green) respective beams, the molecules are oriented by the electric field of the laser pulse before they scatter the electron bunch. The interference pattern is recorded on the screen behind. The fluorescence created by the impact is captured by a CCD camera. Reprinted with permission from [52]. Copyright © 2012 by the American Physical Society.

[11]This process is known as a nonlinear optical conversion. It relies on the possibility that the incident wave generates dipole moments in a particular crystal which are proportional to the magnitude of the electric field but also to its square (and even further powers). If the incident wave frequency is ω, the induced electric current oscillates at ω and 2ω (and possibly further multiples) and, as the incident wave propagates in the crystal, there will be a progressive build-up of secondary waves with higher frequencies 2ω, 3ω, ...

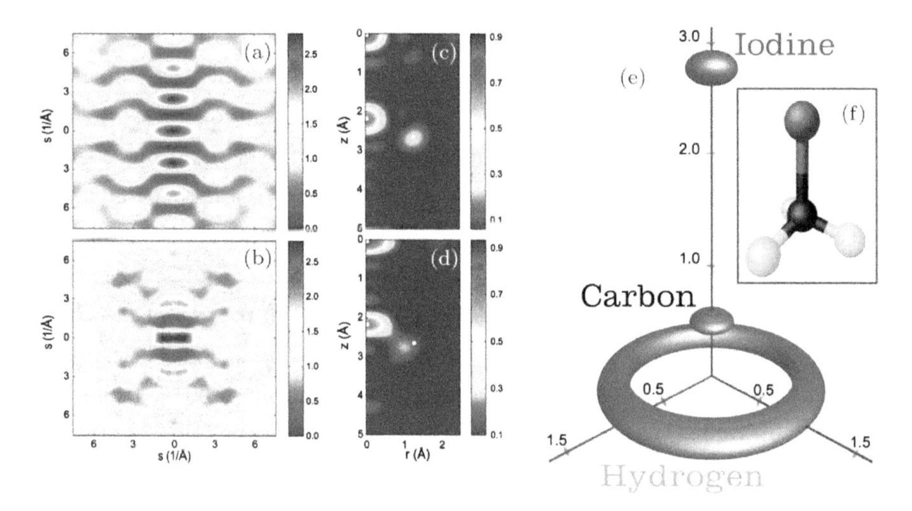

Fig. 2.15. Molecular structure reconstructed from a set of single molecule diffraction patterns. (a) shows the theoretical diffraction image for the atomic distribution displayed in (c). It corresponds to the ball and stick model in (f). The image in (d) is a reconstruction from the experimental interference pattern in (b). It is then possible to infer a global (but cylindrical) molecular structure (e). Reprinted with permission from [52]. Copyright © 2012 by the American Physical Society.

is a well-known mathematical relationship between a diffraction pattern and the shape of the object; Hensley and co-workers could therefore reconstruct a 3D structure of the molecule (Fig. 2.15(d)). Note that, because the only orienting constraint is imposed by the electric field polarization of the laser beam, there still is a rotational degree of freedom along this axis. Hence, molecules in the jet show a random azimuth orientation and the resulting reconstruction inevitably exhibits cylindrical symmetry (Fig. 2.15(e)).

Question 2.4: **Electrons vs. photons**. For a 10-keV energy beam, compare electrons' and photons' resolving power in terms of diffraction spot size.

Answer: The resolving power of radiation on an image is determined by the width of diffraction spots (Rayleigh criterion). Thus, a λ wavelength radiation can correctly distinguish two objects which are approximately λ apart.

A 10-keV X-ray photon beam corresponds to an electromagnetic wave with angular frequency $\omega = \varepsilon/\hbar \approx 10 \times 10^3 \times 1.6 \times 10^{-19}/10^{-34} \approx 1.6 \times 10^{19}\,\mathrm{rad\,s^{-1}}$. The wavelength is $\lambda = 2\pi c/\omega$, so $\lambda_X \approx 18 \times 10^8/(1.6 \times 10^{19}) \approx 10^{-10}\,\mathrm{m} = 1\,\text{Å}$. This radiation is well-adapted to separate out different atoms because it happens to be the approximate distance between neighbouring atoms in a solid, a molecule or a liquid.

Answer: (continued)

When it comes to electrons (or any other particle with a mass), caution is advisable. The wavelength-to-energy relationship is different from the photon case. We first need to compute the momentum, then the wave vector and finally the wavelength:

$$\varepsilon = \frac{p^2}{2m} = \frac{\hbar^2 k^2}{2m} = \frac{1}{\lambda^2}\frac{(2\pi)^2\hbar^2}{2m}$$

We thus obtain:

$$\lambda = 2\pi\hbar\frac{1}{\sqrt{2m\varepsilon}} \approx 6.6\times 10^{-34}\frac{1}{\sqrt{2\times 10^{-30}\times 10\times 10^3\times 1.6\times 10^{-19}}} \approx 10^{-11}\,\text{m}$$

For an identical energy, electrons therefore provide a much better spatial resolution. One great strength of electron microscopy is that it allows high magnification due to the fine details that it can resolve. Note that such a kinetic energy of electrons is obtained by a potential difference of 10 kV. Some electronic microscopes have an accelerating voltage that can commonly reach 500 kV.

Another widely used technique is neutron diffraction. It is the basis for much of our comprehension of the strong magnetism microscopic origin. Coherent inelastic neutron scattering is employed to explain insulators' specific heat and for a better understanding of atomic vibration phenomena in solids (see Volume 3). Neutrons are generally obtained in nuclear reactors during the fission of heavy nuclei (see Volume 2) or as the product of a spallation process in a high energy particle collision. The experiments are therefore conducted in the reactor's immediate environment (tens of meters as in Fig. 2.16).

Question 2.5: **The Kapitza–Dirac effect.** In 1933 Kapitza and Dirac [62] suggested that a diffraction effect of an electron beam from a grating made of standing light waves could be observed. Because it relies on Einstein's stimulated emission (see Volume 3), they warned that the probability of such a phenomenon must be very weak and require a very intense light source to

Fig. 2.16. Inside the neutron reactor at Institut Laue-Langevin in Grenoble (France). The blue light is the Cerenkov radiation created by ultrafast charged particles emitted by nuclear reactions. The purpose of such a reactor is not to produce energy but neutrons. Around it, inside neighbouring buildings, experiments make use of the neutron beams for a variety of scattering experiments, including neutron diffraction. Credits: photo by Jean-Louis Baudet, with kind permission of ILL.

Question 2.5: (continued)

increase the radiation–particle interaction. As a matter of fact, the technical tour de force had to wait for powerful lasers and almost 70 years [41]. Figure 2.17 summarizes the principle of the experiment and its results.

1. The electrons energy is about 380 eV. What is the wavelength λ_e of the incident electron beam?
2. The laser beam has a wavelength $\lambda_\nu = 532\,\text{nm}$. What is the photon energy? To reach a high power, the laser needs to operate with 10 ns pulses, each with 0.2 J. What is the photon flux if the beam is 125 μm in diameter?
3. Justify the presence of the first-order electron diffraction peak at 55 μm using the diffraction angle deviation law $d \sin\theta = n\lambda$, where d is the grating period and λ the wavelength of the incident radiation.

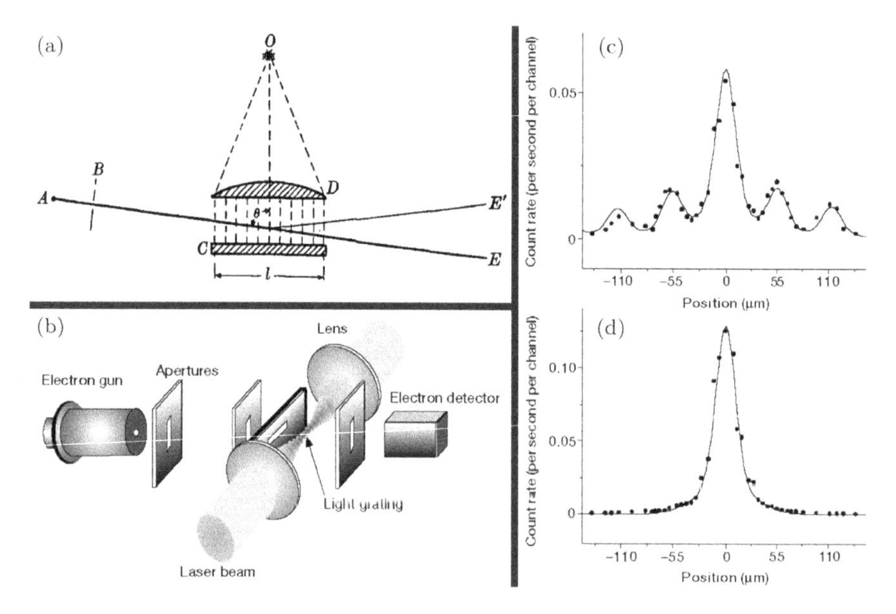

Fig. 2.17. The diffraction of massive particles by electromagnetic standing waves. (a) The principle from Kapitza and Dirac's publication [62]. Reproduced with permission. The electrons are emitted in A and detected in E' when the electromagnetic grating is created between the lens in D and the mirror in C. When the light is turned off, the particles are expected to be detected only in E. (b) The experimental setup by Freimund *et al.* [41]. (c) The interference fringes as seen by the detector 24 cm from the grating. The abscissa refer to the detector's positions. (d) The detected signal when the electromagnetic wave is off. Reprinted with permission from [41]. Copyright © 2001 by Springer Nature.

Answer:

1. The wavelength is $\lambda_e = \frac{2\pi}{k} = \frac{2\pi\hbar}{\sqrt{2m\varepsilon_e}} \approx \frac{6.6\times10^{-34}}{(2\times9\times10^{-31}\times380\times1.6\times10^{-19})^{1/2}} \approx 0.6$ Å.
2. The photon energy is $\varepsilon_\nu = \frac{2\pi\hbar c}{\lambda_\nu} \approx \frac{6.6\times10^{-34}\times3\times10^{8}}{532\times10^{-9}} \approx 3.7\times10^{-19}$ J ≈ 2.3 eV. The laser power density is $\frac{0.2}{10\times10^{-9}\times\pi\times(62.5\times10^{-6})^2} \approx 1.6\times10^{15}$ W/m^2. The photon flux is thus $1.6\times10^{15}/3.7\times10^{-19} \approx 4.4\times10^{33}$ ph/(m^{2}s).
3. The grating is created by the intensity of the radiation. The period is thus half the wavelength $d = \lambda_\nu/2 = 532/2$ nm. Using the grating diffraction law, we find the first-order diffraction peak deviation angle $\theta = \frac{2\times0.6\times10^{-10}}{532\times10^{-9}} \approx 0.226\times10^{-3}$ rad and a first maximum ($n = 1$) for a detector positioned at $0.226\times10^{-3}\times0.24 \approx 54\,\mu$m.

Chapter 2: Nuts & Bolts

- *A matter particle with* ***momentum p*** *and energy* ε *can behave as a wave with frequency* $\nu = \varepsilon/h$ *and* ***wave vector*** $\mathbf{k} = \mathbf{p}/\hbar$.
- *An excited hydrogen atomic gas radiates an* ***electromagnetic spectrum*** *with a set of* ***well-defined wavelengths****. Balmer observed that the visible lines follow the empirical rule* $\frac{1}{\lambda} = R_\infty(\frac{1}{2^2} - \frac{1}{n^2})$ *where* $n \in \mathbb{N}^*$ *is the number of the radiation line and* $R_\infty \approx 1.1\times10^7\ m^{-1}$.

Part II

From Phenomenology to an Axiomatic Formulation of Quantum Physics

Part I described much of the experimental evidence supporting the impression that, contrary to Kelvin's statement, the whole physics story had not been told. Despite its incredible success during the mid-nineteenth century, which led to the industrial and technological massive advances, there were some pending questions. It appeared that the more accurate the measurements were, the higher the electromagnetic frequencies, the lower the temperatures, the lighter and the smaller the objects, the more striking the differences were between experimental results and theoretical expectations based on classical physics models. A paradigm shift had to be conducted: the most disruptive result was probably that particles exhibit wave-like behaviour under particular circumstances. This could not be fitted into Newton's view of mechanics, nor could its equations be distorted to adapt to such a change in the description of matter. This would call for a total upheaval of the physicist's way of describing even the simplest objects.

As Henri Poincaré put it: "Science is built up of facts, as a house is with stones. But a collection of facts is no more a science than a heap of stones is a house." The accumulation of experimental results only pleaded in favour of a change in physical models. It did not provide any. It was a long and winding road to the elaboration of a new proposition that would comply both with the "old" physics and the ever-increasing number of puzzling experimental results. Some daring models were proposed and progressively abandoned when they turned out to only partially embrace experimental reality. Some other models were successful not only at providing explanations for the whole corpus of measurements but also at predicting new effects, as yet not observed, and pointing at fruitful new research areas.

However, it soon appeared that Galileo's statement about the necessity of mastering mathematics because it is the "language of Nature" was reaching an additional degree of truth. Quantum physics looks like a new game that we play with Nature. This new game is based on new rules (written in the language of mathematics) that we need to learn. Playing the new game will often lead to a more intimate acquaintance with the rules which will then unveil their power to interpret and, indeed, predict. Enunciating the rules, known as *the postulates of quantum mechanics*, will be the purpose of Chapter 4 in this second part. However, their mere statement is sometimes seen to lack a direct connection to the experimental results reported in Chapters 1 and 2. Students therefore tend to favour a preliminary qualitative introduction to the matter wave concept that capitalizes on previous knowledge of electromagnetism. It has been generally observed that such an intermediate chapter, using heuristic arguments built upon the similar behaviour of massive particles and photons, would bring more smoothness to the conceptual shock generated by the quantum description of physical behaviours. Hence, the existence of Chapter 3.

3

A Heuristic Approach to Quantum Modelling

The content of this chapter will help you understand ***the lifetime of undetected particles, incompatibilities of position and speed****, and related topics.*

This chapter is a transition. The puzzling experimental results reported in the previous part call for a change in our way of describing how matter behaves and what matter is. Our goal is now to gradually introduce tools and concepts that are new to the physics of massive objects: the wavefunction and Schrödinger equation.

The definition of a wave for particles is something that we have not yet come across. The only dual description that is at our disposal is the awkward case of electromagnetic waves, which we have been familiar with for some time.

It will be seen in the next chapter that the approach proposed here is merely a special (yet very common) case of a general theoretical quantum description.

This chapter has a double goal: while it builds on our familiarity with Maxwell's electromagnetic theory to progressively explain what a particle wave ought to be, it also enriches our physicist's toolbox. The presentation (or reminder) of important concepts in electromagnetism will serve as a means of introducing key mathematical objects such as plane waves, Dirac delta function and others that will prove to be essential in the understanding of many aspects of quantum mechanics.

We are encouraged to consider the similar behaviour of photons and massive particles — in the way they transfer energy and produce interference patterns — to exploit electromagnetic waves as a reference. Their well-known properties will serve as guidelines for inferring a possible mathematical expression for matter waves. Our experience with the dual nature of photons provides hints on what a wave associated with material objects should behave like, what its mathematical representation should be, what kind of equation it should satisfy. The striking resemblance in many aspects encourages us to consider the two types of waves on an equal footing. However, we should keep in mind that photons are known to propagate at light speed c in vacuum. This cannot be the case for massive particles. Therefore, it should be of no surprise that the respective evolution equations exhibit dramatic differences. It is beyond the scope of this book to describe the quantum mechanical treatment of relativistic particles.

3.1. Waves as We Know Them: Let There Be Light

Note: This section can be skipped, at least for the first approach, if the reader is familiar with Maxwell's equations, plane waves, continuity equation, Poynting vectors and much of what constitutes the basics of electromagnetic wave formalism.

We have already made numerous references to electromagnetic waves, but this section will be used to give a few reminders. They will help us to make a transposition to what happens with particles. We will use "light" and "electromagnetic wave" interchangeably on the basis that all our considerations are valid at any frequency (restricted to non-relativistic behaviour of charges).

3.1.1. *The medium*

We know that electromagnetic waves are actually the conjunction of two entities: the electric and the magnetic fields $\boldsymbol{E}$ and $\boldsymbol{B}$. For a given medium, characterized by its dielectric permittivity ϵ, magnetic permeability μ, and its charge and current densities[1] $\rho_q(\boldsymbol{r},t)$ and $\boldsymbol{j}_q(\boldsymbol{r},t)$, compatible fields are given by Maxwell's equations:

$$\vec{\nabla}_{\boldsymbol{r}} \cdot \boldsymbol{B}(\boldsymbol{r},t) = 0 \tag{3.1}$$

$$\vec{\nabla}_{\boldsymbol{r}} \times \boldsymbol{E}(\boldsymbol{r},t) = -\frac{\partial \boldsymbol{B}(\boldsymbol{r},t)}{\partial t} \tag{3.2}$$

$$\vec{\nabla}_{\boldsymbol{r}} \cdot \boldsymbol{E}(\boldsymbol{r},t) = \frac{\rho_q(\boldsymbol{r},t)}{\epsilon} \tag{3.3}$$

$$\vec{\nabla}_{\boldsymbol{r}} \times \boldsymbol{B}(\boldsymbol{r},t) = \mu\left(\boldsymbol{j}_q(\boldsymbol{r},t) + \epsilon\frac{\partial \boldsymbol{E}(\boldsymbol{r},t)}{\partial t}\right) \tag{3.4}$$

At this stage, two points are essential:

- The first two equations do not explicitly refer to any charge or current. The introduction of the vector and scalar potentials, $\boldsymbol{A}(\boldsymbol{r},t)$ and $\Phi(\boldsymbol{r},t)$, is motivated by the possibility of constraining the electric and magnetic fields to a particular form in order to automatically fulfil these two equations. As a matter of fact, from the definition of curl and divergence, the first Maxwell equation implies that there are functions $\boldsymbol{A}(\boldsymbol{r},t)$, named "vector potentials", such that $\boldsymbol{B}(\boldsymbol{r},t) = \vec{\nabla}_{\boldsymbol{r}} \times \boldsymbol{A}(\boldsymbol{r},t)$. Therefore, it becomes possible to forget about $\boldsymbol{B}$ and work only with $\boldsymbol{A}$ so that the first condition is automatically fulfilled. There is a consequence to such a choice: the second Maxwell equation relates the electric field to the vector potential by $\vec{\nabla}_{\boldsymbol{r}} \times \boldsymbol{E}(\boldsymbol{r},t) = -\frac{\partial}{\partial t}\vec{\nabla}_{\boldsymbol{r}} \times \boldsymbol{A}(\boldsymbol{r},t)$ so that the electric field is not determined in a unique manner from a given $\boldsymbol{A}(\boldsymbol{r},t)$. It is

[1] The densities bear a "q" index to specify that what we are dealing with here are the density and current of charges, i.e. the charge or current value per unit volume. Later, we will encounter probability density and probability current density, and will omit the index because they will not necessarily concern charged particles.

always possible to choose any derivable function $\Phi(\boldsymbol{r}, t)$ in order to find the correct electric field: $\boldsymbol{E}(\boldsymbol{r}, t) = -\frac{\partial}{\partial t}\boldsymbol{A}(\boldsymbol{r}, t) + \vec{\nabla}_{\boldsymbol{r}} \cdot \Phi(\boldsymbol{r}, t)$.

- Taking the divergence of the last Maxwell equation yields

$$\vec{\nabla}_{\boldsymbol{r}} \cdot (\vec{\nabla}_{\boldsymbol{r}} \times \boldsymbol{B}(\boldsymbol{r}, t)) = \mu \left(\vec{\nabla}_{\boldsymbol{r}} \cdot \boldsymbol{j}_q(\boldsymbol{r}, t) + \epsilon \vec{\nabla}_{\boldsymbol{r}} \cdot \frac{\partial \boldsymbol{E}(\boldsymbol{r}, t)}{\partial t} \right)$$

The left side is zero by the construction of the differential operators. Permuting the time derivative and the divergence for the electric field on the right-hand side and exploiting Maxwell's third equation leaves us with

$$\vec{\nabla}_{\boldsymbol{r}} \cdot \boldsymbol{j}_q(\boldsymbol{r}, t) + \frac{\partial \rho_q(\boldsymbol{r}, t)}{\partial t} = 0 \tag{3.5}$$

This equation expresses a property which is not related to wave fields but will be useful later. The divergence of a quantity expresses its change between one point and its immediate neighbourhood. Hence, this equation states that, if the electric current is found to exhibit a spatial change, it has to be associated with a time variation of the charge density at this location. It is a **continuity (or conservation) equation** which is understood in terms of fluid mechanics: if the flow changes between two points, there is necessarily a time variation of the particle density in that middle region. We will soon encounter a similar relationship for the flow of energy: if there is a change in energy flux between the entrance and the exit of an elementary volume, it is to be associated with a time variation of the energy stored in the volume.

The electromagnetic wave needs a propagation equation for a given medium. It is readily obtained if we take the curl of the second Maxwell equation. Using $\vec{\nabla}_{\boldsymbol{r}} \times (\vec{\nabla}_{\boldsymbol{r}} \times \boldsymbol{E}(\boldsymbol{r}, t)) = \vec{\nabla}_{\boldsymbol{r}}(\vec{\nabla}_{\boldsymbol{r}}.\boldsymbol{E}(\boldsymbol{r}, t)) - \nabla^2_{\boldsymbol{r}}\boldsymbol{E}(\boldsymbol{r}, t)$, we get:

$$\nabla^2_{\boldsymbol{r}}\boldsymbol{E}(\boldsymbol{r}, t) - \epsilon\mu \frac{\partial^2 \boldsymbol{E}(\boldsymbol{r}, t)}{\partial t^2} = \frac{1}{\epsilon}\vec{\nabla}_{\boldsymbol{r}}\rho_q(\boldsymbol{r}, t) + \mu \frac{\partial \boldsymbol{j}_q(\boldsymbol{r}, t)}{\partial t} \tag{3.6}$$

This is a partial differential equation. Finding the solutions of such an equation is not always easy. Many methods have been developed but none of them is universal.

In most cases, no charge or current is to be accounted for in the propagating medium and the right-hand side is zero. Before seeking solutions for this equation, let us take a brief look at the energetic aspect which will be of paramount importance in the understanding of many facets of wave-based quantum mechanics.

3.1.2. *The energy*

The energy carried by an electromagnetic wave cannot be measured unless there is an interaction with matter at some point. The fields will act on the charges in an elementary volume through the Lorentz force: $\delta\boldsymbol{F} = \sum_i \delta q_i(\boldsymbol{E} + \boldsymbol{v}_i \times \boldsymbol{B})$. The displacement of charges in this volume is induced by the work done on them $\delta W = \sum_i \delta q_i \boldsymbol{r}_i.(\boldsymbol{E} + \boldsymbol{v}_i \times \boldsymbol{B})$. Therefore, the electromagnetic wave transfers some energy to the volume element of matter δV at a rate given by $\frac{\partial \delta W}{\partial t} = \sum_i \delta q_i \boldsymbol{v}_i.\boldsymbol{E}$, where we made use of $\boldsymbol{v}_i \cdot (\boldsymbol{v}_i \times \boldsymbol{B}) = 0$. Hence, since the charge current is $\boldsymbol{j}_q = \sum_i \frac{\delta q_i}{\delta V}\boldsymbol{v}_i$, denoting $w = \delta W/\delta V$ the energy density, the rate of exchange between light and matter is

$$\frac{\partial w}{\partial t} = \boldsymbol{j}_q.\boldsymbol{E} = \frac{1}{\mu}\left(\vec{\nabla}_{\boldsymbol{r}} \times \boldsymbol{B} - \epsilon\mu\frac{\partial \boldsymbol{E}}{\partial t}\right) \cdot \boldsymbol{E}$$

where, for simplicity, we have omitted to write the explicit dependence on space and time, and used the last Maxwell equation to express the current density. By making use of $\boldsymbol{E}.(\vec{\nabla}_{\boldsymbol{r}} \times \boldsymbol{B}) = \boldsymbol{B}.(\vec{\nabla}_{\boldsymbol{r}} \times \boldsymbol{E}) - \vec{\nabla}_{\boldsymbol{r}} \cdot (\boldsymbol{E} \times \boldsymbol{B})$, the rate of energy density exchange becomes

$$\frac{\partial w}{\partial t} = -\frac{1}{\mu}\vec{\nabla}_{\boldsymbol{r}} \cdot (\boldsymbol{E} \times \boldsymbol{B}) - \frac{1}{2\mu}\frac{\partial |B|^2}{\partial t} - \frac{\epsilon}{2}\frac{\partial |E|^2}{\partial t}$$

where the second Maxwell equation was used. This last result can be explained if we express it in the form of a conservation equation similar to (3.5):

$$\frac{1}{\mu}\vec{\nabla}_{\boldsymbol{r}} \cdot (\boldsymbol{E} \times \boldsymbol{B}) = -\frac{\partial}{\partial t}\left(w + \frac{1}{2\mu}|B|^2 + \frac{\epsilon}{2}|E|^2\right) \tag{3.7}$$

The right-hand side corresponds to the rate of change for energy density in matter (first term) and the electromagnetic field (two other terms). Even if there is no matter involved, i.e. no energy transfer to and from the charges ($w = 0$), a change in magnetic and electric energy densities is related to a divergence of the energy density flux represented by the left-hand side part of the equation. We will thus define the Poynting vector as $\mathbf{\Pi} = \frac{1}{\mu}\boldsymbol{E} \times \boldsymbol{B}$. It represents the flux of energy density carried by the electromagnetic field, i.e. an energy current density.

The detection of an electromagnetic signal always involves an energy transfer. Depending on the detector type, the output signal is thus a voltage value (most of the time), a quantity of charges, a colour change (or chemical reaction) or a phase transition (bubbles in particle chambers). All these processes require that the detecting device removes a minimum portion of energy from the incoming Poynting vector. The end result is that one never really measures the electric or magnetic field of light but only, at best,[2] a number of incoming photons.

Where photons are detected, energy is transferred from the light beam to the detector. A photographic picture can be considered as the recording of a photo-detection process. Whether it is a film or a digital photograph (Fig. 3.1), photon energy is converted either to trigger a chemical process

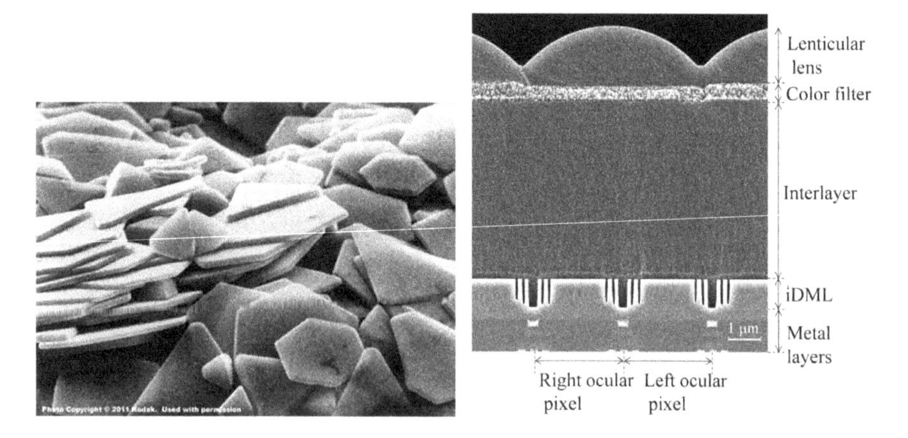

Fig. 3.1. Left: Electron microscopy image of silver halide microcrystals for film photography (used with permission from Eastman Kodak Company). Right: Scanning electron microscopy cut view image of CMOS image sensor. The top microlens directs the light beam towards a detection chip below. Reprinted with kind permission from [64]. Copyright © 2016 by the Optical Society of America.

[2]All photons are not necessarily counted unless the detection yield is perfect.

(in silver iodide particles) or to displace electrons (to a different part of a capacitor). Assuming uniform sensitivity of the detector, a picture can be seen as nothing but the measurement's graphical representation of the spatial photon density distribution, i.e. the number of photons collected per unit area of the detector's surface.

3.1.3. *The waves*

Ubiquitous plane waves

Only the energy of electromagnetic fields is detected. The propagation process nevertheless necessitates the knowledge of mathematical forms of the electric and magnetic fields. The generation process of an electromagnetic wave is not our immediate concern and we will just ignore it for the moment. We consider that the wave has already been emitted and is now propagating in an infinite medium with no charge or current. The corresponding evolution equation is, from (3.6),

$$\nabla_{\boldsymbol{r}}^2 \boldsymbol{E}(\boldsymbol{r},t) - \epsilon\mu \frac{\partial^2 \boldsymbol{E}(\boldsymbol{r},t)}{\partial t^2} = 0 \tag{3.8}$$

Equation (3.8) is a very particular (yet very common) case of partial differential equations. First, it is vectorial. But as there is no mixing between the components, we can decide to project the equation on each cartesian axis. Doing so, we end up with three different problems, each relative to a single dimension of space:

$$\nabla_{\boldsymbol{r}}^2 E_j(\boldsymbol{r},t) - \epsilon\mu \frac{\partial^2 E_j(\boldsymbol{r},t)}{\partial t^2} = 0$$

where $j = x, y, z$. In the remainder of this book, we will separate $\epsilon\mu$ into the vacuum contribution $\epsilon_0\mu_0$ and the medium specific additional dielectric constant ϵ_r and magnetic relative permittivity μ_r. For further convenience, we set the light speed in vacuum $c = 1/\sqrt{\epsilon_0\mu_0}$ and the refractive index $n = \sqrt{\epsilon_r\mu_r}$ so that $\epsilon\mu = n^2/c^2$.

Now, it can be seen that the first term of the propagation equation only operates on the position variables while the second only shows a differentiation with respect to time. This encourages us to seek a solution under the form of a product: $f(\boldsymbol{r})g(t)$. For obvious reasons this trick is called "separation of variables". So far, we have no guarantee that such a solution is correct, that it satisfies the physical constraints, nor that it provides all possible solutions. First, let us look for what kind of solutions this will bring us (if any). We will take care of these details once we are convinced that it is worth the pain.

If we make the substitution in (3.8), it gives

$$g(t)\nabla_{\boldsymbol{r}}^2 f(\boldsymbol{r}) - \frac{n^2}{c^2} f(\boldsymbol{r}) \frac{d^2 g(t)}{dt^2} = 0$$

Wherever none of the two functions is zero, we can divide by $f(\boldsymbol{r})g(t)$ and obtain independent contributions:

$$\frac{1}{f(\boldsymbol{r})}\nabla_{\boldsymbol{r}}^2 f(\boldsymbol{r}) = \frac{n^2}{c^2}\frac{1}{g(t)}\frac{d^2 g(t)}{dt^2}$$

This equality must hold for any position $\boldsymbol{r}$ and time t. As space and time are independent, we are to conclude that both sides must equal a constant. Let us call it $-k^2$ so that it is negative, unless k is complex, and

$$\frac{1}{f(\boldsymbol{r})}\nabla_{\boldsymbol{r}}^2 f(\boldsymbol{r}) = -k^2$$

$$\frac{n^2}{c^2}\frac{1}{g(t)}\frac{d^2 g(t)}{dt^2} = -k^2$$

Each equation has an individual solution that depends on the separation constant $-k^2$.

Let us focus on the time component. The solution comes naturally as

$$g(t) = g_1 e^{i\frac{kc}{n}t} + g_2 e^{-i\frac{kc}{n}t}$$

where g_1 and g_2 are complex constant numbers that depend on the constraints on the field at $t = 0$. This function gives the time evolution of the field. It depends solely on $k/\sqrt{\epsilon\mu}$ which can be interpreted as an angular frequency $\omega_k = kc/n$.

We can now turn to the spatial equation. We will assume that cartesian coordinates are well adapted to the symmetry of the boundary conditions. While the resolution is expected to be very similar to that of the time component, we should keep in mind that we are dealing with a three-dimensional (3D) Laplacian and the spatial equation should thus read

$$\frac{\partial^2}{\partial x^2} f(\boldsymbol{r}) + \frac{\partial^2}{\partial y^2} f(\boldsymbol{r}) + \frac{\partial^2}{\partial z^2} f(\boldsymbol{r}) = -k^2 f(\boldsymbol{r})$$

Again, the method of separation of variables calls for a solution under the form $f(\boldsymbol{r}) = X(x)Y(y)Z(z)$ so that

$$\frac{1}{X(x)} \frac{d^2}{dx^2} X(x) + \frac{1}{Y(y)} \frac{d^2}{dy^2} Y(y) + \frac{1}{Z(z)} \frac{d^2}{dz^2} Z(z) = -k^2$$

wherever none of the functions goes to zero. This equation can also be split into three equations. Setting $k^2 = k_x^2 + k_y^2 + k_z^2$, we are led to solve, for example,

$$\frac{d^2}{dx^2} X(x) = -k_x^2 X(x)$$

the other two equations having the exact same form. Once again, the solution is straightforward: $X(x) = X_1 e^{ik_x x} + X_2 e^{-ik_x x}$. The constants are given by the boundary conditions.

It is now possible to be at ease with the ansatz chosen to separate out the x, y, z and t variables. The product $X(x)Y(y)Z(z)f(t)$ is not the most general solution to the initial second order differential equation. Nevertheless, because (3.8) is a linear homogeneous equation, the most general solution can be expressed as a linear combination of all the possible products.[3]

To further proceed, and without any loss of generality, let us make the choice of simplicity and consider a simple one-dimensional (1D) problem along the x-axis and only one term in $X(x)g(t)$ such as $e^{-i(k_x x - \omega_k t)}$. We see that after a delay δt at the same location x, the phase factor has changed to $e^{-i(k_x x - \omega_k t - \omega_k \delta t)}$. But if we also move the position by δx, it now becomes $e^{-i(k_x x + k_x \delta x - \omega_k t - \omega_k \delta t)}$. The phase factor keeps the same value if the two shifts are such that $k_x \delta x = \omega_k \delta t$. We thereby obtain the velocity at which a constant value of the phase propagates along x. It is called a "phase velocity": $v_\varphi = \omega_k / k_x$. There is of course a similar expression along every space axis. If k_x and ω_k are two positive numbers, then, as time elapses, i.e. t increases, the positions at which $e^{-i(k_x x - \omega_k t)}$ remains constant correspond to an increasing value for x. It thus represents a wave propagating in

[3] For a more detailed treatment see [4] and [28].

the positive direction on the x-axis. This is also true for $e^{+i(k_x x-\omega_k t)}$. The propagation is identical and the two functions only give complex conjugate values for the phase factor. Their sum would just represent the propagation of a real value quantity. This is the case for an electric (or magnetic) field: $E_x(x,t) = E_x(0)\cos(k_x x - \omega_k t)$. Similar arguments will lead us to identify $E_x(x,t) = E_x(0)\cos(k_x x + \omega_k t)$ with a wave propagating in the opposite direction.

Going back to the 3D case, an expression such as

$$\boldsymbol{E}(\boldsymbol{r},t) = \boldsymbol{E}_0 e^{i\boldsymbol{k}\cdot\boldsymbol{r}-i\omega_k t} \tag{3.9}$$

is the complex representation of a wave propagating in the direction of wave vector $\boldsymbol{k}$. A wave front is the location of the points in space that share, at the same time, the same field value. It is a simple matter to show that for waves of the form described by (3.9), the wave front is a plane perpendicular to $\boldsymbol{k}$. This is the reason why they are called "plane waves". They are characterized by the fact that their shape is a function of a unique wave vector.

We can reach another level of intimacy with these plane waves if we apply the second Maxwell equation to them:

$$i\boldsymbol{k} \times \boldsymbol{E}(\boldsymbol{r},t) = i\omega_k \boldsymbol{B}(\boldsymbol{r},t) \tag{3.10}$$

or the last equation:

$$i\boldsymbol{k} \times \boldsymbol{B}(\boldsymbol{r},t) = i\omega_k \frac{n^2}{c^2}\boldsymbol{E}(\boldsymbol{r},t) \tag{3.11}$$

This means that the electric field, the magnetic field and the wave vector are mutually orthogonal. The electric and magnetic fields are thus transverse to the direction of propagation.

Monochromatic plane waves are uniquely characterized by one wave vector $\boldsymbol{k}$ and one angular frequency ω_k. Therefore, according to Planck–Einstein–de Broglie's formulation, they should very naturally represent a photon of momentum $\boldsymbol{p} = \hbar\boldsymbol{k}$ and energy $\varepsilon = \hbar\omega_k$. A given flow of identical photons with celerity c represents a particular flux of electromagnetic energy density which, as we are now well aware, is represented by a Poynting vector. So, what is the Poynting vector for a particular plane wave? If we refer to expression (3.7), we find for volume V in vacuum the total energy $\int_V (\frac{1}{2\mu_0}B_0^2 + \frac{\epsilon_0}{2}E_0^2)d^3\boldsymbol{r}$. But it is possible to use expression (3.10) or

(3.11) to write the magnetic field amplitude as a function of the electric field only, $B_0 = E_0/c$ so that the total energy in the electromagnetic wave is $\int_V \epsilon_0 E_0^2 d^3\boldsymbol{r}$. Since the electric field amplitude E_0 is a constant for a plane wave, this last expression clearly goes to infinity if the integral is over the infinite space volume. It is very unfortunate that the simplest solution to the propagation equation we could come up with does not hold water when it comes to examining its physical content! We can rephrase this result by stating that legitimate wave expressions must be square-integrable, so that there is an upper limit to the total amount of energy available from the field if the integration volume is infinite. A particular condition (necessary but not sufficient) is that such functions belonging to the set of square-integrable functions, $\mathcal{L}_2$, all have a modulus that cancels at infinity. Of course, if the volume is restricted to a finite value, it means that the wave field cancels outside and it thereby satisfies the above condition for square-integrability.

It becomes apparent why, their usefulness notwithstanding, plane waves are not suitable expressions for real physical propagating properties.

Do plane waves really exist? Well, plane waves are natural solutions of the propagation equation in a free medium with no charge, no current and no variation in the dielectric constant or permittivity. One such wave therefore represents a property (electric or magnetic field here) that is constant everywhere in an infinite medium. Of course, such a medium does not exist. Plane waves are thus simply considered as a handy tool to describe what might happen locally if it were possible to forget about boundary conditions.

Question 3.1: **Photons and Poynting.** Consider a 1 mm^2 green laser beam from a pointer ($\lambda \approx 532$ nm) with a mean power of 5 mW. What is the photon flux? What is the intensity of the electric field? What is the magnetic field?

Answer: The angular frequency is $\omega = 2\pi c/\lambda \approx 3.6 \times 10^{15}$ rad/s. The photon number per second and unit of surface is $5 \times 10^{-3}/(10^{-6}\hbar\omega) \approx 1.4 \times 10^{22}$ ph/(s.m^2). The Poynting vector represents the flux of the energy density. Thus, it is the power passing through a unit surface. The given power is a time average. Here, it translates into a time average Poynting vector: $\langle \Pi \rangle = 5 \times 10^{-3}/10^{-6} = 5000$ W/m^2. Approximating the beam locally with plane waves, $E = E_0 \cos(kx - \omega t)$, we find the instantaneous Poynting vector: $\Pi = c\epsilon_0 E^2$. The time average of the Poynting vector is related to the magnitude of the electric field by $\langle \Pi \rangle = c\epsilon_0 E^2/2$ and we get:

$$E = \sqrt{\frac{2\langle \Pi \rangle}{c\epsilon_0}} = 1941 \text{ V/m}$$

and the magnetic field is readily obtained: $B = E/c = 6.5 \times 10^{-6}$ T.

Waves, not so plane

Plane waves are not legitimate mathematical forms to account for the actual propagation of a physical property, but natural solutions to the propagation equation. However, we may well ask what the right functions for the propagation equation are. Since the propagation equation (and Maxwell's equations) is linear, we can consider constructing a solution as a linear combination of plane waves:

$$\boldsymbol{E}(\boldsymbol{r},t) = \sum_{\boldsymbol{k}} \boldsymbol{E}_{\boldsymbol{k}} e^{i(\boldsymbol{k}\cdot\boldsymbol{r}-\omega_k t)} \tag{3.12}$$

where $\boldsymbol{E}_{\boldsymbol{k}}$ is the weight (or amplitude) of a particular plane wave contribution to the total field. To generalize, we can make it a continuous sum

$$\boldsymbol{E}(\boldsymbol{r},t) = \int \widetilde{\boldsymbol{E}}(\boldsymbol{k}) e^{i(\boldsymbol{k}\cdot\boldsymbol{r}-\omega_k t)} d^3\boldsymbol{k} \tag{3.13}$$

If the integral spans all the wave vector space, it is of course just a way of expressing the wave as the Fourier transform of some function $\widetilde{\boldsymbol{E}}(\boldsymbol{k})$. The weight is now represented by a distribution of possible $\boldsymbol{k}$. A broad spectrum wave field would be obtained by a wide distribution while an almost monochromatic and highly collimated beam would require a very narrow distribution $\widetilde{\boldsymbol{E}}(\boldsymbol{k})$.

By means of a superposition of plane waves, it is possible to arrange the respective weights so that the resultant field becomes rather limited in space. Let us assume that the weights are dominated by terms around $\boldsymbol{k}_0$. The sum will be dominated by terms $\widetilde{\boldsymbol{E}}(\boldsymbol{k}_0)$ while the others will play a minor role. Therefore, it is legitimate to limit the wave vector dependence of angular frequency to the first order in a Taylor expansion $\omega_{\boldsymbol{k}} \approx \omega_{\boldsymbol{k}_0} + \vec{\nabla}_k \omega_{\boldsymbol{k}_0} \cdot \boldsymbol{k}$. The linearization of the phase yields

$$\boldsymbol{E}(\boldsymbol{r},t) \approx e^{i(\boldsymbol{k}_0\cdot\boldsymbol{r}-\omega_{k_0} t)} \int \widetilde{\boldsymbol{E}}(\boldsymbol{k}) e^{i\boldsymbol{k}\cdot(\boldsymbol{r}-\vec{\nabla}_k \omega_{k_0} t)} d^3\boldsymbol{k} \tag{3.14}$$

Thus, the field appears as a packet of waves all moving at the same velocity named *group velocity* $\vec{\nabla}_k \omega_{k_0}$. There still is a phase displacement because the packet is multiplied by an overall phase factor given by the dominant wave vector. However, the packet and the phase do not necessarily move at the same speed. The phase velocity is given by $\boldsymbol{v}_\varphi = \frac{\omega_{k_0}}{k_0^2}\boldsymbol{k}_0$.

If the general expression of a legitimate wave is given by (3.13), how can we express a special distribution that yields a perfectly (albeit unrealistic) monochromatic and collimated beam? There is a mathematical object, usually noted $\delta(\boldsymbol{k}-\boldsymbol{k}_0)$, which serves that very purpose. Physicists often call it the "Dirac delta function" while mathematicians prefer to consider such an object within the Theory of Distributions framework. Nevertheless, for us, the Dirac function will represent just the asymptotic form of an extremely narrow distribution[4] with the choice:

$$\int \delta(\boldsymbol{k}-\boldsymbol{k}_0)d^3\boldsymbol{k} = 1$$

so that

$$\boldsymbol{E}(\boldsymbol{r},t) = \int \widetilde{\boldsymbol{E}}(\boldsymbol{k})\delta(\boldsymbol{k}-\boldsymbol{k}_0)e^{i(\boldsymbol{k}\cdot\boldsymbol{r}-\omega_k t)}d^3\boldsymbol{k} = \widetilde{\boldsymbol{E}}(\boldsymbol{k}_0)e^{i(\boldsymbol{k}_0\cdot\boldsymbol{r}-\omega_{\boldsymbol{k}_0} t)} \tag{3.15}$$

It transpires that the Dirac delta function has a wider range of applications than merely selecting one particular plane wave out of a continuous superposition. A more general property, which we will intensively use in quantum physics, is that in the immediate vicinity of a given point x_0, for any "well-behaved" function[5] $f(x)$,

$$\int_{-\infty}^{+\infty} f(x)\delta(x-x_0)dx = f(x_0) \tag{3.16}$$

In our present case, one can thus interpret $\delta(\boldsymbol{k}-\boldsymbol{k}_0)$ as a convenient and compact way to actually write the product: $\delta(k_x-k_{0x})\delta(k_y-k_{0y})\delta(k_z-k_{0z})$.

Note that writing

$$\int \delta(\boldsymbol{k}-\boldsymbol{k}_0)e^{i\boldsymbol{k}\cdot\boldsymbol{r}}d^3\boldsymbol{k} = e^{i\boldsymbol{k}_0\cdot\boldsymbol{r}} \tag{3.17}$$

[4]More details on the Dirac function, including some of its asymptotic forms, are given in Volume 3.

[5]It is probably a characteristic of physicists to ban "bad-behaved" functions from their toolbox. But there seems to be very few exceptions to the implicit statement that Nature abhors discontinuity.

implies, by inverse Fourier transform, that the Dirac delta function can be expressed as

$$\delta(\boldsymbol{k}-\boldsymbol{k}_0) = \frac{1}{(2\pi)^3}\int e^{i(\boldsymbol{k}-\boldsymbol{k}_0)\cdot\boldsymbol{r}} d^3\boldsymbol{r} \tag{3.18}$$

Question 3.2: **Units of the Dirac delta function.**

1. Show that if α is a constant,

$$\delta(\alpha\boldsymbol{q}) = \frac{1}{\alpha^3}\delta(\boldsymbol{q})$$

2. What is the unit of $\delta(\boldsymbol{q})$?

Answer:

1. We use (3.18), changing variables so that $\alpha\boldsymbol{q} = \boldsymbol{k}-\boldsymbol{k}_0$:

$$\delta(\alpha\boldsymbol{q}) = \frac{1}{(2\pi)^3}\int e^{i\alpha\boldsymbol{q}\cdot\boldsymbol{r}} d^3\boldsymbol{r} = \frac{1}{(2\pi)^3}\int e^{i\boldsymbol{q}\cdot\boldsymbol{s}}\frac{1}{\alpha^3} d^3\boldsymbol{s} = \frac{1}{\alpha^3}\delta(\boldsymbol{q})$$

where the change of variable is $\boldsymbol{s} = \alpha\boldsymbol{r}$ and thus, for three dimensions, the infinitesimal volume becomes $d^3\boldsymbol{s} = \alpha^3 d^3\boldsymbol{r}$.

2. Without making any reference to wave vectors or lengths, according to (3.17), $\delta(x)$ is to have the inverse unit of dx. The Dirac delta function is expressed in the inverse units of its argument. We should bear in mind that the argument is never really a vector, even if we often write $\delta(\boldsymbol{k})$. It is always a short notation for the product $\delta(k_x)\delta(k_y)\delta(k_z)$. Since each Dirac delta function possesses the unit of $1/k$, $\delta(\boldsymbol{k})$ should be expressed in the unit of $1/k^3$. This is indeed what we see from the result of question 1.

This property is very convenient as we will now see when computing the electromagnetic energy carried by an electromagnetic wave packet. Let us first note that inserting (3.13) (and the similar expression for the magnetic field) into Maxwell's second equation gives

$$\vec{\nabla}_{\boldsymbol{r}} \times \int \widetilde{\boldsymbol{E}}(\boldsymbol{k}) e^{i(\boldsymbol{k}\cdot\boldsymbol{r}-\omega_k t)} d^3\boldsymbol{k} = -\frac{\partial}{\partial t}\int \widetilde{\boldsymbol{B}}(\boldsymbol{k}) e^{i(\boldsymbol{k}\cdot\boldsymbol{r}-\omega_k t)} d^3\boldsymbol{k}$$

Both differential operators are linear, thus

$$\int \boldsymbol{k}\times\widetilde{\boldsymbol{E}}(\boldsymbol{k}) e^{i(\boldsymbol{k}\cdot\boldsymbol{r}-\omega_k t)} d^3\boldsymbol{k} = \int \omega_k \widetilde{\boldsymbol{B}}(\boldsymbol{k}) e^{i(\boldsymbol{k}\cdot\boldsymbol{r}-\omega_k t)} d^3\boldsymbol{k}$$

which should hold everywhere, at any moment, and therefore yields again

$$\boldsymbol{k}\times\widetilde{\boldsymbol{E}}(\boldsymbol{k}) = \omega_k\widetilde{\boldsymbol{B}}(\boldsymbol{k}) \tag{3.19}$$

The total electromagnetic energy in vacuum calls for the evaluation of

$$\mathcal{E} = \int \frac{1}{2\mu_0} |B|^2 d^3\boldsymbol{r} + \int \frac{\epsilon_0}{2} |E|^2 d^3\boldsymbol{r}$$

or, again,

$$\mathcal{E} = \int \epsilon_0 |E|^2 d^3\boldsymbol{r} = \epsilon_0 \int \left| \int \widetilde{\boldsymbol{E}}(\boldsymbol{k}) e^{i(\boldsymbol{k}\cdot\boldsymbol{r} - \omega_k t)} d^3\boldsymbol{k} \right|^2 d^3\boldsymbol{r}$$

We have thus implicitly assumed that $\boldsymbol{E}(\boldsymbol{r}, t)$ belongs to $\mathcal{L}_2$ so that the transported energy can bear some physical reality. This is now the right moment to make use of the Dirac delta function (3.16) and (3.18), since

$$\begin{aligned}
\mathcal{E} &= \epsilon \iint \widetilde{\boldsymbol{E}}(\boldsymbol{k}) \widetilde{\boldsymbol{E}}^*(\boldsymbol{k}') e^{i(\boldsymbol{k}-\boldsymbol{k}')\cdot\boldsymbol{r}} e^{-i(\omega_k - \omega_{k'})t} d^3\boldsymbol{k} d^3\boldsymbol{k}' d^3\boldsymbol{r} \\
&= \epsilon \int \widetilde{\boldsymbol{E}}(\boldsymbol{k}) \widetilde{\boldsymbol{E}}^*(\boldsymbol{k}') \int e^{i(\boldsymbol{k}-\boldsymbol{k}')\cdot\boldsymbol{r}} d^3\boldsymbol{r} e^{-i(\omega_k - \omega_{k'})t} d^3\boldsymbol{k} d^3\boldsymbol{k}' \\
&= \epsilon \int \widetilde{\boldsymbol{E}}(\boldsymbol{k}) \widetilde{\boldsymbol{E}}^*(\boldsymbol{k}') (2\pi)^3 \delta(\boldsymbol{k} - \boldsymbol{k}') e^{-i(\omega_k - \omega_{k'})t} d^3\boldsymbol{k} d^3\boldsymbol{k}' \\
&= \epsilon \int |(2\pi)^{3/2} \widetilde{\boldsymbol{E}}(\boldsymbol{k})|^2 d^3\boldsymbol{k} \qquad (3.20)
\end{aligned}$$

where the usual notation $\boldsymbol{E}^*(\boldsymbol{r}, t)$ refers to the complex conjugate of $\boldsymbol{E}(\boldsymbol{r}, t)$. The physical interpretation of expression (3.20) is immediate[6]: the total energy results from the sum over the square modulus of each plane wave amplitude (within a $(2\pi)^3\epsilon$ proportionality factor). In Planck–Einstein terms, such a wave can be thought of as an assembly of photons. For each wave vector, the number of photons is thus proportional to $|(2\pi)^{3/2}\widetilde{\boldsymbol{E}}(\boldsymbol{k})|^2$.

To close this very general overview of waves in electromagnetism, we address the following question: is an electromagnetic wave necessarily made from a continuous superposition of plane waves?

The answer comes in two parts. First, plane waves are well adapted to describe the local structure of a freely propagating electromagnetic field far from its source. For example, in the immediate vicinity of an antenna, it

[6]This is often referred to as "Parseval–Plancherel's theorem" which states that the Fourier transform is unitary.

is not a good idea to use a plane wave as we expressed them. Spherical or cylindrical waves would better meet the needs. Secondly, and more importantly, we mentioned that plane waves' amplitude coefficients are subject to boundary and initial conditions. For example, in a finite volume (i.e. a cavity), the electric field must cancel out on the surface of its perfectly reflecting walls. This condition is only satisfied by plane waves which have proper wavelengths so that a standing wave can exist and does not cancel itself after many round trips. For 1D systems, as we have already seen in the case of the black-body cavity, it boils down to the condition: $n\lambda = 2L$, where L is the length of the cavity. Consequently, it is quite clear that not all wave vectors are systematically required to form electromagnetic waves. Depending on the boundary conditions, we may need to express the wave using (3.12) or (3.13). This wave vector sampling is exactly how an optical resonator (or optical cavity) works in conventional laser technology such as the popular Helium–Neon laser sources used in practical classes. The active medium is excited and emits broadband radiation. However, it is placed between two facing mirrors which constitute the cavity. The distance that separates them is adjusted to select the wavelength to be amplified by stimulated emission. In this particular case, the aim is to extract one specific wave vector so that the discrete sum reduces to a single term. Of course, in reality, the mirrors are not absolutely flat[7] and surface imperfection induces a selection which is not perfectly monochromatic but exhibits a very narrow frequency distribution.

Question 3.3: **Dirac's comb.** Use the Dirac delta function to show that (3.13) is merely a particular case of the continuous expansion (3.12).

Answer: If we write the wave vector distribution as $\widetilde{\boldsymbol{E}}(\boldsymbol{k}) = \sum_j \widetilde{\boldsymbol{E}}_{\boldsymbol{k}_j}\delta(\boldsymbol{k} - \boldsymbol{k}_j)$, we find

$$\boldsymbol{E}(\boldsymbol{r}, t) = \int \sum_j \widetilde{\boldsymbol{E}}_{\boldsymbol{k}_j}\delta(\boldsymbol{k} - \boldsymbol{k}_j)e^{i(\boldsymbol{k}\cdot\boldsymbol{r} - \omega_k t)}d^3\boldsymbol{k} = \sum_j \widetilde{\boldsymbol{E}}_{\boldsymbol{k}_j}e^{i(\boldsymbol{k}_j\cdot\boldsymbol{r} - \omega_{k_j}t)}$$

Since it is built from many spiky functions, the discrete distribution that we used here is often called the "Dirac comb" function.

[7]They are actually curved to achieve easier alignment, beam trapping and thereby a better amplification efficiency.

3.2. Matter Wave: Function and Consequences

3.2.1. *A wavefunction to describe particles*

When light strikes a photographic detector, energy transfer occurs and it results in the formation of an image. If the incident wave is uniform, and again assuming a perfect detector, the image will also be uniform. But we can ask ourself what happens if the radiation intensity is reduced to its minimum. The classical representation says that the image will still be uniform. But according to Planck–Einstein's model, for such an interaction, the electromagnetic wave should be represented by its particles: the photons. For a monochromatic field, the minimal possible energy is given by that of a single photon, $\hbar\omega$. This photon cannot be broken into smaller energy pieces. So the interaction with the detector can only take place in a single pixel (or grain if film photography is used). Hence, on the one (classical) hand, we are dealing with a plane wave (of minute intensity) which corresponds to a uniform energy distribution among all pixels and, on the other (quantum) hand, we necessarily detect light in one unique pixel. How this particular pixel is chosen is still an unsolved mystery. There is, of course, always one pixel which happens to be slightly more sensitive than the others but it turns out that, if the exact same experiment is reproduced, the activated pixel is not always the same. So there is another physical process at work. It seems that pixels are randomly picked with no observable correlation between the events. If this is true, repeating the single photon experiment a large number of times is equivalent to having a large number of photons in the beam. We then expect that, in the case of a large number of photons, the image converges to that obtained from classical uniform (monochromatic) lighting. The final image is thus a representation[8] of the probability law at the origin of the random process at work. If the detector's surface corresponds to the (x, y) plane and the emitted Poynting vector is along the z direction, we call $I(x, y)$ the total electromagnetic energy deposited during Δt. It is legitimate to claim that when a large number of photons has been emitted, i.e. for a large enough Δt, $I(x, y)$ is a representation of

[8]The representation is actually sampled in space by pixels.

the probability law for the photon to be found in (x, y, z). The image on the detector should then be considered similar to a bar chart recording the statistics from the outcome of coin tossing or dice throwing: an approximate representation of a probability law.

The above line of reasoning holds no matter what happens to the electromagnetic wave between the source and the detector. An interference pattern is thus a visualization of the probability distribution to detect a photon after it has left the source and has been scattered. The fringes on the screen after Young's slit interferometer clearly show that there are pixels on which there is a negligible probability of detecting a photon, even if the source emits a uniform flux. We can thus safely identify the photon probability distribution with the total monochromatic energy distribution, providing that the latter has been properly normalized. Within a normalization constant, we can then consider that the photon detection probability is given by the modulus-squared of the electric[9] field: $\mathcal{P}(x, y, z) \propto |E(x, y, z)|^2$.

Numerous experiments show that interference patterns can be recorded by all sorts of detectors with a variety of particles (among which are photons, electrons, neutrons, atoms and molecules — see Figs. 3.2 and 3.3).

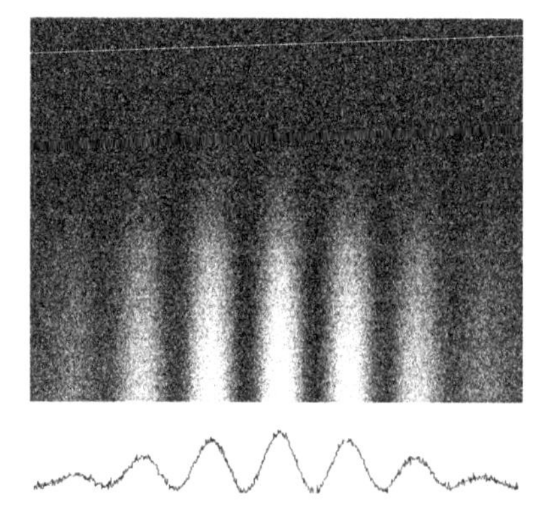

Fig. 3.2. Rueckner and Peidle have designed a pedagogical Young double slit demonstration with single photons. The electromagnetic source (here a blue LED) is made so weak that only one photon impact is recorded per image (0.5 s per exposure). This experiment is an unambiguous demonstration that the interference pattern is not created by the interaction between photons but by an intrinsic property carried by each photon. Each impact corresponds to a realization of the same probability law. Top: Resulting image from a sum of 240 half-second exposures. Bottom: Integrated intensity showing an excellent contrast (the maximum corresponds to 500 counts). Reproduced with permission from [78]. Copyright © 2013 by the American Association of Physics Teachers.

[9]We have mentioned previously that it could also be the magnetic field or the potential vector with no physical difference.

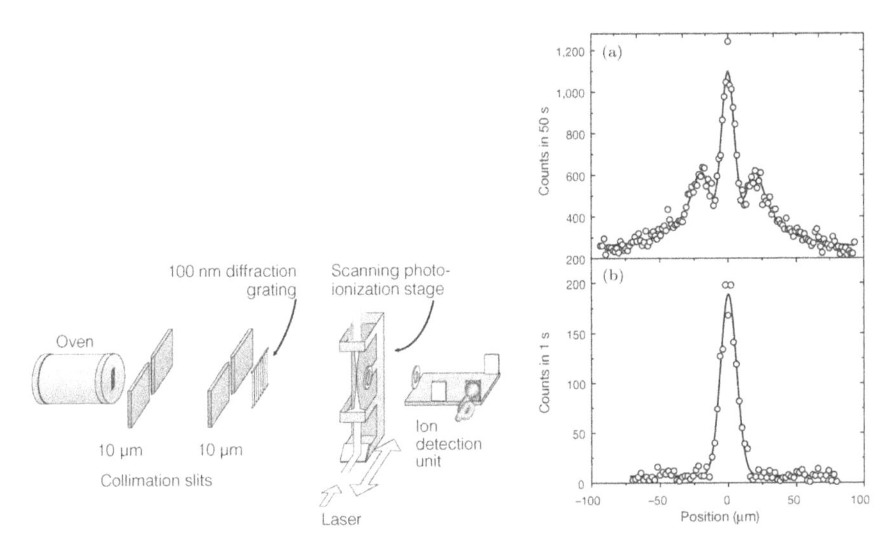

Fig. 3.3. Left: Setup for C_{60} — "bucky balls" — Young slits diffraction experiment. The molecules are emitted from the oven and, after collimation, scattered by the diffraction grating. The molecules are ionized by a laser beam before they can be detected at each transverse position. Right: Number of molecules detected as a function of the detector's transverse position after they have been scattered by the grating (a) or without the grating (b). Within the resolution of the apparatus, interference fringes are clearly visible. Reprinted with permission from [5]. Copyright © 1999 by Springer Nature.

It is then legitimate to postulate the existence of a function describing the wave nature of any object, playing a role similar to the electric field for the photon. There seems to be no straightforward strategy to find out what its mathematical form is like, even though its existence is corroborated by several experimental observations. In Young's experiment, while the field at the detector results from the combination of the two secondary waves emitted through each slit, the detection only reveals the spatial structure of its modulus-squared. The absolute phase information is lost. In a similar manner, we will call "wavefunction", the mathematical expression that describes the wave-like behaviour of particles, but with no hope of achieving better experimental knowledge by particle detection than its mere modulus-squared. The wavefunction has a structure that allows for phase differences, but it is the modulus-squared that can be solely related to a measurable probability distribution. It is postulated that in a "which path experiment", such as Young slits interferometry or diffraction by a crystal, the final wavefunction on the detector is a sum of contributions associated with every possible particle trajectory. The relative weight of each contribution can be inferred from the probability that the particle could follow that particular path. For example, all else being equal, if slit A were twice as large as slit B, it is likely that the wave associated with the path going through A would have a $\sqrt{2}$ larger amplitude.

Consequently, it is hypothesized that there is a "wave function" for each route taken from the source. The wavefunction[10] of the particle, which is finally detected, is made of the superposition (addition) of the functions relating to the various possible paths. The measured signal is, within a scale factor, the modulus-squared of the resultant wavefunction. Note that nothing indicates what the function should be exactly. We just know that it must exist, and its modulus-squared represents a probability distribution of presence. This is what is called the "Born interpretation" of the wavefunction (Fig. 3.4).

In what follows, we will note $\psi(\boldsymbol{r}, t)$ the wavefunction for a single particle at a given instant t. To take the example of Young's slits interferometer, it is natural to think that the final pattern will change with the detector's position, so the probability is expected to depend on $\boldsymbol{r}$ and, because the distance between the slits may vary with time, the wavefunction should also be able to include the variable t. The following table will therefore apply:

wavefunction at t	$\psi(\boldsymbol{r}, t)$
position probability distribution at t	$\lvert\psi(\boldsymbol{r}, t)\rvert^2$
position probability in $d^3\boldsymbol{r}$ at t	$\lvert\psi(\boldsymbol{r}, t)\rvert^2 d^3r$

$|\psi(\boldsymbol{r}, t)|^2$ is called a probability density, or a probability distribution, and noted $\rho(\boldsymbol{r}, t)$. It is of course positive-definite while the wavefunction $\psi(\boldsymbol{r}, t)$ is a complex-valued function. Since $|\psi(\boldsymbol{r}, t)|^2 d^3r$ is the probability that the particle can be found at t in an elementary volume $d^3\boldsymbol{r}$ centred on $\boldsymbol{r}$, we

[10]Wavefunction will now be written as a single word to emphasize the importance of the object.

Fig. 3.4. Max Born (1882–1970) and his wife Hedi. He was awarded the Nobel Prize in 1954. His love for music (playing the piano, he treasured his mother's album with Brahms and Clara Schumann's autographs) made him another sonata companions of Einstein. As a schoolboy, he was enthusiastic when his physics professor reproduced for them Marconi's transmission experiment which had just been demonstrated. Not unusually for the times, he studied between several universities and got to attend lectures by Minkowski and Hilbert (who hired him to prepare his lecture notes) and then Klein, Schwarzschild and later Larmor and Thomson. His long time research, starting with Theodore von Karman, was mostly on the relation between the crystal structures and their physical properties. His Jewish extraction forced him to leave Germany in 1933 (for Italy then Cambridge), the year he published his classic *Principle of Optics*. He later wrote another classic *Atomic Physics* and moved to Edinburgh. He did not participate in the construction of the atomic bomb and after WWII wrote on the social role of scientists. Photo credits: Churchill Archives Centre, The Papers of Professor Max Born, BORN 6/1/22.

can formulate the utmost important "normalization condition":

$$\int_{-\infty}^{+\infty}\int_{-\infty}^{+\infty}\int_{-\infty}^{+\infty} |\psi(\boldsymbol{r},t)|^2\, d^3\boldsymbol{r} = 1 \tag{3.21}$$

This merely means that the sum of all the probabilities of finding the particle in every single volume element of space is 1. In other words, it boils down to the fact that, if the particle exists, it is necessarily somewhere! This allows us to limit the range of possible wavefunctions: they must be **square-integrable functions**, so that (3.21) can be computed. In the preceding section, when we were dealing with photons, we mentioned that to bear some physical reality, the associated electric fields have to belong to $\mathcal{L}_2$. Similarly, we extend here this constraint to wavefunctions of any type of particle and, as a consequence, it will be required to approach zero at the infinity more rapidly than any inverse power of the position coordinates.

Note that if the particle is confined in a box, the probability to detect it anywhere outside is zero. Therefore, the wavefunction should be zero too. Hence, the integration needs to be carried out only over the box volume:

$$\int_{\text{box}} |\psi(\boldsymbol{r},t)|^2 \, d^3\boldsymbol{r} = 1 \tag{3.22}$$

Also note that, at this point, no particular constraint is put on the phase of a wavefunction and a physical interpretation is only given to the modulus-squared wavefunction. Not the wavefunction itself. Furthermore, this point will bear a particular importance in Chapter 5, knowing the position probability distribution does not imply any access to the exact result of a measurement on where the particle would be detected. We all know the probability law of a fair dice. We cannot predict what the outcome of a single roll will be. The Young slit experiment conducted by Tonomura [89] with electrons (Fig. 3.5) is very informative on this issue.[11]

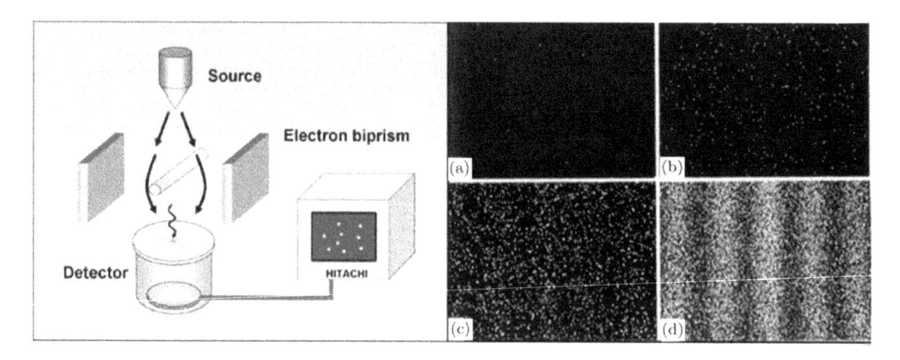

Fig. 3.5. Tonomura and co-workers [89] at Hitachi research laboratories have devised an original experimental setup (left) to demonstrate the quantum origin of electrons interference pattern in the Young double slit experiment. The figures (right) have been obtained after emitting and detecting electrons one by one. Images (a)–(d) are captures of the detector screen at successive moments of the experiment. Obviously, the formation of the interference image is constructed progressively. This shows that the final image on the screen represents a probability distribution and that each electron detection (bright spot) on a given pixel is a mere realization of a possible event from this probability law. Reprinted by courtesy of the Central Research Laboratory, Hitachi, Ltd., Japan.

[11]More details and a very illustrative movie version can be obtained from http://www.hitachi.com/rd/portal/highlight/quantum/.

3.2.2. *Wavefunctions as plane waves or wave packets*

As it was previously emphasized, a monochromatic plane wave describes an electromagnetic field that propagates in a perfectly well-defined direction and with a fixed angular frequency. Its amplitude is a constant complex number and the phase is totally characterized once the wave vector $\boldsymbol{k}$ is known. It is thus associated with a photon for which the momentum is perfectly determined $\boldsymbol{p} = \hbar\boldsymbol{k}$ and the energy is $\varepsilon = \hbar\omega = \hbar kc$.

On the basis of all the similarities between the wave behaviour of photons and other particles, it is safe to propose that plane waves could describe any particle with a perfectly well-defined momentum. Newton's laws would then state that, because the momentum is a constant vector, this corresponds to an isolated particle or with no net force acting on it: such an object will be called a "free particle". The energy of such a particle is of course conserved and is made up of a kinetic component: $\varepsilon = \hbar\omega = \frac{1}{2}mv^2 = \frac{p^2}{2m}$. When the potential energy V is constant, the net forces are also zero. However, the total energy should be written: $\varepsilon_p = \hbar\omega = \frac{p^2}{2m} + V$. Thus, we can write the free particle's wavefunction as

$$\psi(\boldsymbol{r}, t) = Ae^{i(\boldsymbol{p}\cdot\boldsymbol{r} - \varepsilon_p t)/\hbar}$$

Despite its appealing simplicity, we already know that this expression cannot be adopted to describe a particle in an infinite space: it is not square-integrable, so it cannot be normalized. At best, it can be seen as a useful approximation for a free particle in a large volume. If we call V the enclosing volume, the amplitude A is thus found from the normalization condition:

$$\int_V |Ae^{i(\boldsymbol{p}\cdot\boldsymbol{r} - \varepsilon_p t)/\hbar}|^2 d^3\boldsymbol{r} = \int_V |A|^2\, d^3\boldsymbol{r} = 1 \Rightarrow A = \frac{1}{\sqrt{V}}e^{i\varphi} \qquad (3.23)$$

where the constant phase φ cannot be determined here and has no physical meaning in terms of probability. Note that (3.23) describes a particle which would be totally delocalized inside the box.

Particles are seldom thought to be totally delocalized in space. We tend to picture them as objects having at least a minimal localization at a given moment. In terms of probability, we would thus expect that there is a much higher probability of finding the object in a limited region of space and a rather negligible probability of detecting it in other remote positions. This can be expressed by a superposition of plane waves, "wave packet", for example, at $t = 0$:

$$\psi(\boldsymbol{r}, t = 0) = \int c(\boldsymbol{k}) e^{i\boldsymbol{k}\cdot\boldsymbol{r}} d^3\boldsymbol{k} \tag{3.24}$$

This is a simple linear combination of plane waves, whose coefficients are given under the form of the function $c(\boldsymbol{k})$. This addition will give constructive and destructive interferences which modulate the distribution of the probability law.

✍ *Question 3.4:* **Momentum representation of particles from the wave packet distribution.** Using the Dirac delta function and (3.24), show that the normalization condition $\int \rho(\boldsymbol{r}) d^3\boldsymbol{r} = 1$ is equivalent to $\int |\widetilde{\psi}(\boldsymbol{p})|^2 d^3\boldsymbol{p} = 1$, where $\widetilde{\psi}(\boldsymbol{p}) = (2\pi/\hbar)^{3/2} c(\boldsymbol{p}/\hbar)$. This condition implies that $\widetilde{\psi}(\boldsymbol{p})$ must also be square-integrable. Use (3.20) to give an interpretation of $|\widetilde{\psi}(\boldsymbol{p})|^2$ in terms of probability.

Answer: This calculation is totally equivalent to (3.20). From the wave packet expression of the wavefunction, the normalization condition is

$$\iint c^*(\boldsymbol{k}')\ e^{-i\boldsymbol{k}'\cdot\boldsymbol{r}} d^3\boldsymbol{k}' \int c(\boldsymbol{k})\ e^{i\boldsymbol{k}\cdot\boldsymbol{r}} d^3\boldsymbol{k} d^3\boldsymbol{r} = 1$$

Regrouping the terms yields $\iint c^*(\boldsymbol{k}')c(\boldsymbol{k})\ \int e^{i(\boldsymbol{k}-\boldsymbol{k}')\cdot\boldsymbol{r}} d^3\boldsymbol{r} d^3\boldsymbol{k}' d^3\boldsymbol{k} = 1$.
We then make use of Eq. (3.18) and obtain

$$\iint c^*(\boldsymbol{k}')c(\boldsymbol{k})\ (2\pi)^3 \delta(\boldsymbol{k} - \boldsymbol{k}') d^3\boldsymbol{k}' d^3\boldsymbol{k} = 1$$

Hence, from (3.17), it leads to $\int |(2\pi)^{3/2} c(\boldsymbol{k})|^2 d^3\boldsymbol{k} = 1$ It is clearly a normalization condition in the wave vector space. If we express the wave packet in terms of momenta, we write $\widetilde{\psi}(\boldsymbol{p}) = (2\pi/\hbar)^{3/2} c(\boldsymbol{p}/\hbar)$ and the above normalization condition forces us to interpret $|\widetilde{\psi}(\boldsymbol{p})|^2$ as the probability density for a particle described by $\psi(\boldsymbol{r})$ to have a momentum $\boldsymbol{p}$. We will use

$$\psi(\boldsymbol{r}, t = 0) = \frac{1}{(2\pi\hbar)^{3/2}} \int \widetilde{\psi}(\boldsymbol{p})\ e^{i\boldsymbol{p}\cdot\boldsymbol{r}/\hbar} d^3\boldsymbol{p}$$

and

$$\psi(\boldsymbol{r}, t) = \frac{1}{(2\pi\hbar)^{3/2}} \int \widetilde{\psi}(\boldsymbol{p}, t)\ e^{i\boldsymbol{p}\cdot\boldsymbol{r}/\hbar} d^3\boldsymbol{p} \tag{3.25}$$

with

$$\widetilde{\psi}(\boldsymbol{p}, t) = \widetilde{\psi}(\boldsymbol{p}) e^{-i\varepsilon_p t/\hbar} \tag{3.26}$$

Answer: (continued)

Expression (3.25) can be reversed by Fourier transform so that

$$\widetilde{\psi}(\boldsymbol{p},t) = \frac{1}{(2\pi\hbar)^{3/2}} \int \psi(\boldsymbol{r},t)\, e^{i\boldsymbol{p}\cdot\boldsymbol{r}/\hbar} d^3\boldsymbol{r} \tag{3.27}$$

There is bi-univoque relation between $\psi(\boldsymbol{r},t)$, the wavefunction which uses the position as a variable, and its counterpart $\widetilde{\psi}(\boldsymbol{p},t)$. The latter is the wavefunction expressed in terms of momentum. It is also said to be the wavefunction in the "momentum representation". The position and momentum representation are two possible choices for expressing the state of a particle. They will be seen to be special cases of a broader formalism in Chapter 5.

Let us consider a simple 1D example where $c(k)$ takes a gaussian shape ($\alpha > 0$):

$$c(k) = C_0 e^{-\alpha^2 k^2} \tag{3.28a}$$

which yields the 1D wavefunction:

$$\begin{aligned}
\psi(x,t=0) &= \int_{-\infty}^{+\infty} C_0 e^{-\alpha^2 k^2} e^{ikx} dk \\
&= C_0 \int_{-\infty}^{+\infty} e^{-\alpha^2 k^2 + i2\alpha k \frac{x}{2\alpha}} dk \\
&= C_0 \int_{-\infty}^{+\infty} e^{-\left(\alpha k - i\frac{x}{2\alpha}\right)^2 - \left(\frac{x}{2\alpha}\right)^2} dk \\
&= C_0 e^{-\left(\frac{x}{2\alpha}\right)^2} \int_{-\infty}^{+\infty} e^{-y^2} \frac{1}{\alpha} dy \\
&= C_0 \frac{\sqrt{\pi}}{\alpha} e^{-\left(\frac{x}{2\alpha}\right)^2}
\end{aligned} \tag{3.28b}$$

where C_0 is defined by the normalization condition on the wavefunction[12]:

$$\int_{-\infty}^{+\infty} \left| C_0 \frac{\sqrt{\pi}}{\alpha} e^{-\left(\frac{x}{2\alpha}\right)^2} \right|^2 dx = 1$$

[12]Remember that if $I = \int_{-\infty}^{+\infty} e^{-\beta x^2} dx$, obviously the square can be written as

$$\begin{aligned}
I^2 &= \int_{-\infty}^{+\infty} \int_{-\infty}^{+\infty} e^{-\beta(x^2+y^2)} dx dy = \int_0^{2\pi} \int_0^{+\infty} e^{-\beta r^2} d\theta r dr \\
&= 2\pi \int_0^{+\infty} e^{-\beta r^2} \frac{1}{2} d(r^2) = \pi \int_0^{+\infty} e^{-\beta q} dq = \frac{\pi}{\beta}
\end{aligned}$$

Therefore $I = \sqrt{\frac{\pi}{\beta}}$.

so that $C_0 = \sqrt{\frac{\alpha}{(2\pi^3)^{1/2}}}\, e^{i\varphi}$ where φ is some arbitrary real number. The Gaussian wavefunction belongs to $\mathcal{L}_2$ and will prove to be very useful in the particular case of a quantum harmonic oscillator. We can now understand the significance of plane waves on the basis of which it is possible to express wavefunctions. This decomposition operation is known in mathematics under the name "Fourier decomposition".

There is more to the wave packet expression than to merely remark that it is just a Fourier decomposition of a wavefunction: as each plane wave represents a given wave vector $\boldsymbol{k}$, we should note that the Fourier coefficient indicates the importance of a given momentum $\boldsymbol{p} = \hbar\boldsymbol{k}$ in the description of the particle.

This is easily observed in the Gaussian case (Eq. (3.28b)): the larger α gets, the sharper the function $c(k)$, and the more diffuse the wavefunction $\psi(\boldsymbol{r}, t)$. As a consequence, the position probability density $\rho(\boldsymbol{r}, t)$ becomes broader and the particle is more poorly localized. We should consider as a mere consequence of de Broglie's relationship between momentum and wave vector that the better the particle's momentum is known, the sharper the $c(\boldsymbol{k})$ distribution is and the less information the position probability distribution provides about the particle's position.

This observation generalizes what we know about the diffraction of light through a narrow slit. In the Fraunhofer regime,[13] a monochromatic light incident on a narrow slit can locally be approximated by a plane wave with a well-defined wave vector. The light beam collected on a screen after the slit is a spot, the size of which increases as the slit is progressively closed. One could interpret this behaviour in terms of photons: the narrower the slit, the more precise the localization of the photons at that point of their propagation. The inverse Fourier transform gives the momentum distribution in the direction perpendicular to the propagation direction:

$$c(p_x/\hbar) = \frac{1}{2\pi}\int_{-a/2}^{a/2} \psi_0 e^{-ip_x x/\hbar} dx$$

[13]This regime, also known as "far field", corresponds to a distance from the scatterer (or the source) large enough so that the optical path can be supposed to vary linearly with any displacement of the scatterer (or the source). See Volume 3.

where ψ_0 contains the normalization constant and the contribution from the two other space directions. We obtain

$$c(p_x/\hbar) = \psi_0 \frac{a}{2\pi} \frac{\sin\left(\frac{p_x a}{2\hbar}\right)}{\left(\frac{p_x a}{2\hbar}\right)}$$

This is the familiar sinc function. The typical width of the function can be estimated from the value Δp_x between the maximum and the first cancellation so that $\Delta p_x \gtrsim 2\pi\hbar/a$. As the photon keeps the same energy, such a distribution of momentum components along the x direction implies an (half) angular spreading after the slit $\Delta\theta = \Delta p_x/p = \Delta p_x/(\hbar k) \gtrsim \lambda/a$ which is a well-known result from Fraunhofer diffraction.

Question 3.5: **Heisenberg inequality with neutrons.** The curves in Fig. 3.6 represent the distribution measured by Shull [80] of monokinetic neutrons after passing through slits of different widths. Find the incident neutron wavelength. For a sinc θ function, the half-intensity angular position is for $\theta_{\max/2} \approx \pm 1.39$ rad.

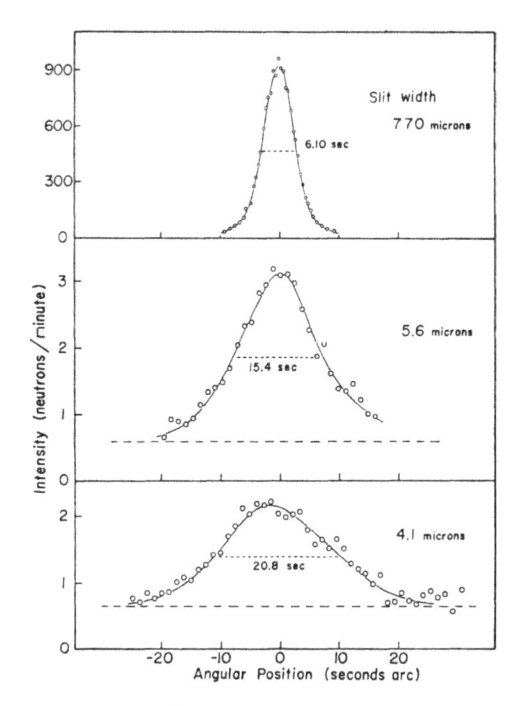

Fig. 3.6. "Long wavelength" neutrons are selected by a monochromator before passing through a slit of adjustable width. The angular distribution is reported for different slit opening widths. Reprinted with permission from [80]. Copyright © 1969 by the American Physical Society.

Answer: The author provides the widths at approximately half the maximum. For the sinc function, with our notations, this roughly corresponds to $\Delta p_x \gtrsim 4\pi/(\hbar a)$ and a full width at half maximum (FWHM) of about $\Delta\theta_{\text{FWHM}} \gtrsim 2\lambda/a$. We thus find the following:

$$a = 770\ \mu\text{m} \quad \text{and} \quad \Delta\theta_{\text{FWHM}} = 3 \times 10^{-5}\ \text{rad} \quad \rightarrow \quad \lambda \approx 220\ \text{Å}$$

$$a = 5.6\ \mu\text{m} \quad \text{and} \quad \Delta\theta_{\text{FWHM}} = 7.46 \times 10^{-5}\ \text{rad} \quad \rightarrow \quad \lambda \approx 4.18\ \text{Å}$$

$$a = 4.1\ \mu\text{m} \quad \text{and} \quad \Delta\theta_{\text{FWHM}} = 10^{-4}\ \text{rad} \quad \rightarrow \quad \lambda \approx 4.13\ \text{Å}$$

The first value is obviously out of range. This distribution is actually the instrumental angular resolution and corresponds to a full beam passing through wide open slits (770 μm). It cannot be used to infer a wavelength from the beam spreading. However, as the FWHM becomes much larger than the instrumental resolution, the wavelength becomes more reliable. The wavelength provided by the author is $\lambda = 4.43$ Å.

As a general result, from the properties of Fourier transforms, we can state that to describe a function with spatial variations of characteristic size Δx, it is necessary to use a wave vector spectrum extending at least over $\Delta k \approx 1/\Delta x$. We thus find the approximate inequality $\Delta k \Delta x \gtrsim 1$ which, in quantum terms (i.e. $p = \hbar k$), results in

$$\Delta p\, \Delta x \gtrsim \hbar \tag{3.29}$$

This relationship brought by the structure of Fourier transforms is a direct consequence of the dual wave and particle descriptions introduced by early experiments of quantum physics. The physical interpretation is as follows: the breadth of the wave packet measures the indeterminate nature of particle position. This is not related to our human incapacity to localize the particle. It is simply that the particle is itself not localized! However, the width Δp (also) indicates that the momentum of this particle is not perfectly well defined either. So, the product of the respective indeterminations is always greater than a value which is not zero. Planck's constant now has a new role: it sets what the typical minimal indetermination product is. As the Planck constant is not zero, a better determination of a particle momentum mechanically increases its delocalization. The extreme case is of course the plane wave description[14] for which the momentum is precisely known, but the position probability distribution is uniform in the infinite universe. Relation (3.29) is called a "Heisenberg inequality" as it

[14]If it had the acceptable mathematical properties.

was first derived[15] by Heisenberg (Fig. 3.8) in 1927 [51]. Occasionally, it is also quoted as an "uncertainty relation" arguing that Δx and Δp are also respectively connected to the minimal simultaneous uncertainties that we have on the particle position and momentum. The term "uncertainty" is however improper because it has nothing to do with our experimental inability to accurately measure both the position and the momentum. It is an intrinsic impossibility of the particle itself to have simultaneously a fixed value for its position and its momentum.

Very similar conclusions can be drawn if the wave is not monochromatic. It is thus represented by an expansion in the frequency domain, which (by $\varepsilon = \hbar\omega$) acknowledges that its energy is not well defined. Still from Fourier's transform properties, the inequality $\Delta\nu\Delta t \gtrsim 1$ translates in quantum language to an additional Heisenberg indetermination

$$\Delta\varepsilon\Delta t \gtrsim \hbar \tag{3.30}$$

The interpretation of this relation has been the subject of much debate into which it is not the purpose of this book to enter. A clarification will be given in volume 3, when time-dependent perturbation will be studied. To put it on the least-controversial level, it can be stated that for a system described by a non-monochromatic wave, the energy is not well defined. If $\Delta\varepsilon$ quantifies the energy indetermination, the particle is not in a "stationary state". It will have a characteristic evolution time given by $\Delta t \gtrsim \hbar/\Delta\varepsilon$. Conversely, a physical system which exhibits a variation of its properties on a time scale of the order of Δt, does not have an energy value defined better than a characteristic range $\Delta\varepsilon \approx \hbar/\Delta t$.

Question 3.6: **The ephemeral particle.** Particle physics makes extensive use of collision experiments. When the system is made of two particles (before and after the collision), the energies and momenta distribution is totally conditioned by the initial positions and momenta. If other particles are created on this occasion, the distribution is consequently modified. As an example, the collision between a meson π^+ and a proton is now considered. The final product is made of the same system with an additional pair of particle–antiparticle

[15] For a much more rigorous and formal derivation than the somewhat qualitative and heuristic explanation given here, see Section 5.2.3 and Question *3.11*.

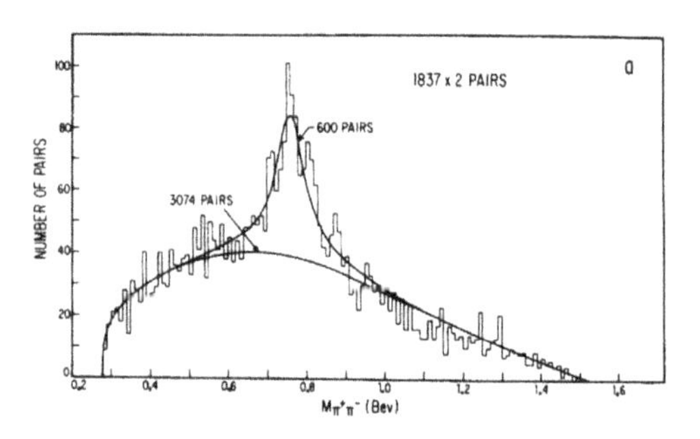

Fig. 3.7. Distribution of possible energies carried by the (π^+, π^-) pair upon the collision of a π^+ meson with a proton. Energies are given in BeV (billions of electron volts), i.e. GeV = 10^9 eV. Reprinted with permission from [1]. Copyright © 1963 by the American Physical Society.

Question 3.6: (continued)

(π^+, π^-). The experiment consists in studying the energy distribution of the energy carried by this particle pair. The result is reported on Fig. 3.7. An analysis of this figure suggests that an intermediate particle has been created with an energy in the order of 770 MeV. What is the lifetime of such a particle? Why is it impossible to directly detect it?

Answer: The pair energy distribution shows that a peak centred at 770 MeV (0.7 GeV) rises above the background noise. From the peak width, the energy spread is estimated to $\Delta\varepsilon \approx 160$ MeV. Such an energy indetermination for the produced pair is the sign of the creation of an intermediate particle with a finite lifetime. It can be estimated to $\Delta t \geq \hbar/\Delta\varepsilon$, thus $\Delta t \geq 10^{-34}/(160 \times 10^6 \times 1,6 \times 10^{-19}) \approx 3,9 \times 10^{-21}$ s.

Such an intermediate particle, if it could move at the speed of light, would not go much farther than $c\Delta t \approx 12 \times 10^{-13}$m, hence a minuscule fraction of an angstrom. The probability of directly detecting such an ephemeral particle is thus very weak.

Question 3.7: **Lifetime vs. Doppler spreading.** When an atomic (or molecular) gas de-excites and emits radiation, the measured wavelength is not unique and the width of the emission line is a function of two phenomena. The first one has a classical origin, it is the Doppler spreading created by the atomic motions. This effect increases with temperature (see Volume 2). The finite lifetime of the atomic (or molecular) excited energy levels is the source of another spreading. We are going to compare the extent of both effects.

1. Among both yellow emissions from a sodium gas, the 589 nm one, named D_2, is the most intense. The excited level is said to bear a finite lifetime because the electron cannot stay in the corresponding state forever. The system returns to the ground state level after an approximate time lapse $\tau \approx 1.625 \times 10^{-8}$ s. Estimate the indetermination on the energy value for this excited level.

2. It will be shown in Volume 2 that, at temperature T, the probability for atoms in a perfect gas to have a speed between v and $v + dv$, is given by $p(v)dv = 4\pi v^2 (\frac{M}{2\pi k_B T})^{3/2} e^{-Mv^2/(2k_B T)} dv$. Compare the amplitude of the Doppler spreading to the emission line, for $T = 10$ K or $T = 300$ K, with that coming from the finite lifetime of the excited level.

Answer:

1. To estimate the lifetime effect, we can simply use the Heisenberg time-energy indetermination relation: $\Delta\varepsilon_\tau \gtrsim \hbar/\tau \approx 4 \times 10^{-8}$ eV.
2. The Doppler shift depends on the source's speed, v, according to $\Delta\varepsilon_D = \varepsilon \times v/c$. The quadratic mean speed is $\overline{v} = \sqrt{3k_BT/M}$, and the Doppler spreading is

$$\Delta\varepsilon_D = \frac{hc}{\lambda} \times \sqrt{\frac{3k_BT}{Mc^2}}$$

so $\Delta\varepsilon_D(T = 10\ \text{K}) \approx 7.3 \times 10^{-7}$ eV or $\Delta\varepsilon_D(T = 300\ \text{K}) \approx 40 \times 10^{-7}$ eV. Obviously, except at very low temperatures, the Doppler contribution amply dominates.

3.3. A Wave Equation: The Schrödinger Equation

The particles are subjected to forces. They can thus be accelerated and have their positions and velocities changed over time. Newton's equation[16] predicts exactly how such a particle must evolve. This change of behaviour upon forces must also affect the wavefunction expression. It is then essential to find the equation that dictates the wavefunction's modification or, for a given environment what the wavefunction should be.[17]

Rather than attempting to demonstrate such an equation, some justifications will be given to facilitate the physical understanding of its use. We will then proceed to describe some basic applications.

3.3.1. *Mean position, mean potential*

A particle, which is described by a wavefunction $\psi(\boldsymbol{r}, t)$, does not have a well-defined position. The experiments reported above show that its interaction with a detector will always indicate a location in a given pixel surface and thickness, thus in a volume δV centred on a position $\boldsymbol{r}$. The experiment can be repeated many times using identical particles described by the same wavefunction. As in the photon case, the detection will seldom be in the same pixel (see Fig. 3.2). From a large sample of detections, however, it will be possible to compute a mean position. The knowledge of the wavefunction

[16]For Lagrange–Hamilton's point of view, see Chapter 6.

[17]When Erwin Schrödinger, in November 1925, was concluding a seminar on de Broglie's hypothesis for a wavefunction, Debye asked him: "You speak of waves. Where is the wave equation?" [87]. This remark was probably an additional incentive for Schrödinger to achieve the result which is exposed here.

Fig. 3.8. Werner Heisenberg (1901–1976) centre front, between Enrico Fermi (left) and Niels Bohr (right). Heisenberg remained close to the latter until his decision to help the Reich in constructing a nuclear reactor during WWII. Heisenberg was awarded a Nobel prize (1932) as well as Arnold Sommerfeld's five other students: P. Debye (1936), M. von Laue (1914), W. Pauli (1945), L. Pauling (1954) and H. Bethe (1967). Sommerfeld was always very supportive to Heisenberg, even during his PhD defence, in the report of which Wien noted that when it came to experiments he had a "bottomless ignorance". Planck advised Heisenberg not to resign when the Nazis took power because the regime was bound to collapse and the country would soon need leaders like him to reconstruct. Heisenberg was not reluctant to serve as a reservist remarking "it is nice not to have to think, for a change, but only to obey". As to whether his mistakes in calculating the necessary critical mass to initiate a nuclear chain reaction were made on purpose are still unclear since he boldly admitted that he wanted Germany to win the war. While he strongly believed that nuclear reactors should be built, he never agreed that the same physical laws should be employed for building a bomb. At the back, from left to right, Peter Debye (see Volume 3), Otto Stern (see Volume 2) and Guglielmo Marconi. Credits: with permission from AIP Emilio Segrè Visual Archives.

allows for the prediction of the mean position $\langle \boldsymbol{r} \rangle = (\langle x \rangle, \langle y \rangle, \langle z \rangle)$. We can even predict the dispersion around this value: for example, the standard deviation in the x direction is $\Delta x = \sqrt{\langle x^2 \rangle - \langle x \rangle^2}$. We will merely need to evaluate the integrals:

$$\langle x \rangle(t) = \int_{-\infty}^{+\infty} x \, |\psi(\boldsymbol{r}, t)|^2 \, d^3\boldsymbol{r}$$

and

$$\Delta x(t) = \left[\int_{-\infty}^{+\infty} x^2 \, |\psi(\boldsymbol{r}, t)|^2 \, d^3\boldsymbol{r} - \left(\int_{-\infty}^{+\infty} x \, |\psi(\boldsymbol{r}, t)|^2 \, d^3\boldsymbol{r} \right)^2 \right]^{1/2}$$

If the particle is subjected to a force field $\boldsymbol{F}(\boldsymbol{r})$, the force is thus not necessarily uniform. As a consequence, the effect on the particle will depend on what its position actually is. It will therefore depend on the wavefunction.

It is possible to compute the mean force acting on the particle, for example, its component along the z-axis:

$$\langle F_z \rangle(t) = \int_{-\infty}^{+\infty} F_z(\boldsymbol{r}) \left|\psi(\boldsymbol{r},t)\right|^2 d^3\boldsymbol{r}$$

Note that, because the wavefunction is time dependent, even if the force field is constant, the mean force will also depend on t. This is also true for the mean position and it is true for every quantum mean quantity: a mean value here is computed over space but is instantaneous. It is a consequence of the time dependence of the probability distribution $\rho(\boldsymbol{r},t) = |\psi(\boldsymbol{r},t)|^2$.

Quantum mechanics is mostly based on energy considerations and forces are scarcely evoked. We will refer in general to *potentials*, which stand for *potential energies* and represent the function $V(\boldsymbol{r})$ from which the force field can be derived:

$$\boldsymbol{F}(\boldsymbol{r}) = -\vec{\nabla} V(\boldsymbol{r}) \tag{3.31}$$

The mean potential acting upon a particle described by $\psi(\boldsymbol{r},t)$ is then:

$$\langle V \rangle(t) = \int_{-\infty}^{+\infty} V(\boldsymbol{r}) \left|\psi(\boldsymbol{r},t)\right|^2 d^3\boldsymbol{r} \tag{3.32}$$

This mean potential should not be confused with the potential at a mean position $V(\langle \boldsymbol{r} \rangle)$. Examples are given in Questions *3.8* and *3.9*.

Question 3.8: **Mean potential in a uniform electric field.** Consider an electron described by a 1D wavefunction: $\psi(x) \propto (x-a)e^{-\alpha(x-a)^2}$. Compute the mean potential energy when it is placed in a uniform electric field E_0.

Answer: To compute a mean value, we need a probability distribution. So far, we only have the general shape of the wavefunction. To have a proper and useful wavefunction, the expression needs to be normalized (so that we make sure that the sum of probabilities is one):

$$\int_{-\infty}^{+\infty} |\psi(x)|^2 dx = 1 = |\mathcal{N}|^2 \int_{-\infty}^{+\infty} (x-a)^2 e^{-2\alpha(x-a)^2} dx$$

Setting $y = x - a$, the Gaussian integral would be:

$$\int_{-\infty}^{+\infty} e^{-2\alpha y^2} dy = \sqrt{\frac{\pi}{2\alpha}}$$

Answer: (continued)

but unfortunately we now have the preceding $y^2 = (x-a)^2$ factor. To compute the requested integral, we can use Feynman's trick of differentiation under the sum

$$\int_{-\infty}^{+\infty} y^2 e^{-2\alpha y^2} dy = \int_{-\infty}^{+\infty} \left(-\frac{1}{2}\right) \frac{\partial}{\partial\alpha} e^{-2\alpha y^2} dy = \left(-\frac{1}{2}\right) \frac{\partial}{\partial\alpha} \int_{-\infty}^{+\infty} e^{-2\alpha y^2} dy$$
$$= \left(-\frac{1}{2}\right) \frac{\partial}{\partial\alpha} \sqrt{\frac{\pi}{2\alpha}} = \frac{1}{4}\sqrt{\frac{\pi}{2\alpha^3}}$$

Thus, the normalization constant is: $\mathcal{N} = (\frac{32\alpha^3}{\pi})^{1/4} e^{i\varphi}$. The mean potential is then:

$$\langle V \rangle = \langle -qE_0 x \rangle = -\left(\frac{32\alpha^3}{\pi}\right)^{1/2} \int qxE_0(x-a)^2 e^{-2\alpha(x-a)^2} dx$$

Setting again $y = x - a$, we then end up with two integrals

$$\langle V \rangle = -\left(\frac{32\alpha^3}{\pi}\right)^{1/2} \left(\int qaE_0 y^2 e^{-2\alpha y^2} dy + \int qE_0 y^3 e^{-2\alpha y^2} dy\right)$$

The second integral is null because the integrand is odd, and

$$\langle V \rangle = -qaE_0$$

In this particular case, the mean potential corresponds to the potential at the mean position $\langle x \rangle = a$.

Question 3.9: **Particle in a gravitational field**. Consider an electron in a gravitational field $\mathcal{G}Mm/|z|$. Compute its mean potential energy when the wavefunction is $\psi(\boldsymbol{r}) \propto z\, e^{-\alpha r^2}$.

Answer: Normalization yields

$$\psi(\boldsymbol{r}) = 2\left(\frac{2}{\pi}\right)^{3/4} \alpha^{5/4} z e^{-\alpha(x^2+y^2+z^2)}$$

The mean potential is expressed as

$$\langle V \rangle = 4\left(\frac{2}{\pi}\right)^{3/2} \alpha^{5/2} \int_{-\infty}^{+\infty} e^{-2\alpha(x^2+y^2)} dx dy \times 2\int_0^{+\infty} \mathcal{G}Mmze^{-2\alpha z^2} dz$$
$$= \mathcal{G}Mm\left(\frac{32\alpha^3}{\pi}\right)^{1/2} 2\int_0^{+\infty} ze^{-2\alpha z^2} dz = \mathcal{G}Mm\left(\frac{32\alpha^3}{\pi}\right)^{1/2} \int_0^{+\infty} e^{-2\alpha z^2} d(z^2)$$
$$= \mathcal{G}Mm2\sqrt{\frac{2}{\pi}}\sqrt{\alpha} \tag{3.33}$$

This result is of course different from the potential at the mean position $\langle \boldsymbol{r} \rangle = 0$, which is infinite.

3.3.2. *Mean momentum, mean kinetic energy*

It is easiest to compute the mean momentum for a particle described by a wavefunction $\psi(\boldsymbol{r},t)$ by first making use of the associated momentum probability density distribution, as described in Question *3.4* $|\widetilde{\psi}(\boldsymbol{p},t)|^2$, so that the expression comes naturally:

$$\langle p_x\rangle(t) = \int p_x|\widetilde{\psi}(\boldsymbol{p},t)|^2 d^3\boldsymbol{p} \tag{3.34}$$

The mean momentum for other directions is obtained with a similar calculation. It is however more informative to figure out how the same result can be reached by using the wavefunction with position coordinates. We can use the inverse Fourier transform from Eq. (3.27):

$$\begin{aligned}
\langle p_x\rangle(t) &= \frac{1}{(2\pi\hbar)^3}\int p_x\int \psi^*(\boldsymbol{r}',t)e^{i\boldsymbol{p}\cdot\boldsymbol{r}'/\hbar}d^3\boldsymbol{r}'\int \psi(\boldsymbol{r},t)e^{-i\boldsymbol{p}\cdot\boldsymbol{r}/\hbar}d^3\boldsymbol{r}d^3\boldsymbol{p}\\
&= \frac{1}{(2\pi\hbar)^3}\iint \psi^*(\boldsymbol{r}',t)e^{i\boldsymbol{p}\cdot\boldsymbol{r}'/\hbar}d^3\boldsymbol{r}'\int \psi(\boldsymbol{r},t)p_x\, e^{-i\boldsymbol{p}\cdot\boldsymbol{r}/\hbar}d^3\boldsymbol{r}d^3\boldsymbol{p}\\
&= \frac{1}{(2\pi\hbar)^3}\iint \psi^*(\boldsymbol{r}',t)e^{i\boldsymbol{p}\cdot\boldsymbol{r}'/\hbar}d^3\boldsymbol{r}'\int \psi(\boldsymbol{r},t)\left(\frac{\hbar}{-i}\right)\\
&\quad\times\left(\frac{\partial}{\partial x}e^{-i\boldsymbol{p}\cdot\boldsymbol{r}/\hbar}\right)d^3\boldsymbol{r}d^3\boldsymbol{p}
\end{aligned}$$

where Feynman's trick was used to extract p_x from the exponential. The last integral can be evaluated by parts. The wavefunction belongs to $\mathcal{L}_2$ and therefore cancels at infinity, so that

$$\begin{aligned}
\langle p_x\rangle(t) &= \frac{1}{(2\pi\hbar)^3}\iint \psi^*(\boldsymbol{r}',t)e^{i\boldsymbol{p}\cdot\boldsymbol{r}'/\hbar}d^3\boldsymbol{r}'\int e^{-i\boldsymbol{p}\cdot\boldsymbol{r}/\hbar}\left(\frac{\hbar}{i}\right)\\
&\quad\times\left(\frac{\partial}{\partial x}\psi(\boldsymbol{r},t)\right)d^3\boldsymbol{r}d^3\boldsymbol{p}\\
&= \frac{1}{(2\pi\hbar)^3}\iint \psi^*(\boldsymbol{r}',t)\left(\frac{\hbar}{i}\right)\left(\frac{\partial}{\partial x}\psi(\boldsymbol{r},t)\right)d^3\boldsymbol{r}\int e^{i\boldsymbol{p}\cdot(\boldsymbol{r}'-\boldsymbol{r})/\hbar}d^3\boldsymbol{p}d^3\boldsymbol{r}'\\
&= \iint \psi^*(\boldsymbol{r}',t)\left(\frac{\hbar}{i}\right)\left(\frac{\partial}{\partial x}\psi(\boldsymbol{r},t)\right)d^3\boldsymbol{r}\delta(\boldsymbol{r}'-\boldsymbol{r})d^3\boldsymbol{r}'\\
&= \int \psi^*(\boldsymbol{r},t)\left(\frac{\hbar}{i}\frac{\partial}{\partial x}\psi(\boldsymbol{r},t)\right)d^3\boldsymbol{r}
\end{aligned} \tag{3.35}$$

where $\int e^{i\boldsymbol{p}\cdot(\boldsymbol{r}'-\boldsymbol{r})/\hbar}d^3\boldsymbol{p} = (2\pi\hbar)^3\delta(\boldsymbol{r}'-\boldsymbol{r})$ was again used, as a 3D extension of (3.18). This expression can be generalized to yield the mean momentum vector:

$$\langle \boldsymbol{p}\rangle(t) = \int \psi^*(\boldsymbol{r},t)\left(\frac{\hbar}{i}\vec{\nabla}_r\psi(\boldsymbol{r},t)\right)d^3\boldsymbol{r} \tag{3.36}$$

Using the exact same procedure, it is straightforward to derive the quadratic mean momentum:

$$\left\langle p_x^2\right\rangle(t) = -\hbar^2\int \psi^*(\boldsymbol{r},t)\left(\frac{\partial^2}{\partial x^2}\psi(\boldsymbol{r},t)\right)d^3\boldsymbol{r} \tag{3.37}$$

and in three dimensions,

$$\left\langle p^2\right\rangle(t) = -\hbar^2\int \psi^*(\boldsymbol{r},t)\left(\nabla_r^2\psi(\boldsymbol{r},t)\right)d^3\boldsymbol{r} \tag{3.38}$$

with the use of the Laplacian $\nabla_r^2 = \vec{\nabla}_r\cdot\vec{\nabla}_r$, i.e. the divergence of the gradient. We can thereby obtain the mean kinetic energy from the wavefunction:

$$\langle K\rangle(t) = \left\langle\frac{p^2}{2m}\right\rangle = -\frac{\hbar^2}{2m}\int \psi^*(\boldsymbol{r},t)\left(\nabla_r^2\psi(\boldsymbol{r},t)\right)d^3\boldsymbol{r} \tag{3.39}$$

Question 3.10: **Gaussian indetermination.** Compute the product of the position and momentum standard deviations $\Delta x\Delta p_x$ if the 1D wavefunction is a Gaussian $\psi(x) = (2\alpha/\pi)^{1/4}e^{-\alpha x^2}$.

Answer: The mean position (and mean momentum) is nil. The quadratic means are:

$$\langle x^2\rangle = \int \psi^*(x)x^2\psi(x)dx = \frac{1}{4\alpha} \text{ and } \langle p_x^2\rangle = \int \psi^*(x)\left(-\hbar^2\frac{\partial}{\partial x^2}\psi(x)\right)dx = \hbar^2\alpha$$

so that the product of standard deviations does not depend on α: $\Delta x\Delta p_x = \frac{\hbar}{2}$.

3.3.3. *Mean total energy*

Each plane wave of the wave packet is associated with a particle in a given constant potential, the energy of which is given by $\varepsilon = \hbar\omega$. The angular frequency depends on the wave vector through the *dispersion relationship*. In the case of photons, we know that $\omega(k) = kc/n$, where n is the index of the medium. In many cases, we already know that the index can also depend on ω. For massive particles, we assume that we can simply write $\varepsilon_{\boldsymbol{p}} = \hbar\omega_{\boldsymbol{p}}$.

This is actually enough to compute the mean total energy[18] as

$$
\begin{aligned}
\langle H\rangle(t) &= \int \hbar\omega_{\boldsymbol{p}} |\widetilde{\psi}(\boldsymbol{p},t)|^2 d^3\boldsymbol{p} \\
&= \int \widetilde{\psi}^*(\boldsymbol{p}) e^{i\omega_{\boldsymbol{p}}t} \hbar\omega_{\boldsymbol{p}} \widetilde{\psi}(\boldsymbol{p}) e^{-i\omega_{\boldsymbol{p}}t} d^3\boldsymbol{p} \\
&= \int \widetilde{\psi}^*(\boldsymbol{p}) e^{i\omega_{\boldsymbol{p}}t} (i\hbar) \frac{\partial}{\partial t} \left(\widetilde{\psi}(\boldsymbol{p}) e^{-i\omega_{\boldsymbol{p}}t} \right) d^3\boldsymbol{p} \\
&= \int \frac{1}{(2\pi\hbar)^3} \int \psi^*(\boldsymbol{r}',t) e^{i\boldsymbol{p}\cdot\boldsymbol{r}'/\hbar} d^3\boldsymbol{r}' (i\hbar) \frac{\partial}{\partial t} \\
&\quad \times \left(\int \psi(\boldsymbol{r},t) e^{-i\boldsymbol{p}\cdot\boldsymbol{r}/\hbar} d^3\boldsymbol{r} \right) d^3\boldsymbol{p} \\
&= \int \psi^*(\boldsymbol{r},t)(i\hbar)\frac{\partial}{\partial t}\psi(\boldsymbol{r},t) d^3\boldsymbol{r}
\end{aligned} \tag{3.40}
$$

where use was made of Eqs. (3.25) and (3.26).

3.3.4. *The Schrödinger equation and its operators*

From the preceding sections, the mean energy could have also been obtained by a mere addition of the mean kinetic and potential energies, so that we obtain the identity:

$$
\begin{aligned}
\int \psi^*(\boldsymbol{r},t)(i\hbar)\frac{\partial}{\partial t}\psi(\boldsymbol{r},t) d^3\boldsymbol{r} &= -\frac{\hbar^2}{2m} \int \psi^*(\boldsymbol{r},t) \left(\nabla_r^2 \psi(\boldsymbol{r},t) \right) d^3\boldsymbol{r} \\
&\quad + \int V(\boldsymbol{r}) \left| \psi(\boldsymbol{r},t) \right|^2 d^3\boldsymbol{r}
\end{aligned} \tag{3.41}
$$

where (3.40), (3.39) and (3.32) were used. This mean energy is obtained by an integration over the possible positions. The total energy is, everywhere, the sum of the kinetic and potential energies. We could also be interested

[18]The symbol H is used for the total energy. It is a very common notation in quantum physics as it is an inheritance from analytical mechanics where the quantity that is conserved because of time invariance is named the "Hamilton function" (see Section 6.2). The quantum mechanics operator associated with the energy will be the Hamiltonian. This is sufficient to adopt the notation from now on.

in the energy in a given elementary volume element centred at any position in space. We would then write

$$\int_{\delta V} \psi^*(\boldsymbol{r},t)\left[i\hbar\frac{\partial}{\partial t}\psi(\boldsymbol{r},t)+\frac{\hbar^2}{2m}\nabla_r^2\psi(\boldsymbol{r},t)-V(\boldsymbol{r})\psi(\boldsymbol{r},t)\right]d^3\boldsymbol{r}=0 \quad (3.42)$$

If this integral is zero for any given volume δV centred on any position $\boldsymbol{r}$, then the integrand must be zero. The (complex conjugate of the) wavefunction cannot be uniformly zero, otherwise the probability density would also be zero and the particle would not exist. We consequently establish, for any position and at any moment

$$-\frac{\hbar^2}{2m}\nabla_r^2\psi(\boldsymbol{r},t)+V(\boldsymbol{r})\psi(\boldsymbol{r},t)=i\hbar\frac{\partial}{\partial t}\psi(\boldsymbol{r},t) \quad (3.43)$$

This partial differential equation relates the time derivative of the wavefunction to its spatial structure and that of the potential acting on the particle. It is obviously an evolution equation for the wavefunction and is named the *Schrödinger equation* after Erwin Schrödinger (Fig. 3.9) who formulated it in 1926.

It should be noted that this equation is not about propagation as it does not contain a second-order derivative with respect to time. It is thus not quite the equivalent to (3.6) or (3.8) and more similar to Fourier's heat equation which is a diffusion equation. As it involves only the first-order derivative with respect to time, knowing the state at time t is enough to infer what the state will be like at any subsequent time. It is thus fully coherent with the statement that a knowledge of the wavefunction at a given moment is sufficient to determine the dynamical state of the system, hence its states in the future (providing that the environment is unchanged). Equation (3.43) can be rephrased in terms of operators acting on the wavefunction. We therefore introduce the Hamiltonian, the momentum and the potential operators. They are mathematical objects that act on a function (a wavefunction, here) and transform it to another function. These three operators are what is needed to compute the mean energy, the mean momentum and the mean potential when one uses the wavefunction and not its plane wave amplitude coefficients:

- The potential operator (from (3.32)) is noted $\widehat{V}(\boldsymbol{r})$ and is simply constructed from the potential acting on the particle. It is assumed to depend

Fig. 3.9. Erwin Schrödinger (1887–1961) was awarded the Nobel Prize in 1933 (with Paul Dirac). Erwin married Annemarie Bertel and she remained by his side as a friend preoccupied with his comfort, occasionally connecting him with other women for his demanding carnal pleasure. While in an alpine sanatorium to cure his tuberculosis (1921), his discovery of *Space, Time and Matter* by H. Weyl triggered his first original ideas in physics. Skiing and romantic intercourses were his necessary fuel for scientific inspiration and it was after one such practice that he produced a set of papers where "the idea springs from pure genius" (Einstein). His view of quantum mechanics was strongly disputed by Born and Heisenberg (even if Hilbert predicted both parties were saying the same thing using different mathematical objects, the reconciliation had to wait for von Neumann's operator approach). Still living with Annemarie, he openly started a second family life (sometimes under the same roof) with a colleague's wife who gave him a daughter. Despite a "repentant confession" which he wrote to satisfy the Nazi Rector of Graz University, he was dismissed from his positions in Vienna and Graz for anti-German activities and his honorary professorship in Vienna for "political unreliability". The double family (and occasional additional fruitful love affairs) ended up in Dublin and had to wait till 1955 to return to Vienna. Photo credits: public domain.

only on the particle's position and acts as a mere multiplication of the wavefunction by the potential expression $\widehat{V}(\boldsymbol{r})\psi(\boldsymbol{r},t) = V(\boldsymbol{r}) \times \psi(\boldsymbol{r},t)$.

- The momentum operator (from (3.36)),

$$\widehat{\boldsymbol{p}} \equiv -i\hbar\vec{\nabla}_r \tag{3.44}$$

is the short notation for three operators[19] (from (3.35)): $\widehat{p}_x = -i\hbar\partial_x$, $\widehat{p}_y = -i\hbar\partial_y$ and $\widehat{p}_z = -i\hbar\partial_z$ so that $\widehat{\boldsymbol{p}} = -i\hbar(\boldsymbol{e}_x\partial_x + \boldsymbol{e}_y\partial_y + \boldsymbol{e}_z\partial_z)$, where $\boldsymbol{e}_x$, $\boldsymbol{e}_y$ and $\boldsymbol{e}_z$ are the units vectors in each direction.

- The Hamiltonian, or total energy operator,

$$\widehat{H} \equiv \frac{1}{2m}\widehat{\boldsymbol{p}}^2 + \widehat{V}(\boldsymbol{r})$$

[19]This expression is arbitrarily given in cartesian coordinates. We will, in due time, have to express it in spherical coordinates. However, the expression lacks the cartesian equivalence between the components and is not useful at this stage.

which is another way of expressing

$$\widehat{H} = -\frac{\hbar^2}{2m}\nabla_r^2 + \widehat{V}(\boldsymbol{r}) \tag{3.45}$$

where the *kinetic energy operator* $\frac{1}{2m}\widehat{\boldsymbol{p}}^2 = -\frac{\hbar^2}{2m}\nabla_r^2$ was defined. We can then summarize the role played by Schrödinger's equation. If the wavefunction $\psi(\boldsymbol{r}, t)$ describes a particle in a given environment, and can consequently be used to predict some (mean) physical properties, then its mathematical structure should obey the following rule: at any position and any moment, the Hamiltonian operator and $i\hbar\partial_t$ modify the wavefunction in the exact same way.

For a given physical system within a given environment, the Hamiltonian is fixed. The Schrödinger equation imposes constraints on possible wavefunctions in much the same way as Maxwell's equations do for the electric and magnetic fields. The latter reveal the electric and magnetic fields compatible with a charge and current density distribution in a medium characterized by a dielectric permeability and a magnetic susceptibility. The Schrödinger equation is central to any quantitative treatment of non-relativistic quantum phenomenon.

Question 3.11: **Heisenberg's inequality.** Below a guide to a mathematical derivation of Heisenberg's inequality in one dimension.

1. Give a physical interpretation of $\widehat{X} = \widehat{x} - \int \psi^*(x,t)\ \widehat{x}\ \psi(x,t)dx$ and $\widehat{P} = \widehat{p} - \int \psi^*(x,t)\ \widehat{p}\ \psi(x,t)dx$.
2. Give a physical interpretation of $\int_{-\infty}^{+\infty} \left|\widehat{X}\psi(x,t)\right|^2 dx$ and

$$\int_{-\infty}^{+\infty} |\widehat{P}\psi(x,t)|^2 dx.$$

3. Let λ be any real number and

$$I_\psi(\lambda) = \int_{-\infty}^{+\infty} \left|\widehat{X}\psi(x,t) + i\lambda\widehat{P}\psi(x,t)\right|^2 dx$$

   What is the sign of $I_\psi(\lambda)$?

Question 3.11: (continued)

4. Develop the computation of $I_\psi(\lambda)$ and regroup terms of same power in λ.
5. Show that the sign of $I_\psi(\lambda)$ imposes a relationship between the quantum mean square values of the operators X and P defined in the first question. Show that this is exactly Heisenberg's inequality.

Answer:

1. $\int \psi^*(x,t)\ \widehat{x}\ \psi(x,t)dx$ and $\int \psi^*(x,t)\ \widehat{p}\ \psi(x,t)dx$ represent the mean value of the position and the momentum, respectively. Therefore, $\widehat{X}$ and $\widehat{P}$ are operators associated with the deviation from a quantum mean value, when the system is represented by wavefunction $\psi(x,t)$. For further convenience, we can write $\widehat{X} = \widehat{x} - \langle x\rangle$ and $\widehat{P} = \widehat{p} - \langle p\rangle$, where $\langle x\rangle$ and $\langle p\rangle$ stand for the quantum mean values for a system described by $\psi(x,t)$.
2. We develop

$$\begin{aligned}\int_{-\infty}^{+\infty} \left|\widehat{X}\psi(x,t)\right|^2 dx &= \int_{-\infty}^{+\infty} \left|(\widehat{x} - \langle x\rangle)\psi(x,t)\right|^2 dx \\ &= \int_{-\infty}^{+\infty} (\widehat{x}^2\rho(x,t) + \langle x\rangle^2\rho(x,t) - 2\widehat{x}\langle x\rangle\rho(x,t))dx \\ &= \langle x^2\rangle - \langle x\rangle^2\end{aligned}$$

where $\int \rho(x,t)dx = 1$ was used. This integral thus yields the quantum quadratic mean deviation in position $\Delta^2 x$ when the particle is described by $\psi(x,t)$.
Similarly, and using $\widehat{p} = \frac{\hbar}{i}\partial_x$:

$$\begin{aligned}\int_{-\infty}^{+\infty} \left|\widehat{P}\psi(x,t)\right|^2 dx &= \int_{-\infty}^{+\infty} \left|(\widehat{p} - \langle p\rangle)\psi(x,t)\right|^2 dx \\ &= \int_{-\infty}^{+\infty} \left|\left(\frac{\hbar}{i}\partial_x\psi(x,t) - \langle p\rangle\psi(x,t)\right)\right|^2 dx \\ &= \int_{-\infty}^{+\infty} (\hbar^2\partial_x\psi^*(x,t)\partial_x\psi(x,t) + \langle p\rangle^2\rho(x,t))dx \\ &\quad - \int_{-\infty}^{+\infty} \left(\frac{\hbar}{i}\langle p\rangle\psi^*(x,t)\partial_x\psi(x,t) + \frac{\hbar}{i}\langle p\rangle\psi(x,t)\partial_x\psi^*(x,t)\right)dx\end{aligned}$$

The first and last terms can be integrated by parts. Since the wavefunction must be square-integrable, $\psi(x,t)$ and $\partial_x\psi(x,t)$ cancel at infinities and

$$\begin{aligned}\int_{-\infty}^{+\infty} \left|\widehat{P}\psi(x,t)\right|^2 dx &= \int_{-\infty}^{+\infty} \Big(-\hbar^2\psi^*(x,t)\partial_x^2\psi(x,t) + \langle p\rangle^2\rho(x,t) \\ &\quad - 2\frac{\hbar}{i}\langle p\rangle\psi^*(x,t)\partial_x\psi(x,t)\Big)dx \\ &= \int_{-\infty}^{+\infty} (\psi^*(x,t)p^2\psi(x,t) + \langle p\rangle^2\rho(x,t) - 2\langle p\rangle\psi^*(x,t)p\psi(x,t))dx \\ &= \langle p^2\rangle - \langle p\rangle^2\end{aligned}$$

This is the quantum quadratic mean deviation in momentum $\Delta^2 p$ with the particle described by $\psi(x,t)$.
3. Whatever the value of λ, $I_\psi(\lambda)$ is the integral of a positive definite function. Hence, $I_\psi(\lambda)$ is positive.

Answer: (continued)

4. We develop $I_\psi(\lambda)$ and use results from question 2:

$$I_\psi(\lambda) = \int_{-\infty}^{+\infty} \left|\widehat{X}\psi(x,t) + i\lambda\widehat{P}\psi(x,t)\right|^2 dx$$

$$= \Delta^2 x + \lambda^2\Delta^2 p + i\lambda \int_{-\infty}^{+\infty} \psi^*(x,t)(x - \langle x\rangle)\frac{\hbar}{i}\partial_x\psi(x,t)dx$$

$$+ i\lambda \int_{-\infty}^{+\infty} \psi(x,t)(x - \langle x\rangle)\frac{\hbar}{i}\partial_x\psi^*(x,t)dx$$

Once more, integration by parts of the last term yields $I_\psi(\lambda) = \Delta^2 x + \lambda^2\Delta^2 p - \lambda\hbar$.

5. $I_\psi(\lambda)$ is positive, for any value of λ, only if $\hbar^2 - 4\Delta^2 x\Delta^2 p \leqslant 0$.

We then obtain Heisenberg's inequality: $\Delta x\Delta p \geqslant \frac{\hbar}{2}$.

3.3.5. *Stationary solutions to Schrödinger's equation*

Finding the possible particle (or assembly of particles) wavefunctions compatible with a given environment requires solving the Schrödinger equation. The natural way to proceed is usually through the method of separation of variables (time and positions) as illustrated in the electromagnetic wave case (3.1.3). This will be done in Question *3.12* (see below). Let us consider what happens if the system is not subjected to any time-dependent force and is given a fixed total energy ε. It can then be described by a wave packet but with the particular constraint that only one angular frequency is activated, so that $\omega_k = \varepsilon/\hbar$. The wave packet expression thus takes the form

$$\psi_\varepsilon(\boldsymbol{r},t) = \int \widetilde{\psi}_\varepsilon(\boldsymbol{k})e^{i\boldsymbol{k}\cdot\boldsymbol{r}}d^3\boldsymbol{k}\ e^{-i\varepsilon t/\hbar} = \phi_\varepsilon(\boldsymbol{r})e^{-i\varepsilon t/\hbar}$$

where the index ε was used to emphasize the fixed energy character of the wavefunction. We set $\phi_\varepsilon(\boldsymbol{r}) = \int \widetilde{\psi}_\varepsilon(\boldsymbol{k})e^{i\boldsymbol{k}\cdot\boldsymbol{r}}d^3\boldsymbol{k}$ to represent the (now separated) spatial component of the wavefunction. By construction, this kind of wavefunction represents a particle with a perfectly well-defined energy. We observe that the associated probability density distribution is time-independent:

$$\rho_\varepsilon(\boldsymbol{r}) = |\psi_\varepsilon(\boldsymbol{r},t)|^2 = |\phi_\varepsilon(\boldsymbol{r})|^2$$

which explains why such wavefunctions are said to represent "stationary states" of a particle.

Plugging the expression for a stationary wavefunction into Schrödinger's equation yields

$$-\frac{\hbar^2}{2m}\nabla_r^2\phi_\varepsilon(\boldsymbol{r})e^{-i\varepsilon t/\hbar}+\widehat{V}(\boldsymbol{r})\phi_\varepsilon(\boldsymbol{r})e^{-i\varepsilon t/\hbar}=i\hbar\phi_\varepsilon(\boldsymbol{r})\frac{\partial}{\partial t}e^{-i\varepsilon t/\hbar}$$

which obviously simplifies to the so-called "stationary Schrödinger equation":

$$-\frac{\hbar^2}{2m}\nabla_r^2\phi_\varepsilon(\boldsymbol{r})+\widehat{V}(\boldsymbol{r})\phi_\varepsilon(\boldsymbol{r})=\varepsilon\phi_\varepsilon(\boldsymbol{r}) \tag{3.46}$$

This equation should hold for any position in space and at any moment. Such stationary wavefunctions are solutions of (3.46) with their corresponding energies. It is possible to rephrase (3.46) in terms of operators to shine a new light on the problem at stake. We use (3.45) so that (3.46) becomes

$$\widehat{H}\phi_\varepsilon(\boldsymbol{r})=\varepsilon\phi_\varepsilon(\boldsymbol{r}) \tag{3.47}$$

This expression shows that looking for stationary wavefunctions and the associated energies is mathematically equivalent to the search for eigenfunctions and eigenvalues of the Hamiltonian, the total energy operator. This result will lead to many explorations in the followings chapters as, to put it quite crudely, the machinery of quantum physics is often reduced to diagonalizing matrices or solving a partial differential equation. Most of the physics is thus in establishing the correct and manageable Hamiltonian for the requested level of accuracy and finding its approximate (or exact in few cases) eigenfunctions and *eigenenergies*, i.e. the eigenvalues of the energy operator. The next chapter will give examples of the quest for stationary wavefunctions in a variety of ideal potentials. More realistic and useful potentials will be studied in the second volume.

✍ *Question 3.12:* **Separation of space and time.** Expression (3.43) is an evolution equation and couples the time and position changes. As many partial differential equations, it is not always easy to find its solutions. However, as in the electromagnetic propagation equation, and as long as the potential acting on the particle is not time dependent, we are glad to notice that spatial and time-dependent terms do not mix. They operate separately on the wavefunction. It was demonstrated that this feature allows for a very fruitful exploitation of the separation of variables method (see Section 3.1.3) with the ansatz: $\psi(\boldsymbol{r},t)=\phi(\boldsymbol{r})g(t)$. Use this expression to split the Schrödinger

Question 3.12: (continued)

evolution equation into one equation concerning only the space component $\phi(\boldsymbol{r})$ and another equation giving an explicit form to the time dependence of the wavefunction $g(t)$.

Answer: Using the proposed ansatz, we find

$$-\frac{\hbar^2}{2m}g(t)\nabla_r^2\phi(\boldsymbol{r}) + V(\boldsymbol{r})\phi(\boldsymbol{r})g(t) = i\hbar\phi(\boldsymbol{r})\frac{\partial}{\partial t}g(t)$$

Everywhere (and at every instant for which) the wavefunction does not cancel, it is possible to divide this expression by $\phi(\boldsymbol{r})g(t)$ so that

$$\frac{1}{\phi(\boldsymbol{r})}\left(-\frac{\hbar^2}{2m}\nabla_r^2\phi(\boldsymbol{r}) + V(\boldsymbol{r})\phi(\boldsymbol{r})\right) = i\hbar\frac{1}{g(t)}\frac{\partial}{\partial t}g(t)$$

Both sides are independent because they refer to different and independent variables (position and time). However, the equality must hold for any position and time. As a consequence, both sides must equal a constant. A dimensional analysis shows that the constant is an energy and we decide to call it ε so that two distinct equations are obtained. The first one is the stationary state Schrödinger equation as in (3.46):

$$-\frac{\hbar^2}{2m}\nabla_r^2\phi(\boldsymbol{r}) + V(\boldsymbol{r})\phi(\boldsymbol{r}) = \varepsilon\phi(\boldsymbol{r})$$

and the second equation only refers to the time evolution of the wavefunction:

$$i\hbar\frac{\partial}{\partial t}g(t) = \varepsilon g(t)$$

which naturally yields $g(t) \propto e^{-i\varepsilon t/\hbar}$. The separation constant is thus the eigenenergy associated with $\phi(\boldsymbol{r})$.

3.3.6. *General solution to Schrödinger's equation*

Schrödinger's equation is constructed from the action of differential (and multiplication) operators. It means that if $(\phi_{\varepsilon_1}(\boldsymbol{r}), \varepsilon_1)$ and $(\phi_{\varepsilon_2}(\boldsymbol{r}), \varepsilon_2)$ are two stationary solutions,

$$\psi(\boldsymbol{r}) = a_1\phi_{\varepsilon_1}(\boldsymbol{r})e^{-i\varepsilon_1 t/\hbar} + a_2\phi_{\varepsilon_2}(\boldsymbol{r})e^{-i\varepsilon_2 t/\hbar}$$

is a solution of the Schrödinger equation (3.43). The most general solution is then constructed as a wave packet involving all possible energies ε. It can be a continuous or a discrete spectrum depending on the nature of the environment and the constraints that the potential imposes on the wavefunctions. In both cases, writing the general solution necessitates the full set of possible stationary solutions. This is why most quantum physics problems start with the search for a full determination of eigenenergies and eigenfunctions of the system's Hamiltonian.

Let us assume that we have found a wavefunction that describes a particle. It thus satisfies the Schrödinger equation and, when divided by $i\hbar$, yields

$$\frac{\partial}{\partial t}\psi(\boldsymbol{r},t) = \frac{i\hbar}{2m}\nabla_r^2\psi(\boldsymbol{r},t) - \frac{i}{\hbar}V(\boldsymbol{r})\psi(\boldsymbol{r},t)$$

and the complex conjugate fulfils

$$\frac{\partial}{\partial t}\psi^*(\boldsymbol{r},t) = -\frac{i\hbar}{2m}\nabla_r^2\psi^*(\boldsymbol{r},t) + \frac{i}{\hbar}V(\boldsymbol{r})\psi^*(\boldsymbol{r},t)$$

We can then compute the time derivative of the probability density distribution $\rho(\boldsymbol{r},t) = \psi^*(\boldsymbol{r},t)\psi(\boldsymbol{r},t)$. The terms involving the potential cancel out and

$$\begin{aligned}\frac{\partial\rho(\boldsymbol{r},t)}{\partial t} &= \frac{\partial}{\partial t}\left[\psi^*(\boldsymbol{r},t)\psi(\boldsymbol{r},t)\right] \\ &= \psi^*(\boldsymbol{r},t)\frac{i\hbar}{2m}\nabla_r^2\psi(\boldsymbol{r},t) - \psi(\boldsymbol{r},t)\frac{i\hbar}{2m}\nabla_r^2\psi^*(\boldsymbol{r},t)\end{aligned}$$

This expression can be rearranged to emphasize the presence of a divergence operator:

$$\frac{\partial\rho(\boldsymbol{r},t)}{\partial t} = \frac{i\hbar}{2m}\vec{\nabla}_r\cdot\left(\psi^*(\boldsymbol{r},t)\ \vec{\nabla}_r\psi(\boldsymbol{r},t) - \psi(\boldsymbol{r},t)\ \vec{\nabla}_r\psi^*(\boldsymbol{r},t)\right)$$

or, using (3.44),

$$\frac{\partial\rho(\boldsymbol{r},t)}{\partial t} = -\vec{\nabla}_r\cdot\left[\psi^*(\boldsymbol{r},t)\ \frac{\widehat{\boldsymbol{p}}}{2m}\psi(\boldsymbol{r},t) + \psi(\boldsymbol{r},t)\ \left(\frac{\widehat{\boldsymbol{p}}}{2m}\psi(\boldsymbol{r},t)\right)^*\right]$$

There is a striking resemblance between this equation and (3.5). This is not an accidental coincidence. They embody the same physical reality: a change of probability in any elementary volume should be related to the difference between the incoming and outgoing of a quantity. Consequently, we call it the *probability current*. It is defined by

$$\begin{aligned}\boldsymbol{j}(\boldsymbol{r},t) &= \frac{i\hbar}{2m}\left(\psi(\boldsymbol{r},t)\ \vec{\nabla}_r\psi^*(\boldsymbol{r},t) - \psi^*(\boldsymbol{r},t)\ \vec{\nabla}_r\psi(\boldsymbol{r},t)\right) \\ &= \mathrm{Re}\left[\psi^*(\boldsymbol{r},t)\ \frac{\widehat{\boldsymbol{p}}}{m}\psi(\boldsymbol{r},t)\right]\end{aligned} \tag{3.48}$$

Important note on the phase of a wavefunction: A wavefunction contains all the important information for predicting quantum mean values. These are the only quantities provided by quantum theory that can be compared to a measurement.[20] Quantum mean values are always obtained by means of an integral involving the wavefunction twice in an expression of the form:

$$\langle Q \rangle = \int \psi^*(\boldsymbol{r},t)\ \widehat{Q}\ \psi(\boldsymbol{r},t)\ d^3\boldsymbol{r}$$

where $\widehat{Q}$ (for example: $\widehat{H}$, $\widehat{x}$, $\widehat{p}$ or $\widehat{V}$) is the operator associated with a measurable quantity Q.

Therefore, an overall *constant* phase factor for the wavefunction can never be expected and it is thus unimportant. It can be set to any convenient value. We will generally choose a setting of 1.

Question 3.13: **Probability current, charge current.** We assume that a free particle can be described by a plane wave in a finite volume.

1. Write the normalized wavefunction of a free particle of charge q, mass m and momentum $\boldsymbol{p}$ in a volume V.
2. What is the probability distribution $\rho(\boldsymbol{r})$ in the volume? What is the charge density $\rho_q(\boldsymbol{r})$?
3. Compute the associated probability current and show that it is similar to the classic charge current density.

Answer:

1. For a free particle of momentum $\boldsymbol{p}$, the energy is $\frac{p^2}{2m}$ and the wavefunction is a plane wave $\psi(\boldsymbol{r},t) = \mathcal{N} e^{i(\boldsymbol{p}\cdot\boldsymbol{r} - tp^2/2m)/\hbar}$. The probability density is normalized by setting

$$\int_V |\psi(\boldsymbol{r},t)|^2 d^3\boldsymbol{r} = V\,|\mathcal{N}|^2 = 1$$

Thus, $\psi(\boldsymbol{r},t) = \frac{1}{\sqrt{V}} e^{i(\boldsymbol{p}\cdot\boldsymbol{r} - tp^2/2m + \phi)/\hbar}$.

2. The probability distribution was obtained while normalizing the wavefunction and is uniform in the volume V: $\rho(\boldsymbol{r}) = 1/V$. The charge density is also uniform: $\rho_q(\boldsymbol{r}) = q/V$.

[20] Relationships between quantum theory results and possible experimental observations will be addressed in Chapter 5.

Answer: (continued)

3. The probability current is readily obtained from (3.48):

$$\boldsymbol{j}(\boldsymbol{r},t) = \frac{i\hbar}{2m}\left(-\frac{i\boldsymbol{p}}{\hbar}|\psi(\boldsymbol{r},t)|^2 - \frac{i\boldsymbol{p}}{\hbar}|\psi(\boldsymbol{r},t)|^2\right) = \frac{\boldsymbol{p}}{m}|\psi(\boldsymbol{r},t)|^2 = \frac{\boldsymbol{p}}{m}\rho(\boldsymbol{r},t)$$

Once the normalized expression of the wavefunction is used, it leads to $\frac{p}{Vm}$. However, the former expression is more interesting as $\boldsymbol{p}/m$ is of course the speed of the free particle and the probability current can thus be compared to the well-known expression of charge current density: $\boldsymbol{j}_q(\boldsymbol{r}) = \rho_q(\boldsymbol{r})\boldsymbol{v}(\boldsymbol{r})$. This classical expression matches exactly the quantum physics expression with $\boldsymbol{j}_q(\boldsymbol{r}) = q\boldsymbol{j}(\boldsymbol{r})$ because, in the free particle case, the momentum (hence the speed) is perfectly determined.

3.4. Stationary States in One Dimension

Solving the Schrödinger equation will be our major occupation for the rest of this volume. For the sake of simplicity, many examples and illustrations will be chosen in an unrealistic 1D world. While several of the results that can be obtained from such problems are directly transferable to many dimensions, it is important to be aware that working in one dimension brings a stringent constraint: whatever the symmetry of the potential, there is a one-to-one correspondence between a stationary state (Hamiltonian eigenfunction) and the eigenenergy. This can be rephrased in mathematical terms: in a 1D world, eigenenergies are non-degenerate. Or, to put it another way, there is a unique wavefunction that can be associated with a given fixed energy value.

The proof is as follows. Assume there are two eigenfunctions, thus representing two different stationary states of the system that share the same eigenenergy. We call them $\phi_1(x)$ and $\phi_2(x)$:

$$-\frac{\hbar^2}{2m}\frac{\partial^2}{\partial x^2}\phi_1(x) + V(x)\phi_1(x) = \varepsilon\phi_1(x) \tag{3.49a}$$

$$-\frac{\hbar^2}{2m}\frac{\partial^2}{\partial x^2}\phi_2(x) + V(x)\phi_2(x) = \varepsilon\phi_2(x) \tag{3.49b}$$

Multiplying Eq. (3.49a) by $\phi_2(x)$ and Eq. (3.49b) by $\phi_1(x)$ and subtracting the resulting expressions leaves

$$-\frac{\hbar^2}{2m}\left(\phi_2(x)\frac{\partial^2}{\partial x^2}\phi_1(x) - \phi_1(x)\frac{\partial^2}{\partial x^2}\phi_2(x)\right) = 0$$

Once the multiplying constant is eliminated, we see that

$$\frac{\partial}{\partial x}\left(\phi_2(x)\frac{\partial}{\partial x}\phi_1(x) - \phi_1(x)\frac{\partial}{\partial x}\phi_2(x)\right) = 0$$

This implies that, if C is a constant,

$$\phi_2(x)\frac{\partial}{\partial x}\phi_1(x) - \phi_1(x)\frac{\partial}{\partial x}\phi_2(x) = C \tag{3.50}$$

However, to be legitimate stationary states wavefunctions, $\phi_1(x)$ and $\phi_2(x)$ need to be square-integrable. As a consequence, they are required to cancel at infinities. This is also true for their first derivatives. Since (3.50) must hold for any value of x, it must be true at $x \to \pm\infty$. This is compatible with a constant value for C if and only if $C = 0$. Therefore, we have the equality:

$$\phi_2(x)\frac{\partial}{\partial x}\phi_1(x) = \phi_1(x)\frac{\partial}{\partial x}\phi_2(x)$$

Wherever the wavefunction is not zero, this can be replaced by[21]

$$\frac{\partial}{\partial x}\frac{\phi_1(x)}{\phi_2(x)} = 0$$

The eigenfunctions can thus only share the same eigenenergy if they only differ by a proportionality constant:

$$\phi_1(x) = \alpha\phi_2(x)$$

The functions need to be normalized, thus α is a mere constant phase factor which, as we have just seen, is not physically meaningful. Consequently, $\phi_1(x)$ and $\phi_2(x)$ refer to the same quantum stationary state and the energy is not degenerate.

It must be emphasized that this result is only valid for the 1D case.

[21] We must refrain from integrating the two functions separately. This will then involve their logarithms which may not be defined because the functions are not positive definite.

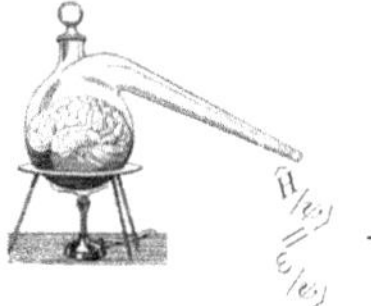

Chapter 3: Nuts & Bolts

- *A quantum particle is described by a continuous function, named a "**wavefunction**" $\psi(\boldsymbol{r},t)$ such that $|\psi(\boldsymbol{r},t)|^2 d^3r$ is the particle's **position probability** in $\boldsymbol{r}$ at time t. Therefore, $\int |\psi(\boldsymbol{r},t)|^2\, d^3r = 1$ must always be satisfied.*
- *The wavefunction is a solution of the **Schrödinger equation** $-\frac{\hbar^2}{2m}\nabla^2\psi(\boldsymbol{r},t) + V(\boldsymbol{r})\psi(\boldsymbol{r},t) = i\hbar\frac{\partial}{\partial t}\psi(\boldsymbol{r},t)$ where $V(\boldsymbol{r})$ is the potential to which the particle is subjected.*
- *The wave-like behaviour of a particle implies the existence of **Heisenberg's inequalities** which relate the respective indetermination on its position and momentum $\Delta x \Delta p_x \geqslant \frac{\hbar}{2}$ (idem in the y and z directions) and on its energy and lifetime $\Delta\varepsilon\Delta t \gtrsim \hbar$.*
- ***Stationary states** are those corresponding to a well-defined energy. The spatial components of such wave functions are solutions of the eigenfunction equation*

$$-\frac{\hbar^2}{2m}\nabla^2\phi_\varepsilon(\boldsymbol{r}) + V(\boldsymbol{r})\phi_\varepsilon(\boldsymbol{r}) = \varepsilon\phi_\varepsilon(\boldsymbol{r})$$

  where ε is the energy associated with the state represented by $\psi_\varepsilon(\boldsymbol{r},t) = \phi_\varepsilon(\boldsymbol{r})e^{-i\varepsilon t/\hbar}$.
- *$\widehat{p}_x = -i\hbar\frac{\partial}{\partial x}$ is the "**momentum operator**" (for direction x). The probability current density (in the x direction) is computed as $j_x(\boldsymbol{r},t) - \frac{1}{m}\mathrm{Re}\left[\psi^*(\boldsymbol{r},t)\widehat{p}_x\psi(\boldsymbol{r},t)\right]$.*

4

Piecewise Constant Potentials

The content of this chapter will help you understand ***quantum dots, scanning tunnelling microscopes, thermoluminescence dating, alpha radioactivity,*** *and related topics.*

In this chapter, we explain how Schrödinger's equation can be used to infer possible wavefunctions for a particle subjected to a variety of ideal potentials. None of these are actually encountered in real problems. However, as examples will show, they can often be invoked to provide a simple and qualitative explanation for a quantum behaviour. These potentials present an idealized reality in the sense that they have constant values in all regions except for discontinuities that are never observed. Even if Nature were to allow such sudden changes in its properties, any instrumental resolution is finite and would smooth out the discontinuity in our observation.

More realistic potentials will be broached in the second volume. Piecewise potentials, which will be studied in this chapter, are frequently used as an approximation of, or a first stage to, a more sophisticated real potential. Techniques that are developed in their treatment will prove to be useful in mastering quantum wave mechanics, before tackling a more delicate and mathematical treatment.

4.1. Potential Jumps and Infinite Forces

This chapter describes particle in ideal potentials. For simplicity, we will restrict ourselves to constant piecewise potentials which have never been proved to exist but bear some pedagogical virtue. We have been taught to work with physical systems by means of Newton's second law of motion, so it is quite natural to primarily think in terms of forces acting on particles instead of potentials. Forces are perfectly legitimate concepts but are not directly accounted for by the Schrödinger equation which imposes an energy-oriented line of reasoning. It is however pertinent to wonder what kind of physical reality is represented by a discontinuous potential: the associated force, derived from $-\vec{\nabla}V(x)$, would be infinite, which does not seem to make much sense.

Let the particle be described by a continuous position probability density $\rho(x)$. One can thus calculate the force endured by the particle when, like in the step potential case described in Section 4.4, the potential is null in the $x < 0$ region and suddenly reaches the constant value $-V_0 < 0$ in the $x > 0$ region. The mean force is obviously found by

$$\langle F \rangle = -\int_{-\infty}^{+\infty} \rho(x)\partial_x V(x)dx$$

This expression can be integrated by parts. Since the potential is finite and the density needs to cancel at both infinities:

$$\langle F \rangle = -\left[V(x)\ \rho(x)\right]_{-\infty}^{+\infty} - V_0 \int_0^{+\infty} \partial_x \rho(x)dx = V_0\ \rho(0)$$

Therefore, as this expression holds for any well-behaving position probability density, the force operator defined from the discontinuous potential $\widehat{\boldsymbol{F}} = -\vec{\nabla}V(x)$ is here given by $\widehat{F}(x) = -\partial_x V(x) = V_0\delta(x)$ where the infinitely "spiky" Dirac delta function provides the ideal spatial behaviour. Therefore, we can rest assured that, because the discontinuity is both finite in amplitude ($V_0 < +\infty$) and occurs at a discrete location, the force is also finite. It is, as one should expect, proportional to the jump amplitude and the probability density for the particle to be located at this specific position. The next section will discuss the rather delicate topic of the continuity of wavefunctions and their derivatives.

4.2. On Wavefunction Continuity

As this chapter will be considering several examples of discontinuous potentials, it is interesting to question the necessity of continuity[1] for the wavefunction and, to some extent, its derivatives.

Since the wavefunction describes the physical system, it is usually required to be single valued. Without any loss of generality, here we will only consider one-dimensional (1D) systems. We have stated that the wavefunction represents all the available knowledge of the system. We usually consider its representation $\phi_\varepsilon(x)$ in the position space. From the preceding section and the physical interpretation of the wavefunction, it appears that choosing such a space is arbitrary and we could just as well have decided to work in the momentum space as there is a one-to-one correspondence with its Fourier transform $\widetilde{\phi}_\varepsilon(p)$ (as stated by Eq. (3.25)). We have also introduced several operators that act on wavefunctions to provide mean values. We will focus here on the kinetic energy operator $\frac{\widehat{p}^2}{2m}$. As it is used to compute the quantum mean kinetic energy, it is required that its action on $\widetilde{\phi}_\varepsilon(p)$ be a mere product

$$\frac{\widehat{p}^2}{2m}\widetilde{\phi}_\varepsilon(p) = \frac{p^2}{2m}\widetilde{\phi}_\varepsilon(p)$$

for any value of p. Application of the kinetic energy operator, like any operator, changes the wavefunction to a different one. However, according to (3.44), the kinetic energy operator acts on a wavefunction so that it becomes another function:

$$\frac{\widehat{p}^2}{2m}\phi_\varepsilon(x) = \frac{-\hbar^2}{2m}\partial_x^2\phi_\varepsilon(x)$$

[1]This section mostly relies on arguments given in [14].

The one-to-one correspondence[2] needs to persist for any such transformation of the wavefunction. Consequently, it is a necessary condition that the resulting wavefunctions in both representations are still related through a Fourier transform:

$$\frac{p^2}{2m}\widetilde{\phi}_\varepsilon(p) = \frac{1}{(2\pi\hbar)^{3/2}}\int_{-\infty}^{+\infty} e^{-ipx/\hbar}\left(\frac{-\hbar^2}{2m}\right)\partial_x^2\phi_\varepsilon(x)dx$$

The following sections will consider cases where the potential shows a sudden, discontinuous jump. For simplicity, we will suppose here that this occurs at $x = 0$. We then split the above integral into two connecting regions:

$$\frac{p^2}{2m}\widetilde{\phi}_\varepsilon(p) = \int_{-\infty}^{0^-}\frac{e^{-ipx/\hbar}}{(2\pi\hbar)^{3/2}}\left(\frac{-\hbar^2}{2m}\right)\partial_x^2\phi_\varepsilon(x)dx + \int_{0^+}^{+\infty}\frac{e^{-ipx/\hbar}}{(2\pi\hbar)^{3/2}}\left(\frac{-\hbar^2}{2m}\right)\partial_x^2\phi_\varepsilon(x)dx$$

The wavefunction, and consequently its derivatives, cancel at infinity. Thus, integration by parts (for both regions) leads to

$$\begin{aligned}\frac{p^2}{2m}\widetilde{\phi}_\varepsilon(p) &= \frac{1}{(2\pi\hbar)^{3/2}}\left(\frac{-\hbar^2}{2m}\right)(\partial_x\phi_\varepsilon(0^-) - \partial_x\phi_\varepsilon(0^+)) \\ &\quad - \left(\frac{ip\hbar}{2m}\right)\frac{1}{(2\pi\hbar)^{3/2}}\int_{-\infty}^{0^-} e^{-ipx/\hbar}\partial_x\phi_\varepsilon(x)dx \\ &\quad - \left(\frac{ip\hbar}{2m}\right)\frac{1}{(2\pi\hbar)^{3/2}}\int_{0^+}^{+\infty} e^{-ipx/\hbar}\partial_x\phi_\varepsilon(x)dx\end{aligned}$$

which can again be integrated by parts and yields

$$\begin{aligned}\frac{p^2}{2m}\widetilde{\phi}_\varepsilon(p) &= \frac{1}{(2\pi\hbar)^{3/2}}\left(\frac{-\hbar^2}{2m}\right)(\partial_x\phi_\varepsilon(0^-) - \partial_x\phi_\varepsilon(0^+)) \\ &\quad - \frac{1}{(2\pi\hbar)^{3/2}}\left(\frac{ip\hbar}{2m}\right)(\phi_\varepsilon(0^-) - \phi_\varepsilon(0^+)) \\ &\quad + \left(\frac{p^2}{2m}\right)\int_{-\infty}^{0^-}\frac{e^{-ipx/\hbar}}{(2\pi\hbar)^{3/2}}\phi_\varepsilon(x)dx \\ &\quad + \left(\frac{p^2}{2m}\right)\int_{0^+}^{+\infty}\frac{e^{-ipx/\hbar}}{(2\pi\hbar)^{3/2}}\phi_\varepsilon(x)dx\end{aligned}$$

[2]This constraint will become even more obvious in the next chapter.

Obviously, this equality holds for any value of momentum if the wavefunction and its first derivative are continuous at $x = 0$ even if the potential is discontinuous.

The single-valued character of the wavefunction is imposed by its physical content. It is not so obvious for its derivatives. However, since the wavefunction must satisfy the stationary Schrödinger equation, we have

$$-\frac{\hbar^2}{2m}\partial_x^2\phi_\varepsilon(x) + V(x)\phi_\varepsilon(x) = \varepsilon\phi_\varepsilon(x)$$

At the discontinuity of the potential, if the Schrödinger equation still holds,

$$\begin{aligned}&-\frac{\hbar^2}{2m}\left(\partial_x^2\phi_\varepsilon(0^+) - \partial_x^2\phi_\varepsilon(0^-)\right) + \left(V(0^+)\phi_\varepsilon(0^+) - V(0^-)\phi_\varepsilon(0^-)\right)\\ &\quad = \varepsilon\left(\phi_\varepsilon(0^+) - \phi_\varepsilon(0^-)\right)\end{aligned}$$

The continuity of the wavefunction imposes cancellation of the right-hand side expression. Thus, two cases need to be considered:

(1) If the potential discontinuity jump is finite, the second derivative also needs to be finite. Hence, the second derivative is not continuous but the first derivative still can be.
(2) If the potential jump is infinite, then the first derivative cannot be continuous. The only remaining continuity condition is on the wavefunction itself.

A more thorough discussion on wavefunction continuity can be found in [6].

4.3. Infinite Well

From a pedagogical point of view, one of the most important examples of piecewise potentials is given by the *infinite one-dimensional quantum well.* It aims to represent that only a limited region in space is accessible to the particle. From this constraint, it will appear clearly that the possible stationary wavefunctions (and the associated energies) constitute a discrete set. As the particle will not be allowed to reach $x \to \pm\infty$, the set of wavefunctions will be said to describe *bound states.* This potential well can be

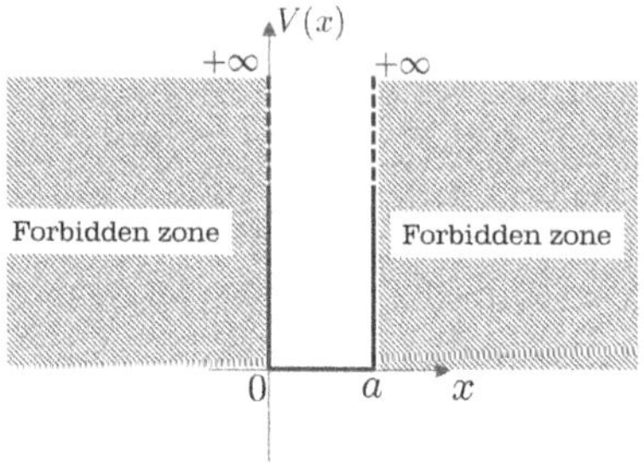

Fig. 4.1. Schematic representation of a 1D infinite quantum well. The particle is allowed only in the $0 < x < a$ region where the potential energy is 0.

thought of as a 1D box with totally impassable walls (see Fig. 4.1) and is modelled by the mathematical expressions:

$$V(x) = \begin{cases} +\infty & \text{if } x \geqslant a \\ +\infty & \text{if } x \leqslant 0 \\ 0 & \text{if } 0 < x < a \end{cases}$$

where a is the width of the potential well.

In this first example, following the description of Section 3.3.5, our aim is to find the Hamiltonian eigenfunctions, i.e. stationary wavefunctions of the system. The particle cannot be found in regions outside $0 < x < a$. Such a configuration is ensured by setting the position probability density, hence the wavefunction, to zero:

$$\phi_\varepsilon(x > a) = \phi_\varepsilon(x < 0) = 0$$

However, since the particle must exist somewhere, we will explore the intermediate zone, where $V = 0$. As the kinetic energy is always positive, we will seek wavefunctions for which the particle has an energy $\varepsilon \geqslant 0$. Inside the well, the potential is constant (and arbitrarily set to zero). The equation for stationary states is then:

$$-\frac{\hbar^2}{2m}\frac{\partial^2}{\partial x^2}\phi_\varepsilon(x) = \varepsilon\phi_\varepsilon(x)$$

Looking for eigenfunctions is thus equivalent to solving the second-order differential equation:

$$\frac{\partial^2}{\partial x^2}\phi_k(x) = -k^2\phi_k(x)$$

where we have changed the index name because $k = \sqrt{2m\varepsilon}/\hbar$. The solutions of the equation take the general form of a linear combination of plane waves:

$$\phi_k(x) = Ae^{ikx} + Be^{-ikx} \tag{4.1}$$

At this point, we still do not know the possible energies (or k values), nor the constants A and B. They will be determined by boundary normalization conditions. The continuity of the wavefunction forces cancellation at $x = 0$. Thus

$$\phi_k(x = 0) = A + B = 0$$

so that

$$\phi_k(x) = A(e^{ikx} - e^{-ikx}) = 2iA\sin(kx)$$

At the other boundary, the wavefunction must also be zero:

$$2iA\sin(ka) = 0$$

We cannot choose $A = 0$ because the particle must exist inside the well. We thus obtain the condition of possible wave vectors:

$$k_n = n\frac{\pi}{a} \quad \text{with } n \in \mathbb{N}^*$$

It is important to note that

- The solution $n = 0$ was rejected because it implies $k = 0$. This solution yields a zero wavefunction, hence the absence of the particle.
- Only positive values for n are considered. Negative values are superfluous because $+k$ and $-k$ have already been taken into account by the combination of two plane waves propagating in opposite directions. Positive and negative n values would thus lead to exactly the same wavefunctions and thus describe the same physical state.
- Cancellation of the wavefunction at the boundary induces an equal contribution of the two plane waves. This means that for a stationary state,

the particle has momenta $p = \hbar k$ and $p = -\hbar k$ with equal weight. The mean momentum (and speed) is then zero.

Since the wavefunction only exists between $x = 0$ and $x = a$, normalization can easily be calculated by restricting the integral:

$$\int_0^a |2iA\sin(k_n x)|^2\,dx = 4|A|^2 \int_0^a \sin^2(n\pi x/a)dx = 1$$

which yields

$$|A|^2 = \frac{1}{2a}$$

Stationary state wavefunctions are then:

$$\phi_n(x) = \sqrt{\frac{2}{a}}\sin(n\pi x/a) \tag{4.2}$$

where the unmeasurable (hence unimportant) phase factor was arbitrarily set to 1.

The possible wave vectors form a discrete set. As a consequence, this is also true for the particle's momentum modulus:

$$p_n = n\frac{\hbar\pi}{a} \quad \text{with } n \in \mathbb{N}^*$$

Since both $+p$ and $-p$ are equally present in the wavefunction, the sign is not defined.

The eigenenergies are uniquely determined by k values and are thus also *quantized*:

$$\varepsilon_n = n^2\frac{1}{2m}\frac{\hbar^2\pi^2}{a^2} \quad \text{with } n \in \mathbb{N}^*$$

As explained in Section 3.3.6, the most general form for a particle wavefunction, at any moment, is then given by a wavepacket built as a linear combination of all the stationary states. In the present case, the packet is constructed from the discrete eigenset:

$$\begin{aligned}\psi(t) &= \sqrt{\frac{2}{a}}\sum_{n\in\mathbb{N}^*} c_n \sin(k_n x)e^{-i\varepsilon_n t/\hbar} \\ &= \sqrt{\frac{2}{a}}\sum_{n\in\mathbb{N}^*} c_n \sin(n\pi x/a)e^{-in^2(\hbar\pi^2)t/(2ma^2)} \end{aligned} \tag{4.3}$$

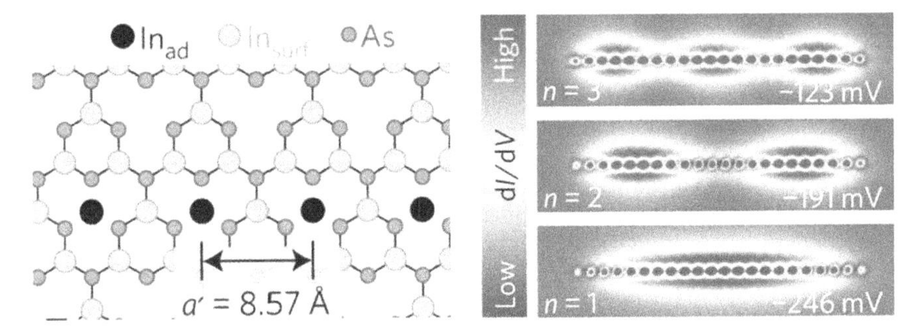

Fig. 4.2. Artificial quantum dot obtained by depositing individual indium atoms on adjacent vacancy sites of InAs surface. Left: Geometry of the surface. Adatoms, which form the quantum dot, are in black. Right: Electron probability density as a function of their energies. Reprinted with permission from [39]. Copyright © 2014 by Springer Nature.

Quantum dots are nanostructures that can be crafted to bear particular properties of an infinite quantum well. Fabrication processes can be as sophisticated as the deposition of one atom at a time thanks to scanning tunnelling microscopy technology (see Section 4.9.2). In the example represented on Fig. 4.2, electron density is measured as a function of STM bias voltage which is proportional to the energy of the probed electron. The attraction potential here is not flat. However, the shape of each displayed density shows that the wavefunction stays very similar to that of the lowest energy states in an infinite quantum well (4.2). The energies also follow approximately the n^2/a^2 rule.

Question 4.1: **Proton or electron in an infinite quantum well. From microscopic to macroscopic systems.** Calculate the eigenenergy values (in electron-volt) for a proton ($m_p \approx 10^{-27}$ kg) in an infinite quantum well when the width is $a = 1$ fermi. What about an electron ($m_e \approx 10^{-30}$ kg) in an $a = 1$ Å infinite well? Can this help us inferring the energy values of a proton in a nucleus or an electron in an atom? What is expected for conduction electrons in a solid?

Answer: If we consider a particle in an infinite quantum well, in 1D, 2D or 3D, the order of magnitude for the energy levels is always given by $\frac{\hbar^2}{2m}\frac{\pi^2}{a^2}$.

The invariant factor is $\hbar^2\pi^2 \approx 10^{-67}$. For a proton in a 1-fm well, $2m_p a^2 \approx 10^{-57}$ whereas for an electron in a 1-Å well, $2m_e a^2 \approx 10^{-50}$. As a result, the typical energies

Answer: (continued)

are of the order of 10 MeV for the proton and 10 eV for the electron. In a nucleus, the strong force restraints the position of the proton in the range of a fermi. So it will be no wonder to find that typical nuclear energies per nucleon (constituent of the nucleus) are of the order of 10 MeV. On the other hand, it will be seen that the typical size of an atom is given by the extension of the electron cloud and of the order of 1 Å. Again, 10 eV gives a fair approximation of the energy of an electron in an atom. Conduction electrons in a solid can be seen as free electrons in a macroscopic box. Consequently, a similar formula can be used, setting a in the millimetre or centimetre range. Therefore, the separation between energy levels is necessarily $(10^{-7})^2$ times smaller and approximately of the order of 10^{-13} eV! Such a value is much smaller than the typical thermal energy which can be absorbed from the environment: it corresponds to a rather atypical equilibrium temperature $T \approx 10^{-13} \times 1.6 \times 10^{-19}/(1.38 \times 10^{-23}) \approx 10^{-9}$ K. The spectrum discreteness for free electrons in a macroscopic solid is thus hardly (never) observable.

✍ *Question 4.2:* **Degenerate states of an electron in a two-dimensional infinite quantum well.**

1. Consider an infinite well identical to the above but in two-dimensional (2D) space. The widths are a and b, along the two respective directions x and y:

$$V(x, y) = \begin{cases} 0 & \text{if } 0 < x < a \text{ and } 0 < y < b \\ +\infty & \text{elsewhere} \end{cases}$$

Show that this time, using the separation of variables method, the wavefunctions of bound states are of the form $B \sin(k_x x) \sin(k_y y)$. Find B, k_x, k_y and the possible energies.

2. Show that in the 2D case, if $b = a$, there are several distinct physical states corresponding to the same energy.

Answer:

1. The wavefunction is null outside $0 < x < a$ and $0 < y < b$.
The Hamiltonian eigenfunction equation is then:

$$-\frac{\hbar^2}{2m}\left(\frac{\partial^2}{\partial x^2} + \frac{\partial^2}{\partial y^2}\right)\phi(x, y) = \varepsilon\phi(x, y) \quad \text{for } 0 < x < a \text{ and } 0 < y < b$$

The solution can be searched in the form: $\phi(x, y) = f(x)g(y)$, and

$$-\frac{\hbar^2}{2m}g(y)\frac{\partial^2 f(x)}{\partial x^2} - \frac{\hbar^2}{2m}f(x)\frac{\partial^2 g(y)}{\partial y^2} = \varepsilon f(x)g(y)$$

The solving technique is the one that makes use of the fact that the two position coordinates are separate and independent variables. This is similar to the considerations that prompted us to split the problem into, respectively, time and position equations

Answer: (continued)

in Section 3.3.5. We thus find

$$-\frac{\hbar^2}{2m}\frac{1}{f(x)}\frac{\partial^2 f(x)}{\partial x^2} - \frac{\hbar^2}{2m}\frac{1}{g(y)}\frac{\partial^2 g(y)}{\partial y^2}\phi(x,y) = \varepsilon$$

Both functions depend on the independent variables x and y. Yet the sum of both terms on the left must always give a fixed value, the total energy.

This is only possible if every term of the sum is itself constant. Thus we find two distinct eigenvalue equations, one for each dimension of space:

$$-\frac{\hbar^2}{2m}\frac{\partial^2 f(x)}{\partial x^2} = \varepsilon_x f(x)$$

$$-\frac{\hbar^2}{2m}\frac{\partial^2 g(y)}{\partial y^2} = \varepsilon_y g(y)$$

with $\varepsilon = \varepsilon_x + \varepsilon_y$. The solutions are now known because, for each dimension, they are identical to the previous 1D problem:

$$\phi(x,y) = A\sin(k_x x) \times B\sin(k_y y)$$

with $k_x = n_x\frac{\pi}{a}$ and $k_y = n_y\frac{\pi}{b}$ and, as previously $(n_x, n_y) \in \mathbb{N}^{*2}$.

Normalization is found by

$$\int_0^a\int_0^b |AB|^2\sin^2(k_x x)\sin^2(k_y y)dxdy = 1$$

so, within a phase factor, $AB = 2/\sqrt{ab}$.

The possible energies are thus given by

$$\begin{aligned}\varepsilon_{n_x,n_y} = \varepsilon_x + \varepsilon_y &= \frac{\hbar^2}{2m}k_x^2 + \frac{\hbar^2}{2m}k_y^2 \\ &= \frac{\hbar^2}{2m}\left(n_x^2\frac{\pi^2}{a^2} + n_y^2\frac{\pi^2}{b^2}\right)\end{aligned}$$

2. If the potential is invariant by a $\pi/2$ rotation then $a = b$. Possible energy levels are

$$\varepsilon_{n_x,n_y} = \frac{\hbar^2\pi^2}{2ma^2}\left(n_x^2 + n_y^2\right)$$

It therefore appears that several wavefunctions correspond to the same energy particle. The energies are said to be "degenerate", i.e. there is no longer a one-to-one relationship between energy and wavefunction. The number of different wavefunctions for a given level of energy is connected to the pairs of numbers (n_x, n_y) so that $n_x^2 + n_y^2 = n \in \mathbb{N}^*$. In particular, due to the $\pi/2$ rotational invariance, the two pairs (n_x, n_y) and (n_y, n_x) systematically correspond to the same energy value.

4.4. Potential Step

The potential step problem is a first opportunity to encounter the important problem of scattering. As it is a 1D system, it is limited to transmission and reflection phenomena.

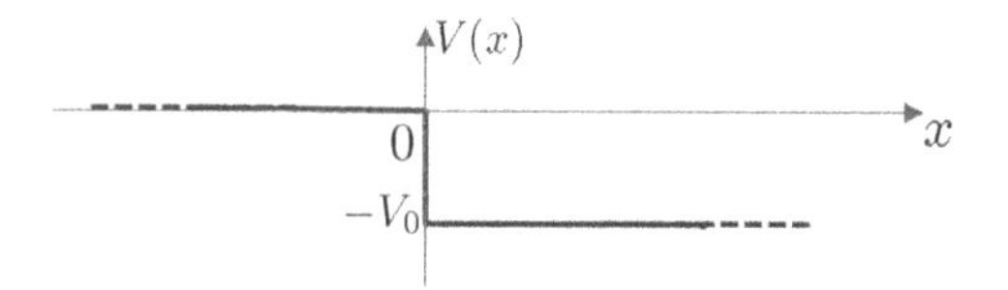

Fig. 4.3. The potential energy changes abruptly at $x = 0$ from 0 to $-V_0$.

4.4.1. *Going down*

A free particle is incident from the $x < 0$ region with kinetic energy ε. We know that this is an idealization since the wavefunction for such a condition cannot be normalized. However, as it will soon become apparent, this is not as important as it seems and we assume that the incoming wavefunction can thus be written $Ae^{i(px-\varepsilon t)/\hbar}$. This function indeed represents a propagation in the positive direction along the x-axis.

The potential has an abrupt (but finite) change at $x = 0$ from $V(x < 0) = 0$ to $V(x > 0) = -V_0 < 0$ (see Fig. 4.3).

Therefore, the stationary state Schrödinger equation for energy ε needs to be separated into

$$\frac{-\hbar^2}{2m}\frac{\partial^2\phi(x)}{\partial x^2} = \varepsilon\phi(x) \quad \text{for } x < 0 \tag{4.4a}$$

$$\frac{-\hbar^2}{2m}\frac{\partial^2\phi(x)}{\partial x^2} - V_0\phi(x) = \varepsilon\phi(x) \quad \text{for } x > 0 \tag{4.4b}$$

For both regions, the most general solutions are superpositions of plane waves

$$\phi(x) = A_1e^{ik_1x} + B_1e^{-ik_1x} \quad \text{for } x < 0 \text{ and } k_1 = \sqrt{2m\varepsilon}/\hbar \tag{4.5a}$$

$$\phi(x) = A_2e^{ik_2x} + B_2e^{-ik_2x} \quad \text{for } x > 0 \text{ and } k_2 = \sqrt{2m(\varepsilon + V_0)}/\hbar \tag{4.5b}$$

Of course, these two expressions pertain to the same wavefunction which is associated with a stationary state of energy ε, so that

$$\psi(x,t) = A_1e^{i(p_1x-\varepsilon t)/\hbar} + B_1e^{-i(p_1x+\varepsilon t)/\hbar} \quad \text{for } x < 0 \tag{4.6a}$$

$$\psi(x,t) = A_2e^{i(p_2x-\varepsilon t)/\hbar} + B_2e^{-i(p_2x+\varepsilon t)/\hbar} \quad \text{for } x > 0 \tag{4.6b}$$

with $p_1 = \hbar k_1 = \sqrt{2m\varepsilon}$ and $p_2 = \hbar k_2 = \sqrt{2m(\varepsilon + V_0)}$. The constants A_1, B_1, A_2 and B_2 are determined from the boundary conditions.

As previously mentioned, A_1 is the amplitude for a wave propagating in the $x < 0$ region towards the step. It can thus be interpreted as describing an incoming particle. However, for a stationary state, the wavefunction cannot be decomposed in such a boldly manner. There is also, in the same region, a second component which propagates counter-wise. If boundary conditions are such that the particle is coming from $x = -\infty$, then the second component can be seen as a wave reflected by the discontinuity in the potential.

In the same line of thought, the first component of (4.6b) is a wave propagating in the positive direction, from $x = 0$. It then describes the particle after it has crossed the discontinuity. Such a function is then associated with a transmitted wave. With the hypothesis of the particle originating from $-\infty$, the last component, with amplitude B_2, describes a wave propagating in the negative direction and coming back from $+\infty$. This last feature is usually discarded because only a change in the potential can create a reflection and it is assumed that no other force than the one at $x = 0$ is acting on the particle. We therefore set $B_2 = 0$.

As stated at the beginning of this chapter (Sections 4.1 and 4.2), the wavefunction and its first derivative are continuous even at the potential jump. This brings the conditions

$$\phi(x = 0) = A_1 + B_1 = A_2 \tag{4.7a}$$

$$\partial_x \phi(x = 0) = i\frac{p_1}{\hbar}(A_1 - B_1) = i\frac{p_2}{\hbar}A_2 \tag{4.7b}$$

These relations show that, even if the wavefunction cannot be normalized, it is possible to seek relations between component amplitudes. For example, taking the incident wave amplitude A_1 as a reference, we have

$$\mathcal{R} = \frac{B_1}{A_1} = \frac{p_1 - p_2}{p_1 + p_2} \tag{4.8a}$$

$$\mathcal{T} = \frac{A_2}{A_1} = \frac{2p_1}{p_1 + p_2} \tag{4.8b}$$

These quantities are not really meaningful unless they are used to express reflection and transmission of probability currents. The reflection coefficient is given by the ratio between the reflected and incident currents:

$$R = \left| \frac{j_r}{j_i} \right| = \frac{\frac{p_1}{m}|B_1|^2}{\frac{p_1}{m}|A_1|^2} = \frac{|p_1 - p_2|^2}{|p_1 + p_2|^2} \tag{4.9}$$

and the transmission coefficient is computed from the transmitted and incident currents:

$$T = \left| \frac{j_t}{j_i} \right| = \frac{\frac{p_2}{m}|A_2|^2}{\frac{p_1}{m}|A_1|^2} = \frac{4p_1 p_2}{|p_1 + p_2|^2} \tag{4.10}$$

where expression (3.48) was used for each component. For stationary states, the probability density does not change over time. Therefore, in one dimension, the continuity equation implies

$$\frac{\partial}{\partial x} j(x,t) = -\frac{\partial}{\partial t}\rho(x,t) = 0$$

The flux should then be conserved everywhere and indeed

$$R + T = \frac{|p_1 - p_2|^2}{|p_1 + p_2|^2} + \frac{4p_1 p_2}{|p_1 + p_2|^2} = 1$$

It is now legitimate to compare this result with what is expected from classical physics considerations. If we omit some technical details, for example those related to the finite dimensions of the particle, what we have been describing here is the case of a marble with an initial kinetic energy which is about to roll down the staircase. As gravity works in its favour, each step of height h will cause the marble to increase its kinetic energy by mgh. There is thus no backward motion of the particle to be expected. With such a configuration of potential, classical physics only predicts a forward motion associated with a transmission of the particle from the $x < 0$ to the $x > 0$ regions. However, the change in the potential is at the origin of a possible reflection of the quantum particle with no classical equivalence. The pertinence of this result can be questioned since we chose to consider a particle with an initially well-defined momentum. Heisenberg's inequality states that, under this particular condition, the particle is totally delocalized which is difficult to relate to any classical behaviour. However, if the wave packet approach is chosen, thanks to the linearity of Schrödinger's equation, we readily obtain a wavefunction as

$$\psi(x,t) = \int \widetilde{\phi}(p)(e^{i(px-\varepsilon_1(p)t)/\hbar} + \mathcal{R}(p)e^{i(px-\varepsilon_1(p)t)/\hbar})dp \quad \text{for } x < 0 \tag{4.11a}$$

$$\psi(x,t) = \int \mathcal{T}(p)\widetilde{\phi}(p)e^{i(px-\varepsilon_2(p)t)/\hbar}dp \quad \text{for } x > 0 \tag{4.11b}$$

The wave packet can be initially built so that the wavefunction is rather well-localized and still exhibits some reflected components.

4.4.2. *Going up*

We now consider, while the particle is still issued from $x = -\infty$, what happens if it encounters a step up (Fig. 4.4).

As the particle is initially free, its energy (which is solely of kinetic origin) is positive. We know what the classical behaviour should be:

(1) If the initial energy is larger than the magnitude of the potential step, the particle will climb the step and continue with a lower speed because some of the kinetic energy has to be converted to potential energy.
(2) If the energy is lower than the potential, the particle will simply bounce back with the same speed modulus.

The first case is not different from what we studied in the "going down" case. We know that the particle will indeed be transmitted, but that quantum physics predicts a possible reflected component.

The second case is interesting because it brings another possibility to our attention. While the mathematical treatment is similar, we are here led

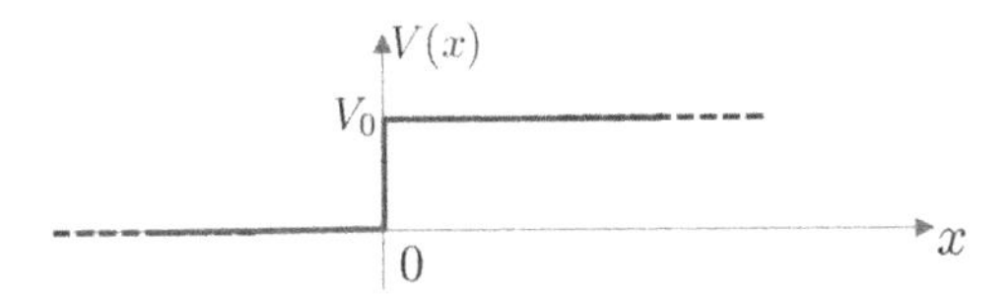

Fig. 4.4. The potential energy changes abruptly at $x = 0$ from 0 to $+V_0$.

to consider $\varepsilon < V_0$ so that in the high-potential region the wave vector is imaginary:

$$k_2 = i\sqrt{2m|\varepsilon - V_0|/\hbar^2} = i\kappa_2$$

with $\kappa_2 = \sqrt{2m|\varepsilon - V_0|/\hbar^2} > 0$. While the wavefunction in the region of negative positions is left unchanged in its form, in the $x > 0$ region, there is no propagation because

$$\psi(x > 0, t) = A_2 e^{(-\kappa_2 x - i\omega t)}$$

The probability density vanishes in an exponential manner as we explore further within the potential step. Indeed, it is still possible to calculate a transmission coefficient, but (3.48) gives:

$$j(x > 0, t) = \text{Re}\left[\psi^*(x,t)\ \frac{\widehat{p}}{m}\psi(x,t)\right] = \frac{1}{m}\text{Re}[i\hbar\kappa_2|A_2|^2 e^{-2\kappa_2 x}] = 0$$

As a result, the presence of a tiny amount of probability density notwithstanding, there is no propagation hence no transmission in the $x > 0$ region. The reflection coefficient is thus maximal:

$$R = \frac{|p_1 - p_2|^2}{|p_1 + p_2|^2} = \frac{|k_1 - i\kappa_2|^2}{|k_1 + i\kappa_2|^2} = 1$$

This phenomenon is the quantum equivalent of the well-known evanescent field in electromagnetism: a light beam cannot penetrate a perfect conductor because the dielectric constant is negative, hence the refractive index is imaginary. However, we are also aware that near-field optical microscopy has proved to be a powerful way to significantly improve imaging resolving power. We will see in Sections 4.7 and 4.8 that quantum physics has its counterpart: the tunnel effect.

4.5. Finite Square Well: Bound and Unbound States

In a way, the finite square well can be considered as a mixed case, halfway between potential steps and the infinite potential well. The potential is

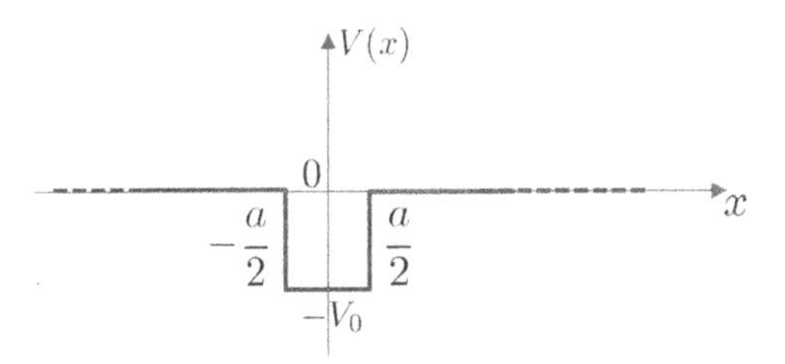

Fig. 4.5. The potential energy describes a finite square well. The potential energy is zero everywhere except between $-a/2$ and $a/2$.

taken to have the form:

$$V(x) = \begin{cases} 0 & \text{if } x \geqslant a \\ 0 & \text{if } x \leqslant 0 \\ -V_0 & \text{if } 0 < x < a \end{cases}$$

As it will be explained in the next chapter,[3] it is always desirable to make use of the symmetry properties of the system to reduce superfluous calculations. We are thus encouraged to shift (see Fig. 4.5) the potential (or change variables) so that it becomes:

$$V(x) = \begin{cases} 0 & \text{if } x \geqslant a/2 \\ 0 & \text{if } x \leqslant -a/2 \\ -V_0 & \text{if } -a/2 < x < a/2 \end{cases}$$

$$-\frac{\hbar^2}{2m}\frac{\partial^2}{\partial x^2}\phi(x) + V(x)\phi(x) = \varepsilon\phi(x)$$

Reversing the position axis changes all x into $-x$, and the equation becomes

$$-\frac{\hbar^2}{2m}\frac{\partial^2}{\partial(-x)^2}\phi(-x) + V(-x)\phi(-x) = \varepsilon\phi(-x)$$

The kinetic energy operator is unchanged and, since the potential is even ($V(x) = V(-x)$), we obtain exactly the same equation for $\phi(x)$ and $\phi(-x)$:

$$-\frac{\hbar^2}{2m}\frac{\partial^2}{\partial x^2}\phi(-x) + V(x)\phi(-x) = \varepsilon\phi(-x)$$

---

[3] But also in Chapter 6 where Noether's theorem expresses the key role played by symmetry invariance in the conservation of observable quantities.

It has been shown (3.4) that, for 1D systems, the energy ε is non-degenerate, i.e. it is associated with a unique wavefunction. Consequently, $\phi(x)$ and $\phi(-x)$ need to be identical within a global phase factor: $\phi(-x) = \alpha\phi(x)$. This argument can be used once more: we start from the equation on $\phi(-x)$, reverse the x-axis and, since the process is exactly the same, arrive at a new relationship $\phi(x) = \alpha\phi(-x)$. But a double inversion brings us back to the initial wavefunction and we are left with $\phi(x) = \alpha^2\phi(x)$. The result is then $\alpha = \pm 1$ so that the wavefunction should have one of the parities:

$$\phi(-x) = \pm\phi(x) \tag{4.12}$$

In an even 1D potential, the wavefunction must be odd or even. This is a very strong restriction on possible wavefunctions and allows for seeking a solution in a reduced region of space. We will then use the symmetry property to reconstruct the wavefunction in its entirety.

In the present case, a symmetric 1D square finite well is considered. We will restrict our quest for eigenfunctions to the $x > 0$ region. The wavefunction in the full position space will then be obtained by folding it to the $x < 0$ region in the two possible ways: $\phi(-x) = \phi(x)$ and $\phi(-x) = -\phi(x)$. For each solution that is found in $x > 0$, we will then generate two possible independent eigensolutions.

The identification of the possible wavefunctions takes place in a manner that is very similar to that seen in the case of the potential step. For each region, we write the Schrödinger equation and its local solution:

- Region 1, $-\infty < x \leqslant -a/2$, the potential is 0

$$-\frac{\hbar^2}{2m}\frac{\partial^2}{\partial x^2}\psi_1(x) = \varepsilon\psi_1(x)$$

$$\psi_1(x) = A_1 e^{ik_1 x} + B_1 e^{-ik_1 x}$$

- Region 2, $-a/2 < x < +a/2$, the potential is $-V_0 < 0$ and

$$-\frac{\hbar^2}{2m}\frac{\partial^2}{\partial x^2}\psi_2(x) - V_0\psi_2(x) = \varepsilon\psi_2(x)$$

$$\psi_2(x) = A_2 e^{ik_2 x} + B_2 e^{-ik_2 x}$$

- Region 3, $+a/2 \leqslant x < +\infty$, the potential is again 0:

$$-\frac{\hbar^2}{2m}\frac{\partial^2}{\partial x^2}\psi_3(x) = \varepsilon\psi_3(x)$$

$$\psi_3(x) = A_3 e^{ik_3 x} + B_3 e^{-ik_3 x}$$

It is immediately found that $k_1 = k_3 = \sqrt{2m\varepsilon}/\hbar$ and $k_2 = \sqrt{2m(\varepsilon + V_0)}/\hbar$. Suppose the particle was initially positioned in the immediate vicinity of $x \approx 0$, there cannot be any wave coming from $-\infty$ nor from $+\infty$. Hence, we can set $A_1 = B_3 = 0$.

As before, the potential jump is finite, which allows for a continuity of both the wavefunction and its first derivative at $x = -a/2$ and $x = +a/2$. Thus:

$$B_1 e^{ik_1 a/2} = A_2 e^{-ik_2 a/2} + B_2 e^{ik_2 a/2}$$

$$-k_1 B_1 e^{ik_1 a/2} = k_2 A_2 e^{-ik_2 a/2} - k_2 B_2 e^{ik_2 a/2}$$

and

$$A_3 e^{ik_1 a/2} = A_2 e^{ik_2 a/2} + B_2 e^{-ik_2 a/2}$$

$$k_1 A_3 e^{ik_1 a/2} = k_2 A_2 e^{ik_2 a/2} - k_2 B_2 e^{-ik_2 a/2}$$

As demonstrated above, given the symmetry of the problem, the wavefunction is even or odd. So we will introduce even functions (denoted by $\psi^+(x)$), for which $B_1 = A_3$ and $A_2 = B_2$, and odd functions (denoted by $\psi^-(x)$), for which $B_1 = -A_3$ and $A_2 = -B_2$. Owing to these parity considerations, only the following continuity conditions at the junction are now useful:

$$B_1 e^{ik_1 a/2} = A_2(e^{-ik_2 a/2} \pm e^{ik_2 a/2})$$

$$-k_1 B_1 e^{ik_1 a/2} = k_2 A_2(e^{-ik_2 a/2} \mp e^{ik_2 a/2})$$

It is thus possible to find B_1/A_2 and the energies ε.

It quickly becomes clear that two classes of energies must be distinguished, $\varepsilon > 0$ and $\varepsilon < 0$. If the energy of the particle is negative, k_1 necessarily becomes imaginary; it can be denoted $i\kappa_1 = i\sqrt{2m|\varepsilon|}/\hbar$. This means, as in the potential step case, that the probability of presence will be exponentially decreasing in the regions outside the well (1 and 3). Hence, for $\varepsilon < 0$, the particle is somehow forced to remain in the well (or very marginally exceeding the limits $\pm a/2$). It is said that the particle is in a *bound state*. Note that for these bound states, in the semi-infinite regions (1 and 3), the wavefunction no longer has a plane wave structure. It is now normalizable.

We now turn to the energy. Depending on whether the wavefunction is even or odd, we must solve

$$\frac{k_2}{\kappa_1} = \frac{1}{\tan(k_2 a/2)} \quad \text{for even functions}$$

and

$$\frac{k_2}{\kappa_1} = -\tan(k_2 a/2) \quad \text{for odd functions.}$$

Recall that $k_2 = \sqrt{2m(\varepsilon + V_0)}/\hbar$ and $\kappa_1 = \sqrt{2m|\varepsilon|}/\hbar$. Unfortunately, there is no explicit analytical solution to this problem and our search for possible energies of a particle contained in a finite depth well V_0 requires a numerical approach (Fig. 4.6). To show better the physical conditions associated with the solutions of this problem, we introduce $\hbar q = \sqrt{2mV_0}$, so that $\kappa_1^2 = q^2 - k_2^2$ for $\varepsilon < 0$, the resolution is reduced to searching:

$$\frac{k_2}{q} = |\cos(k_2 a/2)| \quad \text{for } k_2 \in \left[0, \frac{\pi}{a}\right] \tag{4.13a}$$

for even functions, and

$$\frac{k_2}{q} = |\sin(k_2 a/2)| \quad \text{for } k_2 \in \left[\frac{\pi}{a}, \frac{2\pi}{a}\right] \tag{4.13b}$$

for odd functions.

It can be seen that above a certain slope of k_2/q, i.e. below a given $|V_0|$ value, there is only one bound (even) solution. This corresponds to a particularly shallow well (q, so V_0, is small).

As the well gets narrower ($a \to 0$), the period of $\cos(k_2 a/2)$ or $\sin(k_2 a/2)$ becomes larger and the solutions to (4.13) will correspond to increasingly separate values of k_2. This means that the possible energies will also be

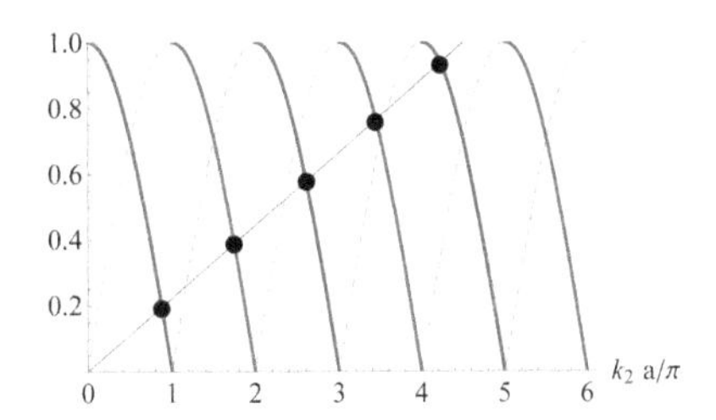

Fig. 4.6. Intersections of the line k_2/q (fine straight line) with the red curves (even wavefunctions) and the blue curves yield a discrete set of possible wave vectors. Only descending parts of the blue or red functions are considered because of the domains where k_2 exists in (4.13a) and (4.13b). The particle energy spectrum, when it is confined ($\varepsilon < 0$), is thus discrete.

separate. Therefore, it is not surprising that soon enough the gap between the first bound state and the next one exceeds the depth of the well. There is then a unique bound state. The next possible stationary state corresponds to a free particle outside the well: an *unbound state*, sometimes called a *scattering state* because it corresponds to the possibility for the particle to go to infinity (Fig. 4.6).

We can also observe what happens when the well becomes infinitely deep ($V_0 \to \infty$). In this case, q becomes infinitely large, the slope of k_2/q becomes negligible and the solutions correspond to the intersections of the sine or cosine with the x-axis. It is then found that the solutions are of the type $k_2 = n\pi/a$ with $n \in \mathbb{N}^*$. This brings us back to the infinite quantum well of Section 4.3.

It should be noted that, very generally speaking, the solutions for which the particle remains bound, confined in the well, correspond to a discrete spectrum of possible energies.

For positive energies, k_1 is real, and the above calculation can be repeated with one important difference: the possible energies form a continuous spectrum. This is obviously due to the fact that for $\varepsilon > 0$ the particle is no longer confined in the well. It can be considered as a free particle whose wavefunction is locally disturbed by the potential change.

4.6. An Application of Quantum Wells: Thermoluminescence and Dating

The use of electron confinement in wells is ubiquitous in nanotechnologies with quantum dots, nanodots, nanowires or nanolayers which yield a variety of potential applications. However, their understanding is often subjected to a knowledge of semi-conductor physics. Therefore, the example given in this section is a simplified version of a more natural type of well as can be found in many ionic crystals.

In 1663, Robert Boyle, an English chemist, reports [79] the case of a diamond that, when exposed to a heat source, shows a faint visible glow. It is now established that, for millions of years, the crystal had been exposed to radioactive elements such as uranium, thorium and potassium. The light was a mere manifestation of the release of such an accumulated energy.

Today, it is possible to acquire sodium chloride crystals, surprisingly coloured in a mixture of yellowish and brown which when the temperature is raised by some hundreds of degrees, emit a sudden visible burst of light and then go back to their familiar white table salt colour. This phenomenon, known as *thermoluminescence*, corresponds to electron migration from a metastable state towards a "standard" valence state. The unusual initial colour of the salt crystal turns out to be the consequence of sustained exposure of ordinary NaCl to an ionizing (i.e. high energy) radiation such as UV, X-rays or even gamma-rays or neutrons. The energy that is thus brought then allows an electron to be ejected from an ion and then captured by a local trap created by a missing chlorine atom.[4] For NaCl, the well ([23] [47] and Fig. 4.7) has an excited state at 2.7 eV. These

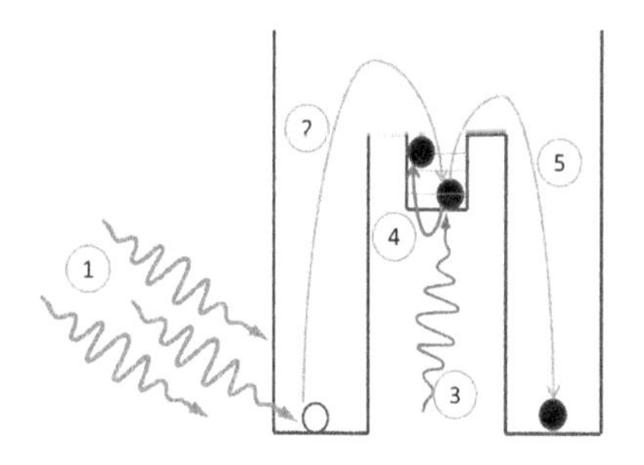

Fig. 4.7. When it is subjected to a highly energetic radiation (1), an electron is released by photoelectric effect. It can then be captured by a vacancy trap (2). In this well, the energy levels are such that the electron can undergo transitions (4) that correspond to light absorption in the visible domain (3). The usual crystal colour then appears to be altered. Heating the crystal at a sufficient temperature then allows the electron to escape the trap and reach a standard stable state (5).

[4]Crystal vacancies sometimes occur naturally or can be created by sodium atmosphere heating in order to increase their proportion against chlorine ions.

transitions in such a quantum well are then responsible for the yellowish-brown colour of the irradiated salt. The term *coloured centres* is therefore legitimately used. Heat energy above 3 eV then allows the electron to be released from the vacancy trap and reach a bound state belonging to a nearby ion. This departure from a metastable state towards a lower energy level is accompanied by a photon emission that is observed in thermoluminescence.

Question 4.3: **Thermoluminescence dating**. Thermoluminescence offers the possibility of dating ancient objects [33]. In that respect, the sample is heated (usually above 450°C and sometimes 600°C for clay) to allow for the system to relax to a stable state. Measuring the intensity of the emitted radiation is used to quantify the number of trapped electrons that are returning to the bound state. It then makes it possible to deduce the cumulative dose of radiation to which the system has been exposed since it was manufactured. This is generally established only after a calibration measurement has been made in which thermoluminescent radiation is quantified from the sample under a calibrated radiation dose. A chemical analysis of the environment (e.g. the amount of radioactive trace elements that are present in the sample or in its immediate vicinity) or an accurate measurement of the ambient radiation provides information on the annual dose to the sample. From the total accumulated energy and the deposited annual dose, the duration of exposure can then be deduced.

Measurement uncertainties are typically of the order of 5–15% but, depending on environmental conditions, it may be as high as 50%. However, this technique was able to clarify an hypothesis concerning human origins in the Middle East [91].

1. From the minimal requested heating temperature, can one estimate the depth of the potential well created by the vacancy trap in sodium chlorine?
2. Why is it necessary that the artefact was initially baked?
3. Even if the object is maintained at a relatively low temperature, the age is sometimes underestimated. Regardless of external phenomena which may have led to a decline (or even vanishing) in background radiation, for example, show that this can be attributed to a quantum phenomenon.
4. Why is it difficult to date very ancient objects, such as rocks as old as several million years?
5. Apart from heating, what other method may be considered for generating the de-excitation radiation? As an example, quartz contains several eV deep traps that would not easily give away their electrons.

Answer:

1. If it is necessary to heat at a minimal temperature of 450°C, this means that the required thermal energy to escape the trap is $E \approx k_B T = 10^{-20}\,\text{J} = 6 \times 10^{-2}\,\text{eV}$.
2. The dating method is based on a measurement of accumulated energy through trapping of electrons on the vacancy sites. It is essential that during the manufacture of the object, if it is for example a pot, the constituent materials have all been discharged from previously accumulated energy. Baking brings the thermal energy needed to "reset the meter to zero".
3. If all the possible causes of external radiation decline are discarded, the age under estimates must be due to an intrinsic and spontaneous process. The underestimation originates in a lower than expected number of trapped electrons on the day of the measurement. This means that electrons were able to escape the trap prior to the measurement. Figure 4.7 explains a possible mechanism. If the trap is not deep enough, weak thermal energy may be sufficient for release. This is named a *fading* process. Another relaxation path is quantum mechanical: if the potential barrier between the vacancy trap and the standard bound site is narrow enough, the electron can tunnel through. This is made possible because the wavefunction does not vanish at the walls but decreases progressively outside the well (see Section 4.7). This is called *anomalous fading*. At this stage, since the shape of the well or the width of the barrier are unknown to us, it becomes difficult to estimate the probability of such a phenomenon.
4. One reason given is that if the irradiation time is too long, there is a high risk that vacancy sites eventually become saturated with trapped electrons. The amount of energy then restored by heating will result in underestimating the age of the sample because it could not store more background radiation energy in the most recent years.
5. Having to heat precious artefacts to very high temperatures requires microscopic sampling. But an alternative is to make use of a short wavelength electromagnetic radiation. This corresponds to an energy high enough to allow for a release of electrons from vacancy traps and a subsequent recombination luminescence. In quartz, such an electron extraction is achieved by exposure to green light and leads to ultraviolet luminescence. The technique is obviously *photoluminescence* and no longer *thermoluminescence*.

4.7. Potential Barrier

Quantum particles exhibit peculiar behaviour when they penetrate a potential barrier. The typical piecewise potential barrier is described as follows (Fig. 4.8):

$$V(x) = \begin{cases} 0 & \text{if } x \geqslant a \\ 0 & \text{if } x \leqslant 0 \\ V_0 > 0 & \text{if } 0 < x < a \end{cases}$$

The potential does not allow any bound state and we will examine, as in the potential step case, what happens when the particle comes from $-\infty$

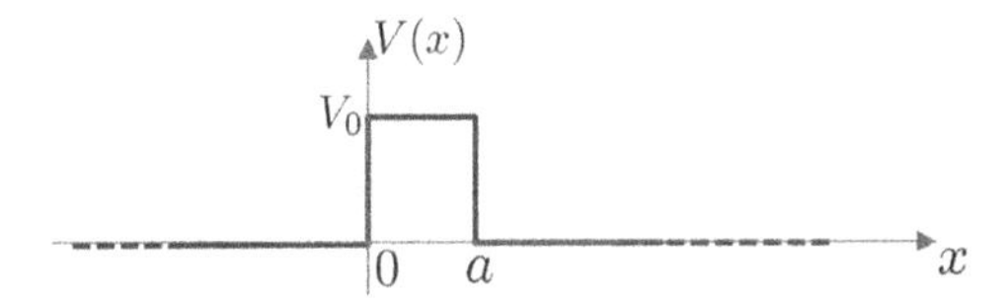

Fig. 4.8. The particle undergoes an abrupt potential energy change at $x = 0$. The potential is constant and equals $V_0 > 0$ until the particle reaches $x = a$ where the potential becomes 0.

and is incident on the barrier. The possible solutions in each part of the position space can be written as follows:

- Region 1, $-\infty < x \leqslant 0$, the potential is zero:

$$-\frac{\hbar^2}{2m}\frac{\partial^2}{\partial x^2}\phi_1(x) = \varepsilon\phi_1(x)$$

$$\phi_1(x) \propto e^{ik_1x} + \mathcal{R}e^{-ik_1x}$$

- Region 2, $0 < x < +a$, the potential is $V_0 > 0$ and

$$-\frac{\hbar^2}{2m}\frac{\partial^2}{\partial x^2}\psi_2(x) + V_0\phi_2(x) = \varepsilon\phi_2(x)$$

$$\phi_2(x) = Ae^{ik_2x} + Be^{-ik_2x}$$

- Region 3, $+a \leqslant x < +\infty$, the Schrödinger equation is identical to region 1 and

$$-\frac{\hbar^2}{2m}\frac{\partial^2}{\partial x^2}\phi_3(x) = \varepsilon\phi_3(x)$$

$$\phi_3(x) = \mathcal{T}e^{ik_1(x-a)}$$

  where the wave returning from $+\infty$ was once more discarded.

The incident wave vector is uniquely determined by the kinetic energy $k_1 = \sqrt{2m\varepsilon}/\hbar$ while $k_2 = \sqrt{2m(\varepsilon - V_0)}/\hbar$.

The scattering case corresponds to the situation that we are now familiar with: $\varepsilon \geqslant V_0$. The wavefunction is a plane wave everywhere and both discontinuities contribute to a reflected component just as in Section 4.4.

The $\varepsilon \leqslant V_0$ case deserves a more detailed study. Classical physics does not allow any particle transmission and, as a consequence, no current probability is expected in the $x > a$ region. Conversely, from a quantum perspective, as broached in the context of the "going up" potential (section 4.4.2), there is a minute possibility for the particle to penetrate the exclusion region. The fact that the $a < x < +\infty$ region corresponds to a zero potential allows for an increase in the possibility for the particle to go through the barrier. To obtain a quantitative result, the continuity conditions need to be exploited:

- continuity of $\phi(x)$ at $x = 0$:

$$1 + \mathcal{R} = A + B \qquad ①$$

- continuity of $\partial_x \phi(x)$ at $x = 0$:

$$\frac{k_1}{k_2}(1 - \mathcal{R}) = A - B \qquad ②$$

- continuity of $\phi(x)$ at $x = a$:

$$Ae^{ik_2 a} + Be^{-ik_2 a} = \mathcal{T} \qquad ③$$

- continuity of $\partial_x \phi(x)$ at $x = a$:

$$k_2(Ae^{ik_2 a} - Be^{-ik_2 a}) = k_1 \mathcal{T} \qquad ④$$

So basic arithmetic is required here. Take ① – ② and ① + ② to obtain B and A respectively as a function of $\mathcal{R}$, k_1 and k_2. Results are inserted into ③ and ④, so that

$$\mathcal{T} = (1 + \mathcal{R}) \cos k_2 a + i\frac{k_1}{k_2}(1 - \mathcal{R}) \sin k_2 a \qquad ③$$

and

$$\mathcal{T} = \frac{k_2}{k_1}\left[i(1 + \mathcal{R}) \sin k_2 a + \frac{k_1}{k_2}(1 - \mathcal{R}) \cos k_2 a\right] \qquad ④$$

It is then possible to extract $\mathcal{R}$:

$$\mathcal{R} = \frac{(k_1^2 - k_2^2) \sin k_2 a}{(k_1^2 + k_2^2) \sin k_2 a + 2ik_1 k_2 \cos k_2 a}$$

and, since $k_1 = k_3$ we have $T = |\mathcal{T}|^2$, and we readily obtain the transmission coefficient:

$$T = |\mathcal{T}|^2 = \left| \frac{2ik_1k_2}{(k_1^2 + k_2^2)\sin k_2a + 2ik_1k_2 \cos k_2a} \right|^2 = \frac{4k_1^2k_2^2}{(k_1^2 + k_2^2)^2 \sin^2 k_2a + 4k_1^2k_2^2 \cos^2 k_2a} \quad (4.14)$$

It is easy to check that the conservation of flux is still valid: $R + T = 1$, whatever the nature of k_2.

If $\varepsilon \leqslant V_0$, k_2 becomes an imaginary number so that $k_2 = i\kappa = i\sqrt{2m|\varepsilon - V_0|}/\hbar$ and using hyperbolic functions $\cosh x = \cos(ix)$ and $i \sinh x = \sin(ix)$, the transmission coefficient becomes

$$T = \frac{1}{\frac{(k_1^2+\kappa^2)^2}{4k_1^2\kappa^2} \sinh^2 \kappa a + 1} \quad (4.15)$$

This expression is better exploited if we introduce two quantities:

- $\Delta V = V_0 - \varepsilon$ is the barrier's height relative to the particle's initial kinetic energy.
- $\ell_\varepsilon = 1/\kappa = \hbar/\sqrt{2m(V_0 - \varepsilon)}$ will be referred to as the *attenuation length.*

For an electron facing a $\Delta V \approx 1$ eV relative barrier, the typical attenuation length is of the order of 2 Å. Consequently, the ratio $2a/\ell_\varepsilon$ is expected to be larger than 1 and the above transmission coefficient can fairly be approximated by

$$T(\varepsilon) \approx \frac{16\varepsilon\Delta V}{V_0^2} e^{-2a/\ell_\varepsilon} \quad (4.16)$$

for $2a/\ell_\varepsilon \gg 1$. Such an expression will prove to be very useful in many applications and is known as a *thick barrier approximation.*[5]

[5] See also the JWKB approximation in Section 4.8.

Question 4.4: **A delta barrier.** A barrier with the shape of a Dirac delta function can conveniently provide a crude model of 1D scattering by a point-like potential. A free incoming particle of mass m is assumed to come from $x = -\infty$ and encounters a potential discontinuity $V(x) = \alpha\delta(x)$, with $\alpha = \frac{\hbar^2}{mb} > 0$.

1. Write the stationary Schrödinger equation in the vicinity of $x = 0$. Compute its integral between $x = -\eta/2$ and $x = \eta/2$ when $\eta \to 0$ to find how the first derivative of the eigen-wavefunction connects between the two regions separated by the potential discontinuity.
2. Compute $\mathcal{R}$ and $\mathcal{T}$, the coefficients for the reflected and transmitted amplitudes, respectively. Deduce the transmission and reflection coefficients.
3. If the potential is now reversed ($\alpha < 0$), what are the possible bound states of the system?

Answer:

1. Integration of Schrödinger's equation in the vicinity of $x = 0$ gives

$$\int_{-\eta/2}^{\eta/2} \frac{-\hbar^2}{2m}\frac{\partial^2}{\partial x^2}\phi(x)dx = \int_{-\eta/2}^{\eta/2} (\varepsilon - \alpha\delta(x))\,\phi(x)dx$$

When $\eta \to 0$, the equation becomes

$$\frac{\hbar^2}{2m}\left(\frac{\partial}{\partial x}\phi_2(0) - \frac{\partial}{\partial x}\phi_1(0)\right) = \alpha\phi_1(0) \tag{4.17}$$

with the continuity of the wavefunction: $\phi_1(0) = \phi_2(0)$. The solution, away from the potential continuity, for a particle originating from $x = -\infty$ gives

$$\phi_1(x) \propto (e^{ikx} + \mathcal{R}e^{-ikx}) \quad \text{and} \quad \phi_2(x) \propto \mathcal{T}e^{ikx}$$

2. The continuity of the wavefunction at $x = 0$ implies $\mathcal{T} = 1 + \mathcal{R}$ while the above relation between the derivatives yields

$$1 + \mathcal{R} = i\frac{k\hbar^2}{2m\alpha}(\mathcal{T} - 1 + \mathcal{R}) = i\frac{kb}{2}(\mathcal{T} - 1 + \mathcal{R})$$

with $b = \frac{\hbar^2}{m\alpha}$. The amplitude coefficients are then

$$\mathcal{T} = \frac{kb}{kb + i} \quad \text{and} \quad \mathcal{R} = \frac{-1}{1 - ikb}$$

Therefore, the current density coefficients are $T = \frac{k^2b^2}{1+k^2b^2}$ and $R = \frac{1}{1+k^2b^2}$. If $kb \ll 1$, no interaction with the potential is detected. This corresponds to an incident wavelength large compared to b that can be interpreted as an interaction distance. In that case, the singularity is undetected. It thus takes $\lambda \approx b$ for the incident particle to be able to probe a short-range effect.

3. Reversing the sign of α gives an infinite quantum well. If it is seen as a limiting case of the infinite potential with $a \to 0$, we would have no negative energy state. However, outside the discontinuity, $\frac{-\hbar^2}{2m}\frac{\partial^2}{\partial x^2}\phi(x) = \varepsilon\phi(x)$, and with $\varepsilon < 0$, the wavefunction needs to take the form $\phi(x) \propto e^{-\kappa|x|}$, with $\kappa > 0$ so that the function still belongs to $\mathcal{L}_2$. The previous conditions at $x = 0$ still hold and replacing $e^{-\kappa|x|}$ into (4.17) yields a single bound state corresponding to $\kappa = m|\alpha|/\hbar^2$. A mere discontinuity of null extension in a universe where the potential is uniformly flat, makes it possible for a particle to be localized with a bound state of energy $\hbar^2/(2mb^2)$.

4.8. The Jeffreys–Wentzel–Kramers–Brillouin Approximation and Non-constant Barriers

The potential barrier treatment, which was explained in the previous section, opens up new possibilities for the study of probability transmission through non-constant potential regions. Consider a potential barrier as pictured in Fig. 4.9. If the monochromatic incident wave phase changes much faster than any characteristic variation of the potential, it is possible to consider that each rectangle is a good local approximation to the potential. This is the basis of a resolution procedure, which makes use of a classical limit identified as $\hbar \to 0$, and developed by H. Jeffreys and later applied to quantum problems by G. Wentzel, H. Kramers and L. Brillouin, hence the acronym JWKB for this method. For a δx wide and $V(x)$ high rectangular potential at x, the Schrödinger equation can be written:

$$-\frac{\hbar^2}{2m}\frac{\partial^2}{\partial x^2}\phi(x) = (\varepsilon - V(x))\phi(x) = -\frac{\hbar^2}{2m}k^2(x)\phi(x)$$

where $k^2(x)$ is a slowly varying function. Section 7.3.2 will identify this condition as the "classical limit", i.e. the regime where the particle wavelength is much shorter than any characteristic spatial variation of the environment.[6] As $k(x)$ is assumed to be a constant over δx, the wavefunction takes the form of a local plane wave:

$$\phi(x) \approx A(x)e^{\pm ik(x)x}$$

where $A(x)$ is assumed to vary much more slowly than the phase $\Phi(x) = k(x)x$. If $\varepsilon < V(x)$, the wave vector is, here also, replaced by $\kappa(x) = ik(x)$.

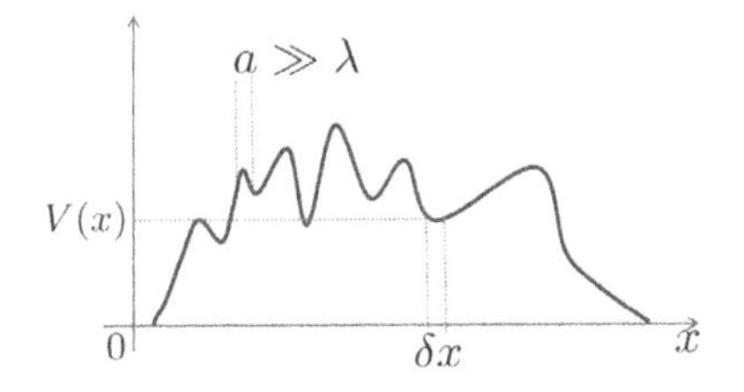

Fig. 4.9. In JWKB approximation, it is assumed that the potential varies much more slowly than the phase of the incoming particle. The potential is partitioned into a succession of barriers which are assumed to be constant over a width δx.

[6]It can already be seen in the case of the infinite quantum well where, as the width is increased, the discrete nature of the particle's energy spectrum becomes less and less observable and, accordingly, the behaviour is closer to the prediction in classical physics.

Setting $\phi(x) = A(x)e^{i\Phi(x)}$ in the Schrödinger equation yields

$$\begin{aligned}
&[\partial_x^2 A(x) - A(x)(\partial_x \Phi(x))^2] + i[A(x)\partial_x^2 \Phi(x) + 2\partial_x A(x)\partial_x \Phi(x)] \\
&\quad = -A(x)\frac{2m}{\hbar^2}(\varepsilon - V(x))
\end{aligned} \tag{4.18}$$

Consider the real part of this equation:

$$\partial_x^2 A(x) - A(x)(\partial_x \Phi(x))^2 = -A(x)\frac{2m}{\hbar^2}(\varepsilon - V(x)) \tag{4.19}$$

As assumed above, the amplitude varies very slowly (compared to $\Phi(x)$) so that $\partial_x^2 A(x) \ll A(x)(\partial_x \Phi(x))^2$ and the equation becomes

$$\frac{\partial \Phi(x)}{\partial x} = \frac{\sqrt{2m(\varepsilon - V(x))}}{\hbar} \tag{4.20}$$

and, assuming that the phase was set to zero at $x = 0$

$$\Phi(x) = \frac{1}{\hbar}\int_0^x \sqrt{2m(\varepsilon - V(x'))}dx' \tag{4.21}$$

Note that the amplitude $A(x)$ is readily found by means of the imaginary part of Eq. (4.18).

It is clear that, for the region where $\varepsilon < V(x)$, the phase function is imaginary and converts into a damping factor so that the transmission coefficient for the probability current is

$$T(x) \approx e^{-2\int_0^x \sqrt{2m(V(x')-\varepsilon)/\hbar^2}dx'}$$

This expression relies on the fact that the wavelength of the local incident plane wave is much smaller than any change in the potential. However, there is an issue when $\varepsilon = V(x)$, i.e. when the initial kinetic energy is not enough for the classical particle to continue through the potential barrier. Such special positions are known as "turning points". While the problem in this region cannot be solved by means of JWKB approximation, it is possible to solve for the wavefunction in both domains $\varepsilon > V(x)$ and $\varepsilon < V(x)$ and then find the missing part by continuity.

A treatment of the JWKB method, more detailed than what we can afford here, is given in classic text books such as [13].

4.9. Applications of the Tunnel Transmission

The possibility for a particle to tunnel through a potential barrier and the exponential character of the transmitted probability current yield numerous applications, ranging from chemistry (Section 5.4.2) and electronics to nuclear physics (Section 4.9.1). The next section gives examples in the form of two questions while Section 4.9.2 explains how the mastering of the tunnel effect has revolutionized microscopy and opened the field of nanotechnologies.

4.9.1. *The tunnel effect at two energy scales*

The following two questions provide examples where the tunnel transmission can be evoked to offer an explanation for the emission of particles, which is impossible in a classical framework. They show that the tunnel effect is not only observed in electronics but also in nuclear physics, therefore in situations where particles have a very significant energy difference.

Question 4.5: **Field emission of (cold) electrons from metal surface.** An important case for which JWKB approximation is directly applicable is the computation of how an emission of electrons from a metal is affected by the application of an electric field. The introduction of the photoelectric effect in Section 1.2 already gave us the opportunity to introduce the work function W. It is the minimal energy that needs to be given to an electron for it to be ejected from a solid. In a simple picture, the electrons in a metal can thus be described as particles in a finite potential well with a depth W. If an electric field is applied, assuming that it is zero in the conductor, the potential seen by an electron thus has a triangular shape as represented in Fig. 4.10. Compute the transmission coefficient for the probability current of one electron as a function of the amplitude of the electric field $\mathcal{E}$.

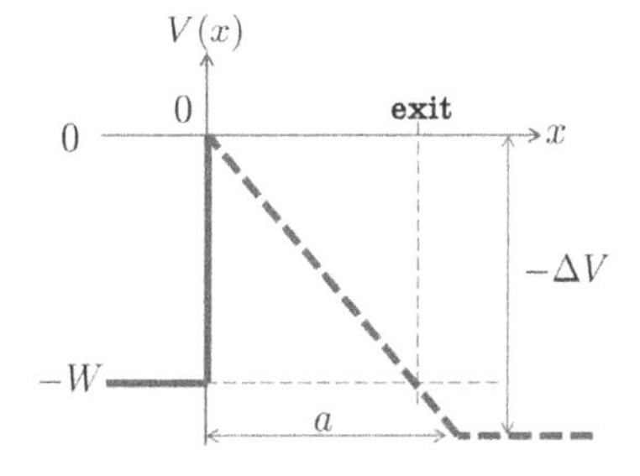

Fig. 4.10. The electron is initially trapped in the solid represented by a $-W$ deep quantum well (thick solid line). Upon the application of a lower electric potential $-\Delta V/q$ at distance $x > a$ from the surface, an electric field is created in the gap region where the potential energy now becomes $-qx\mathcal{E} = -x\Delta V/a$ (dashed thick line). The global potential barrier seen by the particle thus has a triangular shape.

Answer: It is assumed that the electron is initially in a bound state of the metal, i.e. $\varepsilon = -W < 0$. As a consequence, unless high temperature or ultraviolet radiation is used, no energy is given to the electron to allow for an ejection and classical physics states that the electron should remain in the metal.

Application of an electric field induces a change in the shape of the potential experienced by the electron. The position at which the particle exits the barrier is $x = W/(q\mathcal{E})$. If the electron's energy is high enough, i.e. the wavelength of the incident plane wave is much shorter than a, JWKB approximation can be used and

$$
\begin{aligned}
T_{@exit} &\approx \exp[-2\int_0^{W/(q\mathcal{E})} \sqrt{2m(W+V(x))/\hbar^2}dx] \\
&\approx \exp\left[-2\sqrt{2mW}\int_0^{W/(q\mathcal{E})} \sqrt{(1-qx\mathcal{E}/W)/\hbar^2}dx\right]
\end{aligned}
$$

A change of variables $u = 1 - qx\mathcal{E}/W$ gives

$$
T_{@exit} \approx \exp\left[-\sqrt{m}\frac{(2W)^{3/2}}{q\mathcal{E}\hbar}\int_0^1 \sqrt{u}du\right] = \exp\left[-2\sqrt{m}\frac{(2W)^{3/2}}{3q\mathcal{E}\hbar}\right]
$$

This expression was derived in 1928 by Fowler and Nordheim (and is named after them). It has become increasingly important as technology allowed the construction of scanning tunnelling microscopes that rely on the possibility of observing such a transmitted probability (hence electronic) current. The strong dependence of the current on the applied electric field makes it possible to fine-tune the quantity of electrons tunnelling through the gap between the surface of a conductor and the tip of another metallic collecting device. Several aspects of STM are discussed in the next section.

Other applications of this expression are found in transistors and similar devices. Fowler–Nordheim's formula provides an explanation for the observed "gate leakage" in CMOS (Complementary Metal Oxide Semiconductor) devices which becomes a major issue as these components are reduced in size (see for example [76]). On the other hand, electron tunnelling through the oxide layer is used for erasing and consequently reprogramming, the ROM (Read Only Memory) known as Floating-Gate Tunneling Oxide Electrically Erasable Programmable ROM (FLOTOX EEPROM).

Question 4.6: **Gamow's model of alpha decay radioactive lifetime.** Alpha radioactivity is the emission of ^{4}He nuclei by a (generally heavy) nucleus. The name "alpha" particle is justified because this was the first type of radioactivity observed. The "beta" (electrons) and "gamma"(photons) came later, in this order. Alpha decay is thus a spontaneous transmutation of a chemical element as Middle Age alchemists dreamed of (Fig. 4.11).

Alpha particles from this emission were measured to have kinetic energies ε much lower than the height of the potential barrier that keeps the protons and neutrons together in the emitting nucleus. To explain this unexpected phenomenon, Gamow [43] proposed the following rather simple model.[a]

It is assumed that the alpha particle (a cohesive system constituted of two protons and two neutrons) is formed inside the father nucleus prior to the emission. Bethe and Weizsacker's model (see Volume 2) of the nucleus assumes that all nuclei have the same nucleon density so that, if it is made of A nucleons, the radius is estimated to be about $r_0A^{1/3}$ with $r_0 \approx 1.24$ fermi.

[a] Almost simultaneously, Gurney and Condon [46] elaborated a similar model.

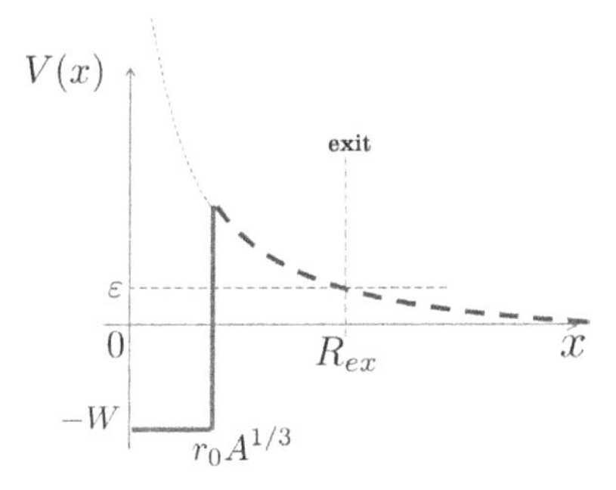

Fig. 4.11. In Gamow's description of alpha emission, the ^{4}He particle pre-exists in the father nucleus. It is trapped in a quantum well ($-W$ deep, represented by the thick line) but modified by the repulsive Coulomb potential (thick dashed line) for $x > r_0 A^{1/3}$, where $r_0 \approx 1.24$ F and A is the mass number of the father nucleus. A particle with initial kinetic energy ε will exit from the barrier at position R_{ex}.

Question 4.6: (continued)

Inside the nucleus, the potential is assumed to be flat and quite deep since the cohesion energy (from the "strong force") per nucleon is of the order of 10 MeV.

However, the strong force has a very limited range of action and outside the $r_0 A^{1/3}$ radius, the Coulomb potential is dominant. This means that, once the alpha particle has reached $x > r_0 A^{1/3}$, it experiences a repulsive potential created by the remaining protons of the nucleus $V(x) = \frac{2(Z-2)q^2}{4\pi\epsilon_0 x}$.

1. What is the height of the Coulomb barrier relative to the zero potential level (the limit value at $x \to +\infty$)?
2. Let ε be the kinetic energy away from the father nucleus. At what distance R_{ex} from the center of the nucleus does the alpha particle exit from the potential barrier?
3. Compute the transmission coefficient for the probability current of the ejected alpha particles. It would be wise to make use of[b]

$$\int_{x_1}^{x_2} \sqrt{\frac{1}{x} - \frac{1}{x_2}}\, dx = \sqrt{x_2}\left[\operatorname{acos}\left(\sqrt{\frac{x_1}{x_2}}\right) - \sqrt{\left(\frac{x_1}{x_2}\right) - \left(\frac{x_1}{x_2}\right)^2}\right]$$

   and then consider that the barrier (as seen by the alpha particle) is thick enough to have $R_{ex} \gg R_0$.
4. If it is assumed that, while trapped in the quantum well, the alpha particle bounces between the walls with a constant energy ε, show that Gamow's model can provide a justification for Geiger–Nuttall's law between the radioactive lifetime τ_α and the energy of the emitted ^{4}He nucleus: $\log_{10}(\tau_\alpha) = -C\frac{Z}{\sqrt{\varepsilon}} + D$, where C and D are constant values.

[b]While particularly cumbersome, the integral offers no difficulty if the change of variable $x_2/x = 1/\cos^2\gamma$ is used. This is made possible because $0 < x < x_2$.

Answer:

1. Coming from infinity and directing towards the centre of the nucleus, a test charge $2q$ will experience a rising Coulomb potential $\frac{2(Z-2)q^2}{4\pi\epsilon_0 x}$ until the strong forces take over. Then the potential suddenly drops to transform (in this model) into a quantum well. The sudden change is at $x = R_0 = r_0 A^{1/3}$. Obviously, at this position, the Coulomb potential is $\frac{A^{-1/3}2(Z-2)q^2}{4\pi\epsilon_0 r_0}$.
2. The depth of the quantum well is not important here. The emitted alpha particle has a kinetic energy far from the nucleus, i.e. where the potential energy can be taken to zero, given by ε. Therefore, assuming that the potential is very deep, it can be stated that ε was also the energy inside the nucleus. Note that the Gamow hypothesis which assumes that the ^{4}He pre-exists inside the father nucleus is not more legitimate than to think that the drip pre-existed in the water tank before it fell from the tap. Since the total energy is conserved, the exit position must satisfy $\frac{2(Z-2)q^2}{4\pi\epsilon_0 R_{ex}} = \varepsilon$ so that $R_{ex} = \frac{2(Z-2)q^2}{4\pi\epsilon_0 \varepsilon}$.
3. The transmission coefficient can be computed using the JWKB approximation so that

$$T \propto e^{-\frac{2}{\hbar}\int_{R_0}^{R_{ex}} \sqrt{2m(V(x')-\varepsilon)}dx'}$$

with $V(x') = \frac{2(Z-2)q^2}{4\pi\epsilon_0 x'}$. We now gratefully make use of the provided integral result to obtain

$$\begin{aligned} &\sqrt{2m\frac{2q^2(Z-2)}{4\pi\epsilon_0}}\int_{R_0}^{R_{ex}}\sqrt{\frac{1}{x}-\frac{1}{R_{ex}}}dx \\ &= \sqrt{\frac{mq^2(Z-2)R_{ex}}{\pi\epsilon_0}}\left[\text{acos}\left(\sqrt{\frac{R_0}{R_{ex}}}\right) - \sqrt{\left(\frac{R_0}{R_{ex}}\right) - \left(\frac{R_0}{R_{ex}}\right)^2}\right] \\ &\approx \sqrt{\frac{mq^2(Z-2)R_{ex}}{\pi\epsilon_0}}\left[\frac{\pi}{2} - \sqrt{\frac{R_0}{R_{ex}}}\right] \end{aligned}$$

where, for the last line, the exit point was assumed to be far enough, i.e. $R_0 \ll R_{ex}$, so that a first-order expansion could be used. It should be emphasized that $R_{ex} \propto 1/\varepsilon$ so that the transmission coefficient $T \propto e^{-\frac{\pi}{\hbar^2}\sqrt{\frac{mq^2(Z-2)R_{ex}}{\pi\epsilon_0}}} e^{-\frac{4}{\hbar^2}\sqrt{\frac{mq^2(Z-2)R_0}{\pi\epsilon_0}}}$ strongly increases with the ejection speed since $R_{ex} \propto 1/v^2$. The first exponential term is often referred to as "Gamow's factor".
4. In this model, the pre-existing alpha particle is assumed to elastically bounce between the two walls inside the quantum well. Since its energy is ε and walls separating distance is $2R_0 = 2r_0 A^{1/3}$, the particle attempts to be transmitted every $\Delta t = \sqrt{m}2R_0/\sqrt{2\varepsilon}$. As a consequence, the decay time, i.e. the mean emission time, is estimated by

$$\tau_\alpha = T\Delta t = \sqrt{m}\frac{2R_0}{\sqrt{2\varepsilon}}e^{-\frac{\pi}{\hbar^2}\sqrt{\frac{mq^2(Z-2)R_{ex}}{\pi\epsilon_0}}} e^{-\frac{4}{\hbar^2}\sqrt{\frac{mq^2(Z-2)R_0}{\pi\epsilon_0}}}$$

This expression clearly shows that there is a strong relation between the alpha decay lifetime of an element and the energy at which it emits the ^{4}He nucleus. It can be seen on a log scale:

$$\log_{10}(\tau_\alpha) = -\frac{1}{2}\log_{10}(A^{-2/3}\varepsilon) - C\frac{(Z-2)}{\sqrt{\varepsilon}} + D\sqrt{(Z-2)A^{1/3}} + \text{Cte} \tag{4.22}$$

where C and D are positive constants. Obviously, the log term is a slowly varying function of ε compared to the second one and it can often be considered as a constant.

Answer: (continued)

This expression explains Geiger–Nuttall's law which states that there is an approximate linear relationship between the half-life of an alpha emitter and $Z/\sqrt{\varepsilon}$, i.e. the ratio between the number of protons in the nucleus and the ejection speed of the ^{4}He (see Fig. 4.12).

The range of application of alpha radioactivity is wide. Americium-241 (half-life 432 years) can be found in ionizing type smoke detectors. Curium-244 (18 years) is used as a source of ^{4}He for Alpha Particle X-Ray Spectroscopy instruments (such as the one on board the Pathfinder mission to Mars in 2004) to perform elemental analysis of rocks by fluorescence. Finally, high energy alpha radiation is known to severely damage organ's DNA when alpha-nuclides are absorbed (making them deadly poisons — see the use of Polonium-210 in Litvinenko's assassination [50]) but is easily stopped by human skin. Recently, nuclides such as Actinium-225 (10 days), Lead-212 (10.6 h), Astatine-211 (7.2 h), Bismuth-213 (45 mn) or Thorium-226 (30.7 mn) have been coupled with antibodies in order to selectively kill tumour cells in the context of ovarian, colon, gastric, blood, breast or bladder cancer therapies.

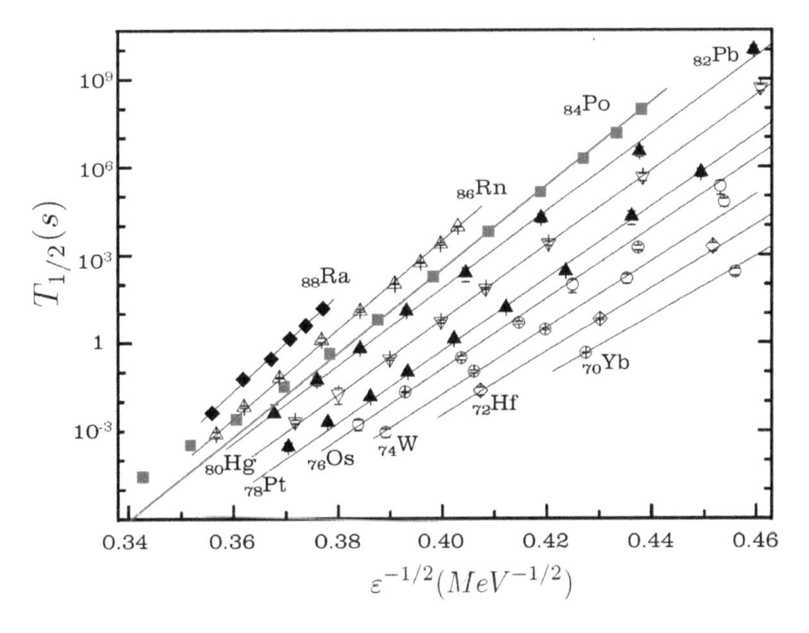

Fig. 4.12. Half-life of some alpha emitters (with even Z and A) (from [74]). The half-life is the duration after which there is a 50% chance that the nucleus has decayed. It is thus $T_{1/2} = \tau_\alpha \log 2$. The graph clearly shows that the linear relation is almost the same for all elements. For a given chemical element (fixed Z value), the slope is constant and the isotopes lie on a common straight line. Isotopic family lines of heavier chemical elements have a stronger slope and are shifted upward in agreement with (4.22). It can be noticed that polonium is one element for which the Geiger and Nuttall law does not hold in the low isotopes region. From Creative Commons CC BY-3.0.

4.9.2. *The scanning tunnelling microscope*

The ability of a quantum particle to penetrate a region where its potential energy is larger than its total energy can be seen as one of the most striking consequences of the existence of a wavefunction. This property is at the origin of unexpected behaviours in chemistry (see the flipping of ammonia molecule in Section 5.4.2), in astrophysics (the formation of interstellar hydrogen), in solid state physics (tunnel diode or gate leakage in CMOS — Question *4.5*) or in nuclear physics (alpha radioactivity — Question *4.6*). However, in the field of microscopy, the tunnel effect has led to a major technological breakthrough; it opened the possibility for observing individual atoms on a surface.

While X-ray, electron or neutron diffraction from atoms at crystal surface give a very accurate and detailed description of each atomic species *on average*, the Scanning Tunnelling Microscope (STM) is able to visualize, and manipulate, each single atom on a surface with no need of periodicity.

Expression (4.16) has proved to be applicable on many practical occasions. As mentioned previously, it works reasonably well for two atoms separated down to a distance of the order of 2 Å. The atomic resolution is then quite straightforward to explain: the tunnel effect allows electrons to flow between two conducting materials separated by a vacuum gap. This electric current is proportional to the transmission coefficient (4.16). Hence, it decreases in an exponential manner with the gap width, represented by distance a. Now, consider the unknown topology of a (macroscopically) flat surface of a solid or liquid. Ideally, if a sharp conducting tip is approached at a short distance from this surface, measuring the tunnel electric current as a function of the tip's transverse position provides a way to recover the structure of the surface at the atomic level. The general principle of such a device is summarized in Fig. 4.13.

Tip

It could be legitimate to question the technical feasibility of such a device. Firstly, contrary to particle diffraction-based techniques, this is a local

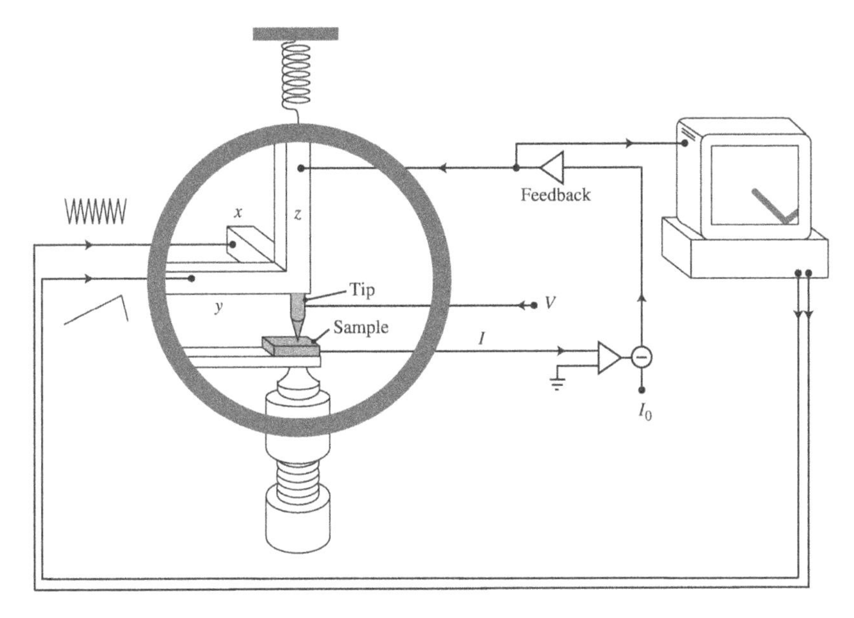

Fig. 4.13. General description of a scanning tunnelling microscope. The tip is brought close to the surface to be explored so that a tunnelling current I_t can flow. As the tip is scanned parallel to the surface (x, y), the current intensity is recorded so that a topographic image of the surface can be reconstructed as $I_t(x, y)$. Reprinted with permission from [84]. Copyright © 2007 by Springer-Verlag Berlin Heidelberg.

probe microscopy. It could be thought of as a blind reader sliding his finger over Braille. Given the distance separating the raised dots of the paper (typically 2 mm), it is advisable to use the part of the finger where touch sensitivity is highest rather than a toe or an elbow. Therefore, the ideal accuracy for an STM tip is obtained with a single atom. Another obstacle is the difficulty of bringing the tip sufficiently close to the surface to be probed. This distance cannot be much longer than the attenuation length ℓ_ε and it is then essential to reduce the barrier's height. This is made possible by using materials with low electron work function (i.e. extraction energy) so that they can easily release the particle into the vacuum gap. Tungsten, long used to construct electron guns in the old television technology and still today in electron microscopes or medical X-ray tubes, is such a material. A typical tungsten tip is showed in Fig. 4.14. If the system is not in high vacuum, the tip may oxidize and create an additional obstacle

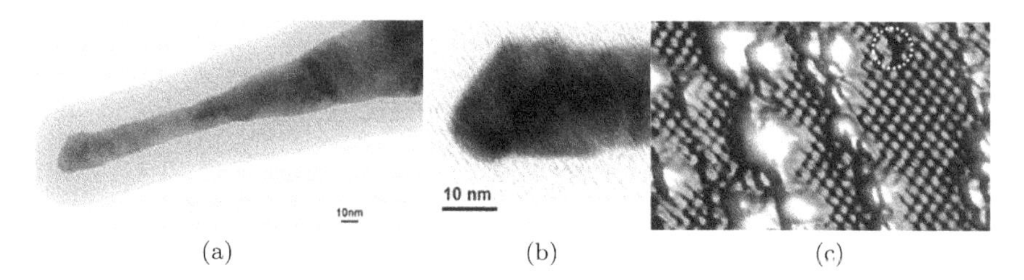

(a) (b) (c)

Fig. 4.14. This tip (a) is made from crystalline tungsten and sharpened by electron beam heating and ion sputtering. The round end shape, with a curvature radius on the order of a few nanometres (b), allows a resolving power such that single atomic defects (c) can clearly be seen in the periodic arrangement of the silicon substrate (circled with a dotted line). Deposited carbon atomic chains are also visible. Reprinted with permission from [18]. Copyright © 2014 by Springer Nature.

to the tunnel electron flow. Other metals, such as platinum–iridium, are then good materials for the tip.

Assuming identical work functions for the tip and the sample, at equilibrium, there is as much current flowing in both directions and consequently no net current is observed. To break the balance, a biased tip-surface voltage from 1 mV to a few volts is applied. Depending on the gap width, a tunnelling current ranging from 0.1 nA to 10 nA is then typically measured.

Subatomic translation

The translation of the tip over the surface needs to be accurately conducted and its position recorded. This cannot be achieved by any sort of motoring device. Piezoelectric tubes are then employed. Piezoelectric materials are crystals (or ceramics) that have the property to shrink or expand when they are subjected to an electric field. Therefore, these crystals cannot have a center of inversion.

At a microscopic level, the application of an electric field induces an opposite displacement of atoms with opposite net charges. This results in a change in inter-atomic distances and consequently to a change in the macroscopic volume. Depending on the material or its orientation, it is possible to induce an adjustable and reversible deformation proportional to the applied voltage. This is known as the *inverse piezoelectric effect.* The direct effect is the appearance of an electric potential between two opposite faces of a squeezed material. Some common materials that exhibit the piezoelectric property are quartz, cane sugar or Rochelle salt. The relation between the deformation and the applied voltage is tensorial, but the major contribution can be estimated as a deformation (relative change in dimension) proportional to the electric field (the voltage V divided by the distance h between the electrodes): $\frac{\Delta x}{x} = d\frac{\Delta V}{h}$. Typical values for the piezoelectric coefficient are a few Å/V. This material property is extensively used in many technological devices, from the simplest such as the piezo igniter (direct effect) to ultrasound transducers for medical echography (inverse effect). The translation technology in STM has progressively evolved from a tripod system to a tube (see Fig. 4.15) allowing very reliable displacement of the tip along the surface and real-time adjustment of the gap.

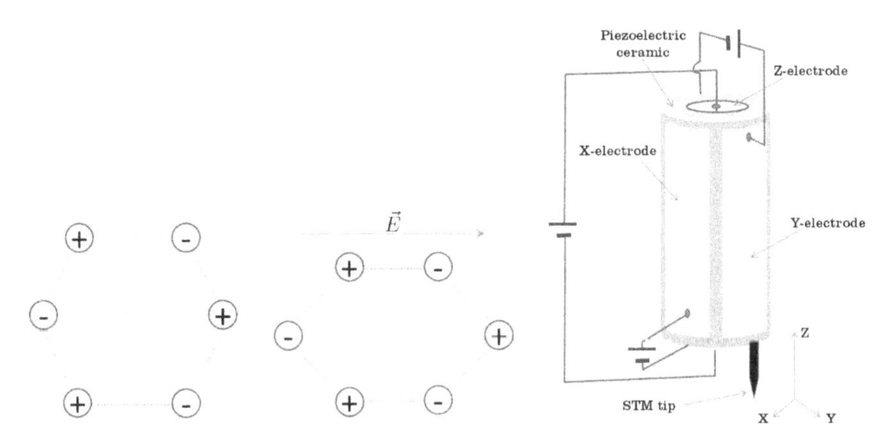

Fig. 4.15. Inverse piezoelectric effect. Left: At equilibrium, the atomic arrangement is such that there is no net dipole moment and no centre of symmetry (the positive and negative atoms pertain to different elements). Middle: when the system is exposed to an electric field, atomic displacements induce a deformation of the system. Right: Displacement of STM tip driven by piezoelectric tube. Applying a voltage on opposite electrodes causes the tube to change its shape (bend in X or Y directions, or expand/contract in Z direction).

Resolution

It is then important to make sure that, during the scan, the system is close enough to the surface to measure enough current, so that its variation can be recorded, and stays a safe distance away to avoid any collision with a sudden and significant variation of the sample's height. Therefore, while a constant height scan mode is adapted to explore flat surfaces, it has become standard procedure to adopt a constant current scan mode. The latter (Fig. 4.13) requires a feedback loop where the tunnelling current variations are directly used to monitor the voltage at the piezoelectric tube that changes the height of the probing tip. As previously mentioned, the resolution power characterizes the ability of the microscope to distinguish between two separate points. The resolution in a direction normal to the surface is basically conditioned by the exponential decrease of the current and can be as good as 0.01 Å. The lateral resolution is a lot more dependent on the shape at the very extremity of the tip. Consider three simple cases: the ideal geometry is when the end of the tip is made up of a line of dangling atoms (Fig. 4.16(a)). Since the tunnel current decreases by a factor 10 for every additional angström in the tip-to-surface distance, a significant drop in intensity should be detected while the tip passes over the region between the two atoms on the surface. The exponential decrease also implies that atom 2 (Fig. 4.16(b)) will bring a negligible contribution to the global current measured by the tip and that atom 1 will be dominant. This inverse pyramidal geometry is of course more likely and still provides an excellent resolution. The last situation (Fig. 4.16(c)) is an issue as atoms 1 and 2 contribute equally. This means that there are two positions for the tip corresponding to a maximal intensity. The lateral resolving power is then deteriorated. We then see that it cannot be significantly better than 1 Å. On 3rd January 1979, just before Rohrer and Binnig submitted the patent that would earn them the Nobel prize (Fig. 4.16), they report on their experimental notebook a possible resolution of 45 Å. It turned out to be much better because of undetected atomic protrusions on the tip.

Today, one way to reach a 1 Å resolution is indeed to have the tip pick up a short molecule and let it hang so that the geometry closely resembles situation (Fig. 4.16(a)).

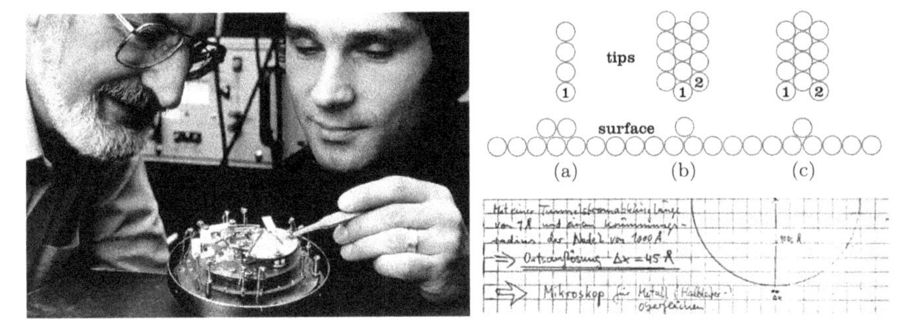

Fig. 4.16. Left: Heinrich Rohrer and Gerd Binnig were awarded half the Nobel Prize in 1986 for the invention of the Scanning Tunnelling Electron Microscope. (Photo credits: with kind permission from IBM Research.) The other half went to Ernst Ruska for his work on electron optics and the design of the electron microscope. Also from the IBM Zurich Research Lab, Christoph Gerber participated in the construction of the first working version of an STM. Top Right: Three typical tip extremities pointing at atoms on a surface. Bottom right: excerpt of G. Binnig's lab notebook, 3rd January 1979 entry. An estimate of 45 Å resolving power is given for a 1000 Å radius tip at 1 Å vacuum from a flat surface. Reproduced with permission from [75]. Copyright © 1986 by the American Institute of Physics.

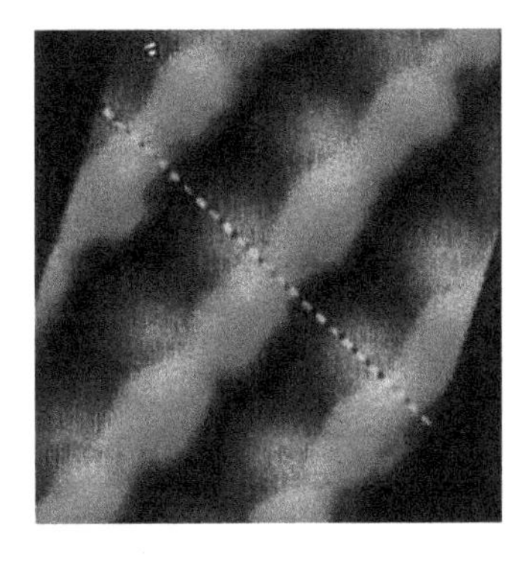

Fig. 4.17. Electron surface states on gallium arsenide. Red and green colours correspond to occupied and unoccupied states, respectively. They are probed by reversing the sample voltage ($\pm 1.9\,\mathrm{V}$) so that the tunnel electron flow is reversed. Reprinted with permission from [34]. Copyright © 1987 by the American Physical Society.

STM does not directly map a distribution of atoms; it is sensitive to the most mobile electrons of the surface and actually measures a signal that is proportional to probability density for electrons with the weakest binding energy. By reversing the bias voltage between the tip and the surface, it is possible to reverse the electron flow. In such a case, electrons from the tip preferentially go on to populate an available surface wavefunction. As in the quantum well, the possible wavefunctions are mutually orthogonal. This is illustrated in Fig. 4.17 which shows the possibility of using an STM to probe both occupied and unoccupied state wavefunctions.

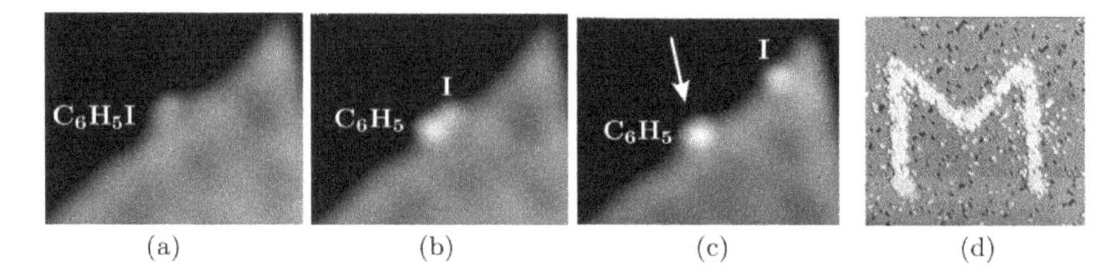

Fig. 4.18. Selective manipulation of atoms. (a)–(c) Iodobenzene dissociation. (a) A $(C_6H_5)I$ molecule is adsorbed on a copper surface. (b) The C–I bond is selectively broken by the tip acting as an electron gun. (c) Additional separation is achieved by using the tip in push mode. Reprinted with permission from [55]. Copyright © 2003 by Elsevier Science B.V. (d) Selective dehydrogenation of a silicon surface as a means to write a letter at the nanoscale. Reprinted with permission from [84]. Copyright © 2007 by Springer-Verlag Berlin Heidelberg.

Seeing can be moving

Finally, the possibility for a precise movement of the STM tip together with the faculty of such a device to map the surface of a solid or a liquid, with atomic resolution, allows for crafting particular "tip-made" objects. When the tip is approached from atoms which are loosely bound, modulating the voltage opens the possibility of modifying the interaction force to pull, push or slide individual atoms. It then becomes possible to position at will atoms on the surface and build nano-sized objects (see Fig. 4.2) for particular fundamental studies, or tailor-made to exhibit specific properties. A demonstration of these capabilities offered to experimentalists by STM is displayed on Fig. 4.18 where each white dot is a unique atom.

Chapter 4: Nuts & Bolts

- *While the* ***wavefunction*** *needs to be* **continuous**, *its derivative ceases to be continuous wherever the potential exhibits an infinite discontinuity.*
- *Two important examples, which are idealizations of real situations, are worth remembering:*

  → *In the* ***infinite 1D quantum well*** *case, the stationary states (at t=0) are* $\phi_n(x) = \sqrt{\frac{2}{a}}\sin(n\pi x/a)$ *and the associated energies are* $\varepsilon_n = n^2 \frac{1}{2m}\frac{\hbar^2\pi^2}{a^2}$ *with* $n \in \mathbb{N}^*$, *where* a *is the width of the well and* m *the particle's mass.*

  → *In the* ***1D square potential barrier*** *case (with thickness* a *and height* V_0*), the eigen-wavefunctions are unnormalizable plane waves but a transmission coefficient, defined as the ratio of the transmitted-to-incident probability current densities* $T(\varepsilon) = \frac{j_t}{j_i}$, *is found to be:* $T(\varepsilon) \approx \frac{16\varepsilon\Delta V}{V_0^2}e^{-2a/\ell_\varepsilon}$ *where* ΔV *is the effective height of the barrier and* ℓ_ε *is the attenuation length.*

5

Quantum Postulates and Their Mathematical Artillery

The content of this chapter will help you understand ***random number generation, quantum cryptography****, and related topics.*

This chapter presents the rules according to which quantum physics can be put into practice in order to explain and predict many experimental results with unprecedented accuracy. One could consider that these rules are the essence of quantum physics. As previously mentioned in this volume (Chapters 1 and 2), during the first decades of the twentieth century, quantum physics was first proposed as a patch to interpret experiments that showed unexpected results to nineteenth century scientific minds. No thorough theory was proposed until Erwin Schrödinger and Max Born, building on Louis de Broglie's idea, put forward the wavefunction and its evolution equation. However, and independently, Werner Heisenberg had his own way of formalizing the new physics. In his opinion, it was all a matter of matrices acting on vectors. He observed that

Fig. 5.1. Paul Adrien Dirac (1902–1984) was awarded the Nobel Prize for physics in 1933 (with E. Schrödinger). Paul's father had French origins and could speak more than eight languages including Esperanto. The rule was that Paul was only permitted to speak French to his father and both shared their meal in a room separated from the rest of the family. Paul explained "... since I couldn't express myself in French, it was better for me to stay silent than talk in English. So I became very silent at that time". As a matter of fact, he remained very laconic for most of his life and even in the friendly atmosphere of the Bohr institute, he remembers "I admired Bohr very much. We had long talks together, very long talks in which Bohr did practically all the talking". Dirac was educated as an electrical engineer but soon decided to go back to Bristol University, then Cambridge, to graduate in mathematics. The year 1926 came as a surprise to Born who remembered that "there appeared a parcel of papers by Dirac, whose name I had never heard. Never have I been so astonished in my life that a completely unknown and apparently young man could write such a perfect paper". Because he hated so much to be in the spotlight, Dirac turned down all honorary degrees and even thought of refusing the Nobel prize until Rutherford suggested that a refusal would create even more curiosity. He asked that only his mother was invited to the ceremony in Stockholm. Photo credits: Paul A.M. Dirac Papers, courtesy of the Florida State University Libraries, Special Collections and Archives.

many aspects could be explained if each physical quantity was described by a matrix. The equivalence between Schrödinger's waves and Heisenberg's matrix mechanics had to wait the distinction between observables and states brought by Dirac and a further mathematical clarification by von Neumann.

Paul Dirac (Fig. 5.1) was interested in global models. He believed that, to give a satisfactory account of reality, a model had to be mathematical and provide a thorough picture of what could be observed. It had to be simple but also able to predict new phenomena. No blind spot was allowed. From his electrical engineer education, with strong emphasis on geometry and technical drawing, he kept his own very visual way of designing a theory. As such, as he boldly admitted, his only interest was to study theoretically what could be observed. He claimed not to worry too much about mathematical refinements and subtleties. He was not primarily interested in mathematical rigour as long as the theory was efficient in describing the physical reality. The mathematical polish had to come in the second stage. It is now obvious that Dirac's imprint on the formalism of quantum mechanics was (and still is) remarkably strong. He proposed a new way of manipulating vector spaces so that the physical meaning would never be overwhelmed by mathematical considerations. We will call this mathematical description the *Dirac formalism* and the new symbols *Dirac's notation.* As Wigner put it [92], we can still be amazed by "The Unreasonable Effectiveness of Mathematics in the Natural Sciences". It not only provides a coherent way of explaining all experimental observations but also predicts yet unobserved phenomena. Einstein, Pauli, Dirac or Higgs were among those who became famous by indicating new directions for experimentalists to focus their investigations.

This chapter proceeds in two stages. The postulates of quantum mechanics will first be presented and accompanied by a brief comment in order to show how they differ from classical physics references. Then, a more in-depth introduction to the necessary mathematical tools will be given.

5.1. New Game, New Rules

Classical physics relies on a limited number of rules. They may differ depending on whether we are eager to adopt Newton's or Lagrange's point of view (see Chapter 6). The former focuses on forces competing to accelerate a massive body while the latter proposes a variational approach to determine the trajectory.

The practice of quantum physics is a completely new game. Players ought to adopt new rules that are best suited to the microscopic world. As we will later see, these rules are not disconnected from the classical world and, if they were not so stringent and — sometimes — cumbersome to manipulate, could very well replace those at work at any scale.

No demonstration of these new rules is proposed. They are presented as the *postulates of quantum mechanics*. As such, their validity only relies on their efficiency to account for the elements of reality that today's instruments allow to measure. If a single piece of data were to contradict these rules, the theory would have to be revamped from scratch.

As Dirac's ambitious goal, it should be clear that this formalism covers the Schrödinger representation described in Chapter 3 and put into practice for the simple examples of Chapter 4. The wavefunction in position or momentum representation will then appear as a mere particular case of Dirac's formalism.

5.1.1. *Representation of a physical state*

Since Galileo, classical physics has relied on a limited representation of the state of a particle. In a three-dimensional (3D) space, it only takes six variables. They represent the position and momentum coordinates. For a particle with no internal degree of freedom, these sole two vectors suffice to entirely describe what a particle is and what the particle will become if Newton's second law of motion is used. An alternative is given by Lagrange or Hamilton equations (see Chapter 6). For a more complex system, composed of N particles, a set of $6N$ variables is used. For a macroscopic object, it is common to limit the number of internal degrees of freedom assuming that a large number of parts are frozen for the usual forces acting on it and the subsequent displacements. When such an approximation is not possible, a thermodynamical approach is then chosen: the system state is then given by the six numbers describing its center of mass. All other values pertain to its internal degrees of freedom and are limited to *state functions*. Among those functions that are intensive, we can mention the pressure, the temperature or the chemical potential. Extensive state functions are among the volume, the entropy, the number of particles, the dipole or magnetic moment.

First postulate: The state of a quantum system at a moment t is described by the *state vector* noted $|\psi(t)\rangle$.

The reason for this state vector to be written differently than a usual vector[1] is somewhat arbitrary and could be considered as a legacy from Dirac's graphical mind. It will prove to be a convenient choice when scalar products will be considered.

A state vector contains every possible information on the system. Like any other kind of vector, it is difficult to really describe it unless it is represented by its components in a given reference frame (a set of axes). The choice for a given reference frame is arbitrary and often depends on particular properties that are being studied.

One essential property of state vectors, which should already be noted at this point, is that every state vector is requested to have norm 1. This is required by the probabilistic signification of state vectors as imposed by the fourth postulate. As a consequence, different states of the system merely differ by vectors pointing in different directions. A vector direction is not significant on its own unless it is compared to another vector or relative to a reference frame. Comparison between states will be carried out by computing scalar products, i.e. projections of a state vector onto the other one. Therefore, state vectors are described in an abstract complex inner-product space named a *Hilbert space*.

A linear combination of state vectors is also a state vector. This is known as the *superposition principle*.

5.1.2. *Physical quantities and operators*

In classical physics, all quantities that can be measured are real-valued scalars. Complex numbers, vectors or tensors are mere groups of scalars with a particular relationship. They all require to have each of their components measured separately and are obtained as real scalar.

[1] In bold letter (like in this book) or with an arrow on top (pointing to the right).

Second postulate: For quantum physics, any physical quantity that can be measured is described by an *observable*. An observable is associated to an operator which is linear and Hermitian.[a]

[a]For mathematical details, see Section 5.2.

It is essential to note that there is a clear distinction between observables and quantum states. A state is represented by a vector, also named[2] a "ket", while an observable is given by an operator. There is thus an action of the observable onto the ket that transforms it to another ket. The ket, in reverse, does not modify the observable. We will evoke the mean value of an observable of the system while it is in a given state described by a particular ket. A state gathers information of all sorts about the system. On the contrary, an observable associated to a given physical property is a fixed mathematical object: it merely operates on kets.

5.1.3. *Results of measurements*

Whether a particle is trapped in a box or moves freely in interstellar void, its physical properties described by classical physics always belong to a continuous set of possible values.

Third postulate: If a measurement of a particular physical quantity is performed on a quantum system, the outcome necessarily belongs to the eigenvalues set of the associated observable.

Consequently, for an observable described by operator $\widehat{A}$ with a set of eigenvalues and associated eigenvectors $\{(a_n, |\phi_n\rangle)\}$, the result of an ideal experimental measurement should give only one of the eigenvalues a_n.

As an experiment always gives real numbers, it is thus coherent with the previous requisite that the observables should be described by Hermitian operators (postulate 2).

We are thus led to consider different sorts of observables: those which have a continuous spectrum and those which have a discrete spectrum. The

[2]The reason for such a name will become apparent in a next section.

latter correspond to physical quantities for which a measurement yields numerical values that can be clearly distinguished. For example, it is not possible to obtain an experimental result lying between two discrete eigenvalues a_n and a_{n+1}; the spectrum is quantized. We have already encountered such a case with the bound state energy of a particle in a quantum well (see Section 4.3) but this postulate is primarily supported by the observation of electromagnetic spectra from excited atoms. They are a clear-cut manifestation that not all energies are accessible to an electron bound to its nucleus. Detection of these emission lines can be considered as one possible means to measure the accessible energies of a trapped electron.

5.1.4. *Probability of a measurement outcome*

The precision associated with a measurement outcome is only limited by the resolution power and sensitivity of the experimental setup. Classical physics is deterministic in the sense that, the precision of the measurement process notwithstanding, for a given measurement, it is admitted that there exists only one possible "exact" result which solely depends on the state of the system. All fluctuations about this ideal value are attributed to the experimentalist's skill, instrumental bias and resolution limits.

Fourth postulate: At time t, the quantum system is in a state described by the ket $|\psi(t)\rangle$. An observable is represented by an operator $\widehat{A}$, the eigenspectrum of which is $\{a_n\}$. A measurement of the property will yield one of the possible results a_n with a probability p_n given by

$$p_n = |\langle\phi_n||\psi(t)\rangle|^2 = \langle\phi_n||\psi(t)\rangle\langle\psi(t)||\phi_n\rangle$$

This is the modulus square of the scalar product[a] $\langle\phi_n||\psi(t)\rangle$ between the current state vector of the system $|\psi(t)\rangle$ and the eigenvector $|\phi_n\rangle$ of $\widehat{A}$ that corresponds to the measurement outcome a_n.

[a] Mathematical details are given in Section 5.2.

It should be noted that if the system is in *any* quantum state, then the measurement of an observable can give any of the expected results, i.e. *any* of the eigenvalues of the associated operator. Some results are simply more likely than others.

It is essential to emphasize that what is determined in this way is the probability of an expected result. Quantum physics (in its so-called *Copenhagen version*) does not specify the result of the measurement nor does it define the mechanism by which a particular result has finally materialized. This is a kind of random drawing from several possible outcomes for which we admit not knowing the realization process. It is this indeterminism against which Einstein (not to mention de Broglie or Erhenfest) struggled until the end of his life.[3] Referring to the excitation of an electron subjected to a stream of photons, Einstein said in a letter to Max and Hedwig Born in 1924: "Bohr's opinion about radiation is of great interest. But I should not want to be forced into abandoning strict causality without defending it more strongly than I have so far. I find the idea quite intolerable that an electron exposed to radiation should choose of its own free will, not only its moment to jump off, but also its direction. In that case, I would rather be a cobbler, or even an employee in a gaming-house, than a physicist. Certainly, my attempts to give tangible form to the quanta have foundered again and again, but I am far from giving up hope." [11]

5.1.5. *Collapse of the wave packet*

Classical physics admits that the process of a good measurement does not interfere with the system under study. The system is thus thought to remain in the same state before and after the measurement. While this is obviously not the case for the so-called "destructive measurements", such as those for determining the tensile strength of a material, we choose the ideal voltmeter with high impedance to avoid disturbance of the electric circuit. It is generally accepted that regardless of any salutary caution, a police radar speed detector does not change the car velocity.

[3]Einstein was convinced that this quantum theory was not complete, that it lacked some possibly hidden variables in order to more satisfactorily explain the random behaviour confirmed by the most accurate measurements.

Fifth postulate: If a measurement's outcome of an observable gives the eigenvalue a_n of its associated operator $\widehat{A}$, then immediately after the measurement, the system is in the corresponding eigenstate described by the eigenket $|\phi_n\rangle$.

If the eigenvalue is degenerate, the system will be projected onto the degeneracy eigensubspace. It will therefore be described as a linear combination of the eigenkets that share the resulting eigenvalue a_n: $\{|\phi_{n,1}\rangle, |\phi_{n,2}\rangle, \ldots, |\phi_{n,g}\rangle\}$, where g is the order of degeneracy.

This postulate therefore acknowledges that measurement disturbs the state of the system. Whatever its initial state, it can be completely changed after measurement. Unless, of course, the system was already in an eigenstate of the operator corresponding to the measured observable.

The consequences of this postulate are of the utmost importance. It is known as the *reduction of the wave packet* because, as it has already been stated, the initial state can be written as a superposition of the eigenstates of the observable:

$$|\psi(t)\rangle = \sum_n c_n(t)\ |\phi_n\rangle \tag{5.1}$$

As detailed later in this section, each vector can be represented by a particular wavefunction. Therefore, expression (5.1) can be seen as a mere (weighted) sum of wavefunctions similar to what was introduced in Section 3.2.2. Then it is postulated that the measurement has the effect of reducing this development to a limited number of functions (or states): those corresponding to the eigenvalue which came out as the result of this measurement. While, prior to the measurement, the possible results were associated with a set of probabilities (postulate 4), once the measurement has been performed, all the probabilities become zero except that which corresponds to the measured value, which is of course 1. The state is thus projected onto the subspace of the states that are associated with the numerical result of the measurement.

Question 5.1: **A many-measurement problem.** Consider a system whose states are described in a 2D Hilbert space. Let $\widehat{A}$ represent an observable in this space and $\{|a\rangle, |b\rangle\}$ an orthonormal eigenbasis of $\widehat{A}$ with associated eigenvalues $\{\alpha, \beta\}$. This could, for example, be photon polarization states.

1. Why does a measurement of the physical quantity represented by $\widehat{A}$ not necessarily reveal what the state was prior to the measurement?
2. Suppose that it is possible to make N identical copies of the system so that they are all prepared in the same quantum state

$$|\psi\rangle = a_1|a\rangle + b_1|b\rangle.$$

We set a_1 and b_1 to be real positive numbers. If N is arbitrarily large, how can we find out with measurable precision the values for a_1 and b_1?

Answer:

1. The measurement outcome of the physical quantity represented by $\widehat{A}$ can only be one of the two eigenvalues of $|a\rangle$ or $|b\rangle$. If polarization states are taken as an example, the polarizer axes need to be oriented so that polarization a can be tested. If the photon does not go through, a polarization b can be stated since the scalar product $\langle a||b\rangle$. Any other polarization state would yield a transmission because $\langle a||\psi\rangle \neq 0$. Actually, any result could be obtained. What changes as a function of the initial state is the probability of one of the two possible results.
2. If we could obtain a large number of copies for the system in the exact same state, it becomes possible to perform the measurement of the same physical property described by $\widehat{A}$ on every copy and record the results. The ratio of the number of occurrences for each of the two possible results will get closer to $(a_1/b_1)^2$ as the sampling increases. Since both quantities are positive and normalization imposes $a_1^2 + b_1^2 = 1$, it will be possible to determine the values for a_1 and b_1 with increasing precision.

5.1.6. *Time evolution of a state vector*

Time evolution in classical physics is given by Newton's second law or, equivalently, the Lagrange equations. For a description in the phase space, Hamilton's equations will be equally efficient (see Section 6). All approaches are based on a possible trajectory which implies the simultaneous existence of a position and momentum for each constituent of the system. The wave behaviour of particles forbids such an approach to a quantum system and as the state is described by a vector in an abstract vector space, the only evolution that is relevant concerns the vector itself.

Sixth postulate: The temporal evolution of a quantum state described by the vector $|\psi(t)\rangle$ is given by the time-dependent Schrödinger equation:

$$i\hbar\frac{\partial}{\partial t}|\psi(t)\rangle = \widehat{H}|\psi(t)\rangle$$

where $\widehat{H}$ is the Hamiltonian. This is a Hermitian operator associated to the observable "total energy" of the system under study.

The Hamiltonian is thus a key quantity of quantum mechanics, as it is its action on state vectors that will allow the prediction of any evolution. In this respect, it can be seen that the eigenvectors of the Hamiltonian operator, the vectors $|\phi_j\rangle$ for which $\widehat{H}|\phi_j\rangle = \varepsilon_j|\phi_j\rangle$, evolve as $|\phi_j(t)\rangle = e^{-i\varepsilon_j t/\hbar}|\phi_j(0)\rangle$. Consequently, the eigenvectors of the Hamiltonian operator have a particular property: their time evolution does not modify their orientation in the Hilbert space (nor their norm), i.e. they change by a mere phase factor. The eigenvectors of the Hamiltonian are said to be the "stationary states" of the problem. This is a generalization of what has been enunciated in Section 3.3.5.

5.2. The Mathematical Artillery

The postulates of quantum mechanics rely on some basic mathematical concepts. They mostly involve linear algebra (vector spaces and matrices) and mathematical analysis. Both aspects are usually presented as disjoint ways to put quantum mechanics into practice. However, it was Dirac's and von Neumann's great achievement to show that Heisenberg's and Schrödinger's interpretations could be reunited into a unique and all-encompassing mathematical formalism.

It cannot be denied that all the postulates rely on a great deal of mathematics that might appear to blur the physical concepts and make it look a little too abstract to be useful. Nothing could be more wrong. A mere enunciation of the postulates keeps the engineers and the scientist to a light conversation, which certainly raises interesting philosophical questions but hardly helps at understanding quantitative breakthrough in technology. Great dishes are not served if the cook only reads the recipe. He needs to patiently boil, fry or stir. He needs to sharpen his knives and learn the tricks of the trade. This section is devoted to some technical aspects of practical quantum mechanics without lingering to unnecessary mathematical considerations. A more detailed account can for example be found in [58].

5.2.1. *State space and kets*

Not just any vector space is suitable to represent the possible state vectors. We will only be concerned with *Hilbert spaces.* They allow a definition of scalar products with a result in the complex numbers space $\mathbb{C}$. In such a complex space (i.e. a space where the vector components are complex numbers) computing a scalar product calls for a little care. The scalar product will be used to compare different states. For two different state vectors[4] $|\psi_1\rangle$ and $|\psi_2\rangle$, the scalar product can be written as the projection of $|\psi_1\rangle$ on $|\psi_2\rangle$ and writes

$$\langle\psi_2||\psi_1\rangle$$

Therefore, the norm of a state vector $|\psi\rangle$ is computed as: $||\psi\rangle| = \sqrt{\langle\psi||\psi\rangle}$.

Reduction of state space

A particle state vector encapsulates all the possible information that are needed for a thorough definition of what the particle is and how it behaves. Some of these data are constant and do not need to be attached to the state once the kind of particle has been specified. For example, the mass or the charge of the electron are not subject to variation and all electrons bear the same charge and mass values under any circumstance.[5] We are not necessarily aware of all the parameters needed to unambiguously define a state. We can even presume that it takes an infinite number of parameters

[4]For simplicity, the time parameter is temporally omitted.

[5]We refer here to the rest mass. The relativity theory tells a different story for the mass of a moving particle.

to distinguish particular states and the Hilbert space should therefore be of infinite dimension. However, in practice, there is no need to consider the complete space and it is advisable to limit its dimension. As a first example, let us consider photons passing through a system of circular polarizers. As transverse electromagnetic waves can only be right-handed or left-handed polarized, a two-dimensional (2D) state space should be enough. However, we are aware that photons can have different wave vectors and they usually belong to an infinite and continuous set. But for this particular experiment, we do not really care about what the wave vector (or the momentum) actually is as it does not make much difference in a polarization measurement. We can thus happily restrain the state space to two dimensions and the basis vectors could for example correspond to the clockwise $|+\rangle$ and anti-clockwise $|-\rangle$ state vectors respectively.

Another example can be taken from the Young slits experiment carried out with C_{60} molecules ejected from an oven. An individual molecule possesses a large number of internal degrees of freedom. But, at least as a first approximation, they are not relevant to explain the interference patterns that are observed by the detector when it is positioned at a large distance after the slits (as in Fig. 3.3). Assuming that all molecules have similar speeds, the relevant information, thus once more, belongs to a 2D state space the basis of which can be chosen to be $\{|\text{slit } 1\rangle, |\text{slit } 2\rangle\}$, corresponding to the passage through slit 1 or 2, respectively.

Components and scalar products

If a set of state vectors $\{|\phi_j\rangle\}$ is chosen as a basis of the state space, we can then express any state vector $|\psi\rangle$ as

$$|\psi\rangle = \sum_j c_j |\phi_j\rangle \tag{5.2}$$

The complex-valued components c_j are the projections of vector $|\psi\rangle$ on each basis vector. This is thus obtained by computing the scalar product:

$$\langle\phi_k||\psi\rangle = \langle\phi_k| \sum_j c_j |\phi_j\rangle = \sum_j c_j \langle\phi_k||\phi_j\rangle$$

where the linearity of the scalar product was used. If the basis vectors have a norm 1, i.e. $\langle\phi_j||\phi_j\rangle = 1$, and if they are mutually orthogonal, i.e. $\langle\phi_k||\phi_j\rangle = 0$ for $j \neq k$, this can be expressed using the Kronecker symbol, $\delta_{j,k}$:

$$\langle\phi_k||\phi_j\rangle = \delta_{j,k}$$

with $\delta_{j,k} = 0$ for $j \neq k$ and $\delta_{j,k} = 1$ for $j = k$. We then have

$$\langle\phi_k||\psi\rangle = \sum_j c_j\ \delta_{j,k} = c_k$$

If the convention is such that $|\psi_1\rangle$ is represented by a column vector, $\langle\psi_2|$ is a row vector. Therefore, $\langle\psi_1|$ could be considered as the transposed of $|\psi_1\rangle$. However, for a complex vector space, there is one additional requirement. One primary interest of a scalar product is to allow for the computation of the norm of a state vector. As previously mentioned, the norm is defined as the square root of the scalar product of a vector by itself, $\sqrt{\langle\psi||\psi\rangle}$. The self-scalar product thus ought to be positive definite. This is possible only if, going from $|\psi\rangle$ to $\langle\psi|$, the transposition of coordinates is accompanied by taking the complex conjugate of each component of the vector. There is a one-to-one correspondence between the Hilbert space of state vectors of the (column) form $|\psi\rangle$ and the space of state vectors of the (row) form $\langle\psi|$. Each vector of one space is said to be the *Hermitian conjugate* of its counterpart in the other space. The two spaces are said to be their reciprocal *dual spaces* and one can summarize the above descriptions as

$$|\psi\rangle \quad \text{described by} \quad \begin{pmatrix} c_1 \\ c_2 \\ c_3 \\ \vdots \\ c_N \end{pmatrix}$$

transforms to its Hermitian conjugate:

$$\langle\psi| \quad \text{described by} \quad (c_1^*\ c_2^*\ c_3^*\ \cdots\ c_N^*)$$

Such a transformation is sometimes written as follows:

$$(c_1^* \ c_2^* \ c_3^* \ \cdots \ c_N^*) = \begin{pmatrix} c_1 \\ c_2 \\ c_3 \\ \vdots \\ c_N \end{pmatrix}^{t*}$$

where the superscript t stands for "transposed" and means that the conversion from column to line (or the reverse) is carried out while '$*$' indicates, as previously, a complex conjugation. This combination of transformations defines a *Hermitian conjugation* and is denoted by the *dagger symbol* '$\dagger$' so that

$$\langle\psi| \equiv (|\psi\rangle)^\dagger$$

or

$$(c_1^* \ c_2^* \ c_3^* \ \cdots \ c_N^*) = \begin{pmatrix} c_1 \\ c_2 \\ c_3 \\ \vdots \\ c_N \end{pmatrix}^{\dagger}$$

and conversely $|\psi\rangle = (\langle\psi|)^\dagger$ or

$$\begin{pmatrix} c_1 \\ c_2 \\ c_3 \\ \vdots \\ c_N \end{pmatrix} = (c_1^* \ c_2^* \ c_3^* \ \cdots \ c_N^*)^\dagger$$

As a consequence, the norm of a vector in a Hilbert space is

$$||\psi|| = \sqrt{\langle\psi||\psi\rangle} = \left[(c_1^* \ c_2^* \ c_3^* \ \cdots \ c_N^*) \begin{pmatrix} c_1 \\ c_2 \\ c_3 \\ \vdots \\ c_N \end{pmatrix}\right]^{1/2}$$

$$= \sqrt{\sum_j c_j^* c_j} = \sqrt{\sum_j |c_j|^2}$$

where it was assumed that the vector is expressed in an orthonormal basis set. If the basis is not orthonormal, it will soon appear that Dirac's notation provides a very efficient means to handle any case.

Scalar product is generally used to express the projection of a vector onto another one. In a complex space, the result depends on which one is projected and which one is the reference since, for two state vectors in Hilbert space, the projection of $|\psi_\alpha\rangle$ on $|\psi_\beta\rangle$ is chosen to be

$$\langle\psi_\beta||\psi_\alpha\rangle = \sum_j c^*_{\beta,j} c_{\alpha,j}$$

which is obviously different from $\langle\psi_\alpha||\psi_\beta\rangle = \sum_j c^*_{\alpha,j} c_{\beta,j}$ since $\langle\psi_\alpha||\psi_\beta\rangle = (\langle\psi_\beta||\psi_\alpha\rangle)^\dagger = (\langle\psi_\beta||\psi_\alpha\rangle)^*$. In the following sections, the scalar product will play a central role. Its notation $\langle\psi_\alpha||\psi_\beta\rangle$ is often abbreviated to $\langle\psi_\alpha|\psi_\beta\rangle$; with a fusion of the double vertical bars. Such a notation is actually inherited from the scalar product of complex functions (here in 1D):

$$(\phi_\alpha, \phi_\beta) = \int_{-\infty}^{+\infty} \phi^*_\alpha(x)\phi_\beta(x)dx$$

It bears Dirac's signature and ability to graphically manipulate mathematical objects. The shape of the brackets has been changed to highlight the particular physical nature of state vectors. Obviously, these are no ordinary vectors. Consequently, the left part of the bracket, which represents the dual space state vector, is often called *state bra* while the right part is of course named *state ket*. State vectors will thus be often invoked as "kets". Both denominations are synonymous and interchangeable.

State representation of an ensemble of particles

The states of a system $\mathcal{S}_a$ are represented by kets in a Hilbert space $\mathcal{H}_a$ and denoted by $\{|a_i\rangle\}$. Another system $\mathcal{S}_b$ has its kets $\{|b_i\rangle\}$ taken from a Hilbert state space $\mathcal{H}_b$. It is legitimate to ask how states of a composite system built from the union of $\mathcal{S}_a$ and $\mathcal{S}_b$ can be represented. States for this new system, let it be called $\mathcal{S}_c$, ought to render all the possible degrees of freedom of each of its constituents. A state space is thus built from the conjunction of the two individual state spaces. This construction is usually referred to as a *tensor product* of spaces $\mathcal{H}_a$ and $\mathcal{H}_b$ and written $\mathcal{H}_c = \mathcal{H}_a \otimes \mathcal{H}_b$. The elements of this new state space are named *state tensors* or, preferably, *tensor products of states* $\{|c_{i,j}\rangle = |a_i\rangle \otimes |b_j\rangle\}$. This notation will often be shortened to $\{|a_i, b_j\rangle\}$.

Tensor product is distributive with respect to the linear combination of kets, hence

$$|a\rangle \otimes (\beta_1|b_1\rangle + \beta_2|b_2\rangle) = \beta_1|a\rangle \otimes |b_1\rangle + \beta_2|a\rangle \otimes |b_2\rangle \tag{5.3}$$

Question 5.2: **Tensor product of states. Factorizable or entangled states.**

1. Assume system $\mathcal{S}_a$ can be described in a 2D Hilbert space. The basis kets for its state space are $\{|a_1\rangle, |a_2\rangle\}$. System $\mathcal{S}_b$ is described in a 2D state space supported by $\{|b_1\rangle, |b_2\rangle\}$. Write the expression of a state describing the composite system $\mathcal{S}_c = \mathcal{S}_a \bigcup \mathcal{S}_b$ as a tensor product of any two states respectively associated to $\mathcal{S}_a$ and $\mathcal{S}_b$.
2. Show that not all states of the composite system $\mathcal{S}_c$ are necessarily factorizable; in other words, they cannot be written as a tensor product of states of $\mathcal{S}_a$ and $\mathcal{S}_b$. Non-factorizable states are said to be *entangled states*.
3. Give an example of an entangled state for $\mathcal{S}_c$.

Answer:

1. Any state taken from state space $\mathcal{H}_a$ of $\mathcal{S}_a$ is: $|u_a\rangle = \alpha_1|a_1\rangle + \alpha_2|a_2\rangle$. In a similar manner, we write for $\mathcal{S}_b$: $|u_b\rangle = \beta_1|b_1\rangle + \beta_2|b_2\rangle$. We thus find the tensor product of these two states:

   $$|v\rangle = |u_a\rangle \otimes |u_b\rangle = \alpha_1\beta_1|a_1\rangle \otimes |b_1\rangle + \alpha_1\beta_2|a_1\rangle \otimes |b_2\rangle + \alpha_2\beta_1|a_2\rangle \otimes |b_1\rangle + \alpha_2\beta_2|a_2\rangle \otimes |b_2\rangle$$

   or, as a more compact expression

   $$|v\rangle = |u_a, u_b\rangle = \alpha_1\beta_1|a_1, b_1\rangle + \alpha_1\beta_2|a_1, b_2\rangle + \alpha_2\beta_1|a_2, b_1\rangle + \alpha_2\beta_2|a_2, b_2\rangle$$

2. Any given state describing the composite system can be expressed on the basis constructed upon the tensor product of the two respective bases. Hence, as a general result and for any state, we have

   $$|w\rangle = \gamma|a_1, b_1\rangle + \delta|a_1, b_2\rangle + \eta|a_2, b_1\rangle + \theta|a_2, b_2\rangle$$

   For this state to be factorizable, it is necessary to satisfy some conditions: those that were obtained while constructing the state vector $|v\rangle$. Therefore, there are some states obtained from the construction $\mathcal{H}_a \otimes \mathcal{H}_b$ that cannot be brought back to a mere juxtaposition of the two respective state spaces.

   In the present case, it can be shown that a necessary and sufficient condition for the state to be factorizable is that $\gamma \times \theta = \delta \times \eta$.

Answer: (continued)

3. We can mention a state corresponding to $\gamma = 0$. For such a state to be factorizable, it takes α_1 or β_1 to be zero. This necessarily implies that δ or η is also zero. In the opposite case, the state is not factorizable and it cannot be written in the form $|u_a\rangle \otimes |u_b\rangle$. Thus, the ket

$$|w\rangle = \frac{1}{\sqrt{2}}(|a_1, b_2\rangle + |a_2, b_1\rangle)$$

obviously represents an entangled state.

5.2.2. *Operators*

It may be useful to remind some key definitions and properties:

- An operator is a mathematical object that, broadly speaking, acts on a function to give another (or the same) function or acts on a ket to transform it to a different (or the same) ket. For instance, $\widehat{p}_x = -i\hbar\, \partial_x$ is an operator that acts on wavefunctions with a position variable x. The result is another wavefunction. In matrix terms, $\begin{pmatrix} 0 & 1 \\ 1 & 0 \end{pmatrix}$ is an operator that acts on vectors in a 2D space.

  In general, we will write $\widehat{A}|\psi_1(t)\rangle = |\psi_2(t)\rangle$ which should be interpreted as the action of operator $\widehat{A}$ on ket $|\psi_1(t)\rangle$, transforming it to ket $|\psi_2(t)\rangle$.
- If $\widehat{A}$ is an operator which acts on kets so that: $\widehat{A}|\psi_1(t)\rangle = |\psi_2(t)\rangle$, its *adjoint*, which is also called its *Hermitian conjugate*, is denoted by a dagger symbol $\widehat{A}^\dagger$ and acts according to

$$\langle\psi_1(t)|\widehat{A}^\dagger = \langle\psi_2(t)| = (\widehat{A}|\psi_1(t)\rangle)^\dagger$$

  Consequently, if the original operator acts on the ket on its right, its conjugate will act on a bra on its left.
- A *Hermitian operator* is a *self-adjoint operator*, i.e. the operator is its own adjoint $\widehat{A}^\dagger = \widehat{A}$.

  A Hermitian operator can thus act on its left or its right with no special care. This is very convenient since, as stated in the second postulate, all measurable quantities will be described by self-adjoint operators.

  Some intermediate operators, which do not pertain to any measurable quantity, will need to be constructed. They are not required to be Hermitian. Conversely, not all Hermitian operators correspond to a (known) measurable physical quantity.

Question 5.3: **Hermitian or not Hermitian? (part I).** Consider an operator acting on kets in a 2D state space:

$$\widehat{m} = \begin{pmatrix} 0 & 1+i \\ -1-i & 0 \end{pmatrix}$$

and a state ket represented by a column vector (in a given representation):

$$|\psi\rangle \propto \begin{pmatrix} i \\ 1 \end{pmatrix}$$

Is the operator $\widehat{m}$ self-adjoint? Check that $(\widehat{m}|\psi\rangle)^\dagger = \langle\psi|\widehat{m}^\dagger$.

Answer: The operator $\widehat{m}$ is not self-adjoint because

$$\widehat{m}^\dagger = (\widehat{m}^t)^* = \begin{pmatrix} 0 & -1+i \\ 1-i & 0 \end{pmatrix} \neq \begin{pmatrix} 0 & 1+i \\ -1-i & 0 \end{pmatrix}$$

Let us consider how $\widehat{m}$ acts on the given ket: $\widehat{m}|\psi\rangle = \begin{pmatrix} 1+i \\ 1-i \end{pmatrix}$ hence $(\widehat{m}|\psi\rangle)^\dagger = (1-i, 1+i)$. We also get

$$\langle\psi|\widehat{m}^\dagger = (-i, 1)\begin{pmatrix} 0 & -1+i \\ 1-i & 0 \end{pmatrix} = (1-i, 1+i)$$

- An operator $\widehat{A}$ possesses a set of eigenkets and eigenvalues so that

$$\widehat{A}|\phi_n\rangle = a_n|\phi_n\rangle$$

A Hermitian operator's eigenvalues $\{a_n\}$ are necessarily real. They constitute its *spectrum*. The eigenkets associated to distinct eigenvalues are orthogonal:[6]

$$\langle\phi_m|\phi_n\rangle = \delta_{m,n}$$

where $\delta_{m,n}$ is the Kronecker symbol. It can be proven by considering any two eigenkets of the Hermitian operator:

$$\langle\phi_m|\widehat{A}|\phi_n\rangle = \langle\phi_m|(\widehat{A}|\phi_n\rangle) = \langle\phi_m|a_n|\phi_n\rangle = a_n\langle\phi_m|\phi_n\rangle \quad (5.4)$$

$$\langle\phi_m|\widehat{A}^\dagger|\phi_n\rangle = (\langle\phi_m|\widehat{A}^\dagger)|\phi_n\rangle = a_m^*\langle\phi_m|\phi_n\rangle = a_m\langle\phi_m|\phi_n\rangle \quad (5.5)$$

[6]Remember the Kronecker symbol $\delta_{m,n}$ is 1 if $m = n$ and 0 if $m \neq n$.

because the eigenvalues are real. Since $\widehat{A}$ is Hermitian, i.e. $\widehat{A} = \widehat{A}^\dagger$, the two equations are identical, and their subtraction yields

$$(a_n - a_m)\langle\phi_m|\phi_n\rangle = 0$$

It can thus be concluded that, if the eigenvalues are different, the eigenkets need to be orthogonal.

If the eigenvalues are identical (degenerate case), kets from the degenerate subspace can be used to reconstruct (by linear combination) a set of orthogonal kets, which will of course share the same eigenvalue.

By definition, an observable is represented by a Hermitian operator, the eigenkets of which can be used as a basis for the Hilbert space of the system's possible states.

Consequently, any state ket can be expressed in this basis and we have

$$|\psi(t)\rangle = \sum_n c_n(t)\ |\phi_n\rangle \tag{5.6}$$

with $c_n(t) = \langle\phi_n|\psi(t)\rangle$. This constitutes the *spectral decomposition principle*.

Question 5.4: **Hermitian or not Hermitian? (part II).** Consider the following operator: $\widehat{m} = \begin{pmatrix} 0 & 1+i \\ -1-i & 0 \end{pmatrix}$. Are the eigenvalues real?
What about the eigenvalues of $\widehat{q} = \begin{pmatrix} 0 & -1+i \\ -1-i & 0 \end{pmatrix}$? Is one of these operators Hermitian?

Answer: As it was seen in the previous question, $\widehat{m}$ is not Hermitian. Consequently, its eigenvalues are not real.
Operator $\widehat{q}$ is Hermitian. Its eigenvalues are: $\pm\sqrt{2}$.

Probabilities for eigenvalues

If we consider the decomposition (5.6), the probability of finding a system described by $|\psi\rangle$ in the state $|\phi_n\rangle$ is also given by $|c_n|^2$. This coefficient is the projection of ket $|\psi(t)\rangle$ (that describes the state of the system at the moment the measurement is performed) onto the eigenket associated to the eigenvalue a_n:

$$p_n = \left|\langle\phi_n|\sum_m c_m(t)|\phi_m\rangle\right|^2 = \left|\sum_m c_m(t)\langle\phi_n|\phi_m\rangle\right|^2$$
$$= \left|\sum_m c_m(t)\delta_{mn}\right|^2 = |c_n(t)|^2$$

where we made use of the orthonormality of the kets in the eigenbasis set of $\widehat{A}$.[7]

If the expected eigenvalue a_n has g-fold degeneracy, i.e. it corresponds to g distinct eigenstates $\{|\phi_{n,1}\rangle, |\phi_{n,2}\rangle, \ldots, |\phi_{n,g}\rangle\}$, the associated probability is naturally given by a sum over all the degeneracy subspace: $p_n = \sum_g |\langle\phi_{n,g}|\psi(t)\rangle|^2$.

Question 5.5: **A quantum random number generator** can now be bought on shelves. They rely on a very stable and unbiased source of single photons (i.e. emitted one by one) all prepared in a given polarization state

$$\propto |\uparrow\rangle + |\rightarrow\rangle,$$

where the basis kets represent vertical and horizontal polarization states, respectively (Fig. 5.2). Show that a measurement of polarization states for a set of such photons can provide a set of perfect random bits and thereby a bias-free random number generator.

Answer: The source sends one photon at a time to a half-reflecting polarizer. If the photon is measured to be in the horizontal polarization state $|\rightarrow\rangle$, it is transmitted and counted in the "1" bit counter. If it is measured to be in the vertical polarization state $|\uparrow\rangle$, it is reflected and detected by the "0" bit counter. Since all photons emitted by the source are in the $|\psi\rangle = \frac{1}{\sqrt{2}}(|\uparrow\rangle + |\rightarrow\rangle)$ normalized state, the probability for the polarization measurement to be, say, $|\rightarrow\rangle$ is $|\langle\rightarrow|\psi\rangle|^2 = \frac{1}{2}$. For each photon, either results are equally probable and quantum physics ensures that the result from a set of eight successive photons provides a random number coded on eight bits.

Projectors

Let $\widehat{A}$ be an operator with eigenvectors $\{|\phi_n\rangle\}$ and eigenvalues $\{a_n\}$. Assume that the eigenkets span all the Hilbert space. It is possible to use

[7]Remember that $\widehat{A}$ has to be Hermitian.

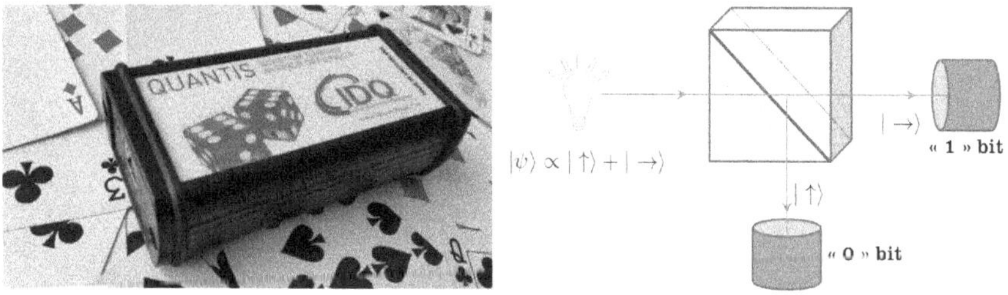

Fig. 5.2. Left: An example of a commercial quantum-based random number generator (QRNG) (courtesy of ID Quantique SA). Right: Physical principle of a QRNG (see answer to question).

the spectral decomposition to express any state ket on the eigenset of $\widehat{A}$:

$$|\psi\rangle = \sum_n c_n |\phi_n\rangle$$

where $c_n = \langle\phi_n|\psi\rangle$ is a scalar. The action of $\widehat{A}$ on the state ket gives

$$\begin{aligned} \widehat{A}|\psi\rangle &= \widehat{A}\sum_n c_n|\phi_n\rangle = \sum_n c_n \widehat{A}|\phi_n\rangle = \sum_n a_n|\phi_n\rangle c_n \\ &= \sum_n a_n|\phi_n\rangle\langle\phi_n|\psi(t)\rangle = \left(\sum_n a_n|\phi_n\rangle\langle\phi_n|\right)|\psi\rangle \end{aligned} \tag{5.7}$$

The operator can thus be written in terms of its eigenvalues and eigenvectors as

$$\widehat{A} = \sum_n a_n|\phi_n\rangle\langle\phi_n| \tag{5.8}$$

Another way to look at this representation is a sum, weighted by the eigenvalues, of operators which take the form: $\widehat{\mathcal{P}}_n = |\phi_n\rangle\langle\phi_n|$. Operator $\widehat{\mathcal{P}}_n$ is called *projector* because, quite obviously, its action on a given vector is to project it onto the eigenvector $|\phi_n\rangle$:

$$\widehat{\mathcal{P}}_n|\psi\rangle = (|\phi_n\rangle\langle\phi_n|)\ |\psi\rangle = |\phi_n\rangle\langle\phi_n|\psi\rangle = c_n|\phi_n\rangle$$

Whatever the starting vector, the action of $\widehat{\mathcal{P}}_n$ results into a vector which is colinear to $|\phi_n\rangle$.

Question 5.6: **Idempotent projectors.** Show that if ket $|\psi_j\rangle$ represents a quantum state, the projector represented by $\widehat{\mathcal{P}}_j = |\psi_j\rangle\langle\psi_j|$ is idempotent, i.e. $\widehat{\mathcal{P}}_j^2 = \widehat{\mathcal{P}}_j$.

Answer: $\widehat{\mathcal{P}}_j = |\psi_j\rangle\langle\psi_j|$ so $\widehat{\mathcal{P}}_j^2 = |\psi_j\rangle\langle\psi_j||\psi_j\rangle\langle\psi_j|$. Since ket $|\psi_j\rangle$ represents a quantum state, its norm must be 1: $\langle\psi_j||\psi_j\rangle = 1$. We hence find: $\widehat{\mathcal{P}}_j^2 = |\psi_j\rangle\langle\psi_j||\psi_j\rangle\langle\psi_j| = |\psi_j\rangle 1\langle\psi_j| = |\psi_j\rangle\langle\psi_j| = \widehat{\mathcal{P}}_j$.

The sum of the projectors that share the same eigenvalues a_n, $\sum_g |\phi_{n,g}\rangle\langle\phi_{n,g}|$ is a projector on the degeneracy subspace.

Question 5.7: **Pure state projector.** Check that if kets $|\psi_i\rangle$ and $|\psi_j\rangle$ represent any mutually orthogonal quantum states, operator

$$\widehat{\mathcal{P}} = \alpha_i|\psi_i\rangle\langle\psi_i| + \alpha_j|\psi_j\rangle\langle\psi_j|$$

is idempotent if $\alpha_i, \alpha_j \in \{0, 1\}$ (assuming they are real).

Answer: We compute

$$\widehat{\mathcal{P}}^2 = (\alpha_i|\psi_i\rangle\langle\psi_i| + \alpha_j|\psi_j\rangle\langle\psi_j|)\,(\alpha_i|\psi_i\rangle\langle\psi_i| + \alpha_j|\psi_j\rangle\langle\psi_j|) = \alpha_i^2|\psi_i\rangle\langle\psi_i| + \alpha_j^2|\psi_j\rangle\langle\psi_j|$$

This is identical to the initial operator only if $\alpha_i^2 = \alpha_i$ and $\alpha_j^2 = \alpha_j$. These conditions imply that $\alpha_i, \alpha_j \in \{0, 1\}$.

It is possible to go further by examining the action of an operator expressed as a sum of all projectors on the eigenkets of an observable. Since they constitute a complete basis set of the state space,

$$\begin{aligned}\left(\sum_n |\phi_n\rangle\langle\phi_n|\right)\ |\psi\rangle &= \left(\sum_n |\phi_n\rangle\langle\phi_n|\right)\ \left(\sum_m c_m|\phi_m\rangle\right)\\ &= \sum_{n,m} c_m\ |\phi_n\rangle\langle\phi_n|\ |\phi_m\rangle\\ &= \sum_{n,m} c_m\ |\phi_n\rangle\delta_{m,n} = \sum_n c_n\ |\phi_n\rangle\\ &= |\psi\rangle \end{aligned} \tag{5.9}$$

We hence obtain what is known as the *closure* or *completeness relation*. It states that the mere sum of all projectors, relative to *all* the kets of the eigenbasis, yields the identity operator (Fig. 5.3):

$$\sum_n |\phi_n\rangle\langle\phi_n| = \widehat{\mathbb{1}}$$

This relation is very useful for conducting many of the calculations to come and, in particular, allows a straightforward change of representation, i.e. reference frame.

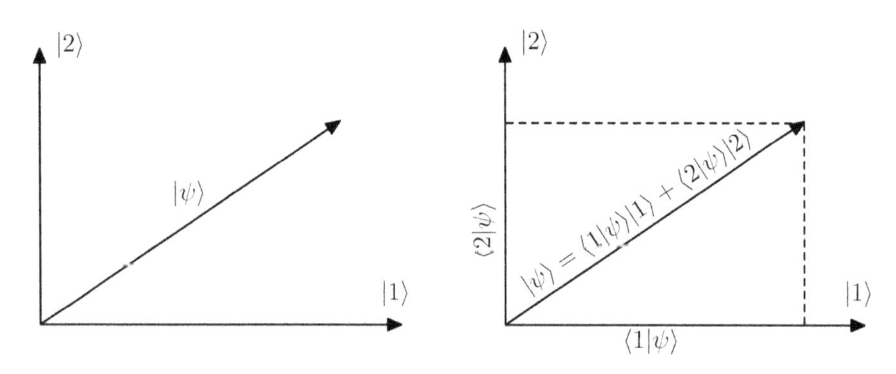

Fig. 5.3. The completeness relation is illustrated in the case of a basis built from two orthogonal kets $|1\rangle$ and $|2\rangle$. If this basis is sufficient for expressing $|\psi\rangle$, the basis is then complete, we have $|\psi\rangle = \langle 1|\psi\rangle|1\rangle + \langle 2|\psi\rangle|2\rangle$ where $\langle i|\psi\rangle$ is a mere scalar. Consequently, it is possible to change its position and factorize $|\psi\rangle = (|1\rangle\langle 1| + |2\rangle\langle 2|)\,|\psi\rangle$ so that $|1\rangle\langle 1| + |2\rangle\langle 2| = \widehat{\mathbb{1}}$.

Consider any operator $\widehat{B}$ to be expressed in the eigenbasis of the Hermitian operator $\widehat{A}$. When $\widehat{B}$ acts on an eigenket of $\widehat{A}$, it transforms it to a new ket which can be expressed in the eigenbasis:

$$\widehat{B}\,|\phi_n\rangle = \sum_m b_{m,n}\,|\phi_m\rangle$$

where $b_{m,n}$ are complex numbers.

If $\widehat{B}$ acts on any vector $|\psi\rangle$, using the closure, it is found that

$$\begin{aligned}\widehat{B}|\psi\rangle &= \widehat{B}\,\widehat{\mathbb{1}}|\psi\rangle \\ &= \widehat{B}\left(\sum_n |\phi_n\rangle\langle\phi_n|\right)|\psi\rangle = \left(\sum_n \widehat{B}|\phi_n\rangle\langle\phi_n|\right)|\psi\rangle \\ &= \left(\sum_n\sum_m b_{m,n}\,|\phi_m\rangle\langle\phi_n|\right)|\psi\rangle \end{aligned} \tag{5.10}$$

This is valid for any vector $|\psi\rangle$, therefore operator $\widehat{B}$ can be written in the form

$$\widehat{B} = \sum_{m,n} b_{m,n}\,|\phi_m\rangle\langle\phi_n| \tag{5.11}$$

Question 5.8: **Adjoint of a product.** Consider any two operators $\widehat{A}$ and $\widehat{B}$. Show that $(\widehat{A}\widehat{B})^\dagger = \widehat{B}^\dagger\widehat{A}^\dagger$.

Answer: Take any vector $|\psi\rangle$. We have

$$(\widehat{A}\widehat{B}|\psi\rangle)^\dagger = \langle\psi|(\widehat{A}\widehat{B})^\dagger$$

Let us set $\widehat{B}|\psi\rangle = |\phi\rangle$ so that $\langle\psi|\widehat{B}^\dagger = \langle\phi|$. Therefore, we also have

$$(\widehat{A}|\phi\rangle)^\dagger = \langle\phi|\widehat{A}^\dagger = \langle\psi|\widehat{B}^\dagger\widehat{A}^\dagger$$

These relations hold whatever $\widehat{A}$, $\widehat{B}$ and $|\psi\rangle$. We then indeed find $(\widehat{A}\widehat{B})^\dagger = \widehat{B}^\dagger\widehat{A}^\dagger$.

Function of operator

From the few operators at our disposal, it is possible to construct new and more sophisticated operators. This is achieved by means of constructing *functions of operators*. By definition, these functions are built using the Taylor expansion of the "normal" function:

$$f(x) = \sum_n \frac{1}{n!}\left(\frac{\partial^n f(x)}{\partial x^n}\right)_{x=0} x^n$$

which yields, for an operator $\widehat{A}$, a new operator which is a function of operator $\widehat{A}$:

$$f(\widehat{A}) = \sum_n \frac{1}{n!}\left(\frac{\partial^n f(x)}{\partial x^n}\right)_{x=0} \widehat{A}^n \tag{5.12}$$

Question 5.9: **Function of operator.** Use (5.8) to show by recurrence that if $\{a_n\}$ and $\{|\phi_n\rangle\}$ are the eigenvalues and eigenkets of operator $\widehat{A}$:

$$f(\widehat{A}) = \sum_n f(a_n)|\phi_n\rangle\langle\phi_n|$$

Answer: Obviously, relation (5.8) holds for $\widehat{A}^2$

$$\widehat{A}\widehat{A} = \sum_{n,n'} a_n a_{n'}|\phi_n\rangle\langle\phi_n||\phi_{n'}\rangle\langle\phi_{n'}| = \sum_n a_n^2|\phi_n\rangle\langle\phi_n|$$

where the orthonormality of eigenkets was used. Then, if the relation holds for $\widehat{A}^{\ell-1}$, it is easy to see that

$$\widehat{A}^{\ell-1}\widehat{A} = \sum_{n,n'} a_n^{\ell-1} a_{n'}|\phi_n\rangle\langle\phi_n||\phi_{n'}\rangle\langle\phi_{n'}| = \sum_n a_n^\ell|\phi_n\rangle\langle\phi_n|$$

Answer: (continued)
This result can be applied to each term in the linear combination which defines the function of operators:

$$f(\widehat{A}) = \sum_\ell \frac{1}{\ell!} \left(\frac{\partial^\ell f(x)}{\partial x^\ell} \right)_{x=0} \widehat{A}^\ell = \sum_\ell \frac{1}{\ell!} \left(\frac{\partial^\ell f(x)}{\partial x^\ell} \right)_{x=0} \sum_n a_n^\ell |\phi_n\rangle\langle\phi_n|$$
$$= \sum_n \left[\sum_\ell \frac{1}{\ell!} \left(\frac{\partial^\ell f(x)}{\partial x^\ell} \right)_{x=0} a_n^\ell \right] |\phi_n\rangle\langle\phi_n| = \sum_n f(a_n)|\phi_n\rangle\langle\phi_n|$$

Unitary operator

Among the operators that we will frequently encounter, some are not associated with measurable quantities. They merely modify kets, transforming one quantum state to another. If $\widehat{U}$ is such an operator and if $|\psi_1\rangle$ represents any given state, $\widehat{U}|\psi_1\rangle$ represents another quantum state $|\psi_2\rangle = \widehat{U}|\psi_1\rangle$. This new ket should be of norm 1, thus

$$\langle\psi_2||\psi_2\rangle = \langle\psi_1|\widehat{U}^\dagger \widehat{U}|\psi_1\rangle = 1$$

with $\langle\psi_1\rangle||\psi_1\rangle = 1$. Hence, $\widehat{U}$ is a unitary operator: $\widehat{U}^\dagger\widehat{U} = \widehat{\mathbb{1}}$. This last expression implies $\widehat{U}^{-1} = \widehat{U}^\dagger$ (and necessarily $\widehat{U}^\dagger\widehat{U} = \widehat{U}\widehat{U}^\dagger$).

✍ *Question 5.10:* **Unitary linear operator**. Show that a unitary operator $\widehat{U}$ is necessarily linear, i.e.

$$\widehat{U}(a|\psi_1\rangle) = a\,\widehat{U}|\psi_1\rangle$$

Hint: A good way to start is to use the linearity of the scalar product.

Answer: To compute the norm of $\widehat{U}(a|\psi_1\rangle) - a\widehat{U}|\psi_1\rangle$ we consider the scalar product:

$$(\widehat{U}(a|\psi_1\rangle) - a\widehat{U}|\psi_1\rangle)^\dagger(\widehat{U}(a|\psi_1\rangle) - a\widehat{U}|\psi_1\rangle) = \langle\psi_1|a^*\widehat{U}^\dagger\widehat{U}a|\psi_1\rangle + \langle\psi_1|\widehat{U}^\dagger a^* a\widehat{U}|\psi_1\rangle$$
$$- \langle\psi_1|\widehat{U}^\dagger a^*\widehat{U}a|\psi_1\rangle - \langle\psi_1|a^*\widehat{U}^\dagger a\widehat{U}|\psi_1\rangle$$

The linearity of the scalar product implies $\langle\psi_1|a|\psi_2\rangle = a\langle\psi_1||\psi_2\rangle$. Therefore,

$$(\widehat{U}(a|\psi_1\rangle) - a\widehat{U}|\psi_1\rangle)^\dagger(\widehat{U}(a|\psi_1\rangle) - a\widehat{U}|\psi_1\rangle)$$
$$= |a|\psi_1\rangle|^2 + |a\widehat{U}|\psi_1\rangle|^2 - (\langle\psi_1|\widehat{U}^\dagger a^*\widehat{U})a|\psi_1\rangle - \langle\psi_1|a^*(\widehat{U}^\dagger a\widehat{U}|\psi_1\rangle)$$

Answer: (continued)

$$= |a|^2||\psi_1\rangle|^2 + |a|^2|\widehat{U}|\psi_1\rangle|^2 - (\langle\psi_1|\widehat{U}^\dagger a^*\widehat{U})a|\psi_1\rangle - \langle\psi_1|a^*(\widehat{U}^\dagger a\widehat{U}|\psi_1\rangle)$$
$$= 2|a|^2||\psi_1\rangle|^2 - a(\langle\psi_1|\widehat{U}^\dagger)a^*(\widehat{U}|\psi_1\rangle) - a^*(\langle\psi_1|\widehat{U}^\dagger)a(\widehat{U}|\psi_1\rangle)$$
$$= 2|a|^2||\psi_1\rangle|^2 - 2|a|^2||\psi_1\rangle|^2 = 0$$

The ket has thus a zero norm. This implies that $\widehat{U}(a|\psi_1\rangle)$ and $a\widehat{U}|\psi_1\rangle$ are identical kets.

Continuous spectra

The position is certainly a property of particles that quantum mechanics should describe. It is generally accepted that localization can be measured by means of detectors. Whatever the type of detector being used, none records particle positions in an actual continuum space. For some very sensitive but large-window detectors, the position resolution can be on the order of a centimetre while scanning tunnelling microscopes or low-temperature neutron diffractometers can measure a position at sub-angström resolution.

In any practical case, no actual measurement renders the actual continuous nature of space-time. So, to begin, we will introduce a discrete localization Hermitian operator $\widehat{X}$. It is associated with the result of the detection of a particle position using some ideal experimental device (Fig. 5.4) which makes use of an array of pixels along the x-axis. All pixels have the same dimension Δ_x and the pixel position along the array is thus given by its index j so that pixel j is at $j \times \Delta_x$. When a position is measured, it does

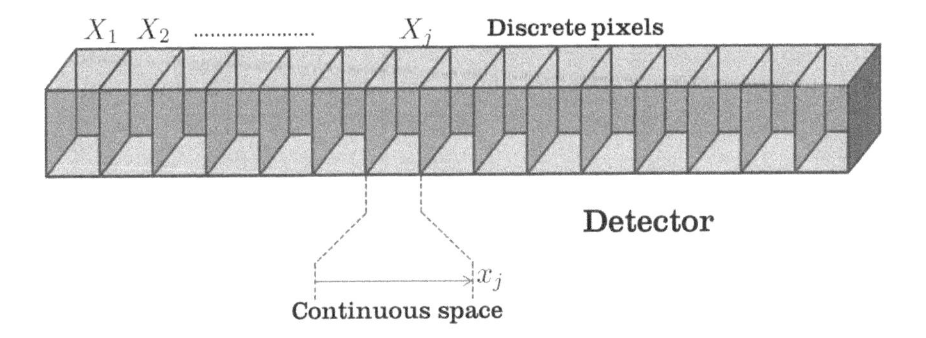

Fig. 5.4. From a discrete pixel scheme to continuous bounded space.

not matter whether it has been caught by pixel j at its centre or on its side; the recorded position is merely attributed to pixel j.

The ensemble of the possible detections $\{X_j = j \times \Delta_x\}$ are eigenvalues of the discrete localization operator $\widehat{X}$. Additionally, we denote by $|X_j\rangle$ the eigenvector associated with position $j \times \Delta_x$ so that it is possible to claim that a particle detected by the jth pixel is in a state approximately described by ket $|X_j\rangle$:

$$\widehat{X}|X_j\rangle = X_j|X_j\rangle$$

Provided that the $\{|X_j\rangle\}$ form a finite set, any ket state can be expressed in this basis:

$$|\psi\rangle = \sum_j c_j|X_j\rangle$$

Once a particle is detected by a given pixel, there is no chance that it can be detected by the neighbouring one, i.e. the discrete position states are mutually exclusive (Fig. 5.5). In mathematical terms, this means that the eigenkets are orthonormal: $\langle X_i|X_j\rangle = \delta_{i,j}$ and the coefficients are given by $c_j = \langle X_j|\psi\rangle$.

Ideally, it would be convenient to find a way of expressing the eigenstates of a continuous position operator $\widehat{x}$ so that it would allow a continuous spectrum of eigen-positions and eigenvectors:

$$\widehat{x}|x\rangle = x|x\rangle$$

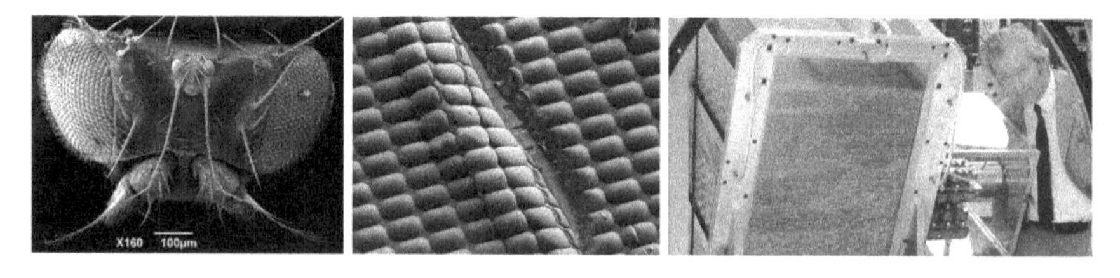

Fig. 5.5. Sensors of a drosophila eye (left) or lenses on a CCD chip (centre) discretize the position space for photon detection. The multiwire proportional chamber (which earned the Nobel Prize to G. Charpak in 1992) also provides a track record as a discrete set of points for particle positions over time. Photo credits: with kind permission of Bioimaging Centre, University of Exeter (left), Image courtesy to Carl Zeiss Microscopy GmbH (middle), and CERN (right).

However, unless the space domain is limited, these continuous state space vectors cannot have a unit norm and they are said to form an *improper basis.* Thus, within a single pixel j, the discrete eigenstate can conveniently be expressed as

$$|X_j\rangle = \int_{\Delta_x(j)} \alpha |x_j\rangle dx_j$$

where the integral is carried out in the limited domain of the jth pixel denoted by $\Delta_x(j)$. This means that if the pixel is assumed to be perfect, any position within its surface contributes equally, and the weight (chosen to be a real number) is denoted by α. Pixel discrete states are normalized, so

$$\begin{aligned}\langle X_j|X_j\rangle =& 1\\ =& \int_{\Delta_x(j)} \alpha\langle x'_j|dx'_j \int_{\Delta_x(j)} \alpha|x_j\rangle dx_j\\ =& \alpha^2 \int_{\Delta_x(j)} \int_{\Delta_x(j)} \langle x'_j|x_j\rangle dx_j dx'_j \end{aligned} \tag{5.13}$$

Even in a bounded domain of positions, it is not trivial to express the fact that a particle located at x_j cannot be described by $|x'_j\rangle$, no matter how small the separation is. This is expressed by the Dirac function: $\langle x'_j|x_j\rangle = \delta(x_j - x'_j)$. Obviously, this expression shows that $\langle x_j|x_j\rangle \neq 1$ so the continuous eigenvectors cannot represent real states. The orthonormality of discrete eigenstates (5.13) yields $\alpha^2\Delta_x(j) = 1$ and, assuming that all pixels have equal size $\Delta_x(j) = \Delta_x$, it results that $\alpha = (\Delta_x)^{-1/2}$.

This result is interesting because as the pixel size reduces, the continuity of position space becomes better rendered and, now assuming a constant state, it is possible to write

$$|x_j\rangle \overset{\Delta_x \to 0}{\approx} \frac{1}{\sqrt{\Delta_x}}|X_j\rangle$$

and the expansion $|\psi\rangle = \sum_j \langle X_j|\psi\rangle|X_j\rangle$ becomes $|\psi\rangle \approx \sum_j \langle x_j|\psi\rangle|x_j\rangle\Delta_x$ which, at the continuous limit $\Delta_x \to 0$, is equivalent to

$$|\psi\rangle \approx \int \langle x|\psi\rangle|x\rangle dx$$

In the case of a continuous position space, and actually any continuous variable, we will then make the following substitutions:

$$\langle X_j|X_i\rangle = \delta_{i,j} \longrightarrow \langle x|x'\rangle = \delta(x - x') \tag{5.14}$$

and for closure,

$$\sum_j |X_j\rangle\langle X_j| = \widehat{\mathbb{1}} \longrightarrow \int |x\rangle\langle x|dx = \widehat{\mathbb{1}} \tag{5.15}$$

If such a pseudo-continuous eigenbasis is then used for the position observable, any quantum state will then be written as

$$|\psi\rangle = \int_{-\infty}^{+\infty} |x\rangle\langle x|\psi\rangle dx \tag{5.16}$$

where the scalar products $\langle x|\psi\rangle$ are the continuous components of state-ket $|\psi\rangle$ on the continuous position basis. Furthermore, the modulus square of this component is interpreted as a probability density that a particle, described by ket $|\psi\rangle$ is found in a position eigenvalue x. Therefore, $|\langle x|\psi\rangle|^2 dx$ is the probability for the particle to be measured between x and $x + dx$. Consequently $\langle x|\psi\rangle$ bears the exact same significance as the wavefunction introduced in Section 3.2. The wavefunction should thus be seen as a particular representation of a quantum state: the projection of its state-ket onto the position (pseudo-continuous) eigenbasis. We will write

$$|\psi\rangle = \int_{\infty}^{+\infty} \psi(x)|x\rangle dx \tag{5.17}$$

with $\psi(x) = \langle x|\psi\rangle$. Position is not the sole physical quantity that is associated with a continuous spectrum. Energy or momentum of ideally unbound states suffer from the same pathology. Position is essentially discrete from an experimental perspective because any localization relies on a discretization of space but, as far as we are aware, it is usually admitted that the position space possesses an intrinsic continuous structure. We will make the same (convenient) assumption for the momentum space. We will then write $|p\rangle$ for the eigenvectors of the momentum operator (here in 1D) $\widehat{P}$ associated with a continuous eigenvalue spectrum. Any state-ket can thus be written in such a basis:

$$|\psi\rangle = \int_{-\infty}^{+\infty} \psi(p)|p\rangle dp \tag{5.18}$$

with the momentum representation of the wavefunction given by $\psi(p) = \langle p|\psi\rangle$. Swapping from a position to a momentum representation of a given ket is achieved by a judicious insertion of the closure:

$$\begin{aligned}
|\psi\rangle &= \int_{-\infty}^{+\infty} \psi(p)|p\rangle dp \\
&= \int_{-\infty}^{+\infty} \psi(p) \left[\int_{-\infty}^{+\infty} |x\rangle\langle x|dx\right] |p\rangle dp \\
&= \int_{-\infty}^{+\infty} \int_{-\infty}^{+\infty} \psi(p)\langle x|p\rangle dp|x\rangle dx
\end{aligned}$$

where $\langle x|p\rangle$ is the position representation of a momentum eigenstate. It is thus the wavefunction of a particle which has a fixed momentum. As mentioned in Section 3.2.2, this is given by a plane wave $Ae^{ipx/\hbar}$ and since, still using the closure:

$$\begin{aligned}
\langle p'|p\rangle = \langle p'| \left[\int_{-\infty}^{+\infty} |x\rangle\langle x|dx\right] |p\rangle &= \int_{-\infty}^{+\infty} \langle p'|x\rangle\langle x|p\rangle dx \\
&= \int_{\infty}^{+\infty} A^* e^{-ip'x/\hbar} A e^{ipx/\hbar} dx = |A|^2 \int_{-\infty}^{+\infty} e^{i(p-p')x/\hbar} dx \\
&= |A|^2 (2\pi\hbar)\delta(p-p')
\end{aligned} \tag{5.19}$$

we find that (within an insignificant constant phase factor) $A = 1/\sqrt{2\pi\hbar}$. This confirms that a change from position to momentum representations is carried out by a mere Fourier transform:

$$\psi(x) = \frac{1}{\sqrt{2\pi\hbar}} \int_{-\infty}^{+\infty} \psi(p) e^{ipx/\hbar} dp \tag{5.20}$$

Operators acting on a continuous eigenbasis

In this part, the continuous position eigenspace is used but every result and remark will remain valid in any other continuous space.

If the system is described by any ket $|\psi\rangle$, we can be led to consider its representation on a continuous basis for two principal reasons:

- This is the most convenient representation for the quantum state under study. As a practical example, quantum chemistry is elaborated in terms of chemical bonds or atomic sites which are generally best described in position space terms.
- The action of an operator is only known on components of the ket in this particular representation.

Thus, an operator acts on a state-ket $|\psi_1\rangle$ and transforms it to $|\psi_2\rangle$ according to

$$\widehat{A}|\psi_1\rangle = |\psi_2\rangle$$

$$\int \widehat{A}|x\rangle\langle x|\psi_1\rangle dx = \int |x\rangle\langle x|\psi_2\rangle dx$$

$$\int \widehat{A}\psi_1(x)|x\rangle dx = \int \psi_2(x)|x\rangle dx$$

Consequently, in a continuous representation, operator $\widehat{A}$ transforms the wavefunction from $\psi_1(x)$ to $\psi_2(x)$. As expressed by (3.44), the momentum operator takes the form $-i\hbar\partial_x$. This is thus an example of such an operator, the action of which is only in position or momentum representations:

$$\widehat{p}\widetilde{\psi}(p) \text{ or, equivalently, } \langle p|\widehat{p}|\psi\rangle = p\langle p|\psi\rangle$$

$$\widehat{p}\psi(x) \text{ or, equivalently, } \langle x|\widehat{p}|\psi\rangle = \int \langle x|p\rangle\langle p|\widehat{p}|\psi\rangle dp$$

✍ *Question 5.11.* **Hermiticity of $\widehat{p}$.** Show that the momentum operator acting on wavefunctions $\widehat{p} = -i\hbar\frac{\partial}{\partial x}$ is Hermitian.

Answer: We need to show that $\widehat{p} = \widehat{p}^\dagger$. Take any two state-kets $|\psi_1\rangle$ and $|\psi_2\rangle$. We have to show that $\langle\psi_1|\widehat{p}^\dagger|\psi_2\rangle = \langle\psi_1|\widehat{p}|\psi_2\rangle$. We have previously seen that $\langle\psi_1|\widehat{p}^\dagger = (\widehat{p}|\psi_1\rangle)^\dagger$. Therefore,

$$\begin{aligned}\langle\psi_1|\widehat{p}^\dagger &= \left[\int_{-\infty}^{+\infty} \widehat{p}|x\rangle\langle x|\psi_1\rangle dx\right]^\dagger = \left[\int_{-\infty}^{+\infty} \left(\frac{\hbar}{i}\frac{\partial}{\partial x}\psi_1(x)\right)|x\rangle dx\right]^\dagger \\ &= \int_{-\infty}^{+\infty} \langle x|\left(-\frac{\hbar}{i}\frac{\partial}{\partial x}\psi_1^*(x)\right)dx\end{aligned}$$

Answer: (continued)

We can then write the matrix element:

$$\begin{aligned}\langle\psi_1|\widehat{p}^\dagger|\psi_2\rangle &= \int_{-\infty}^{+\infty} \langle x| \left(-\frac{\hbar}{i}\frac{\partial}{\partial x}\psi_1^*(x)\right) dx|\psi_2\rangle \\ &= \int_{-\infty}^{+\infty} \langle x|\psi_2\rangle \left(-\frac{\hbar}{i}\frac{\partial}{\partial x}\psi_1^*(x)\right) dx \\ &= \int_{-\infty}^{+\infty} \psi_2(x) \left(-\frac{\hbar}{i}\frac{\partial}{\partial x}\psi_1^*(x)\right) dx\end{aligned}$$

which can be integrated by parts. Using the cancellation of wavefunctions at infinities, we find

$$\langle\psi_1|\widehat{p}^\dagger|\psi_2\rangle = \int_{-\infty}^{+\infty} \psi_1^*(x) \left(\frac{\hbar}{i}\frac{\partial}{\partial x}\psi_2(x)\right) dx = \int_{-\infty}^{+\infty} \langle\psi_1|x\rangle\langle x|\widehat{p}\psi_2\rangle dx = \langle\psi_1|\widehat{p}|\psi_2\rangle$$

Commutators

Operators act on kets, or wavefunctions, and transform them into vectors, or functions. Take the successive actions of $\widehat{B}$ and $\widehat{D}$. We consider here to what extent the order of their actions may matter. Let us express them with $\{|\phi_n\rangle\}$, the eigenbasis formed by the eigenkets of $\widehat{A}$. We consider how the successive actions of these two operators transform any ket $|\psi\rangle$. Using (5.11),

$$\begin{aligned}\widehat{B}\widehat{D}|\psi\rangle &= \left(\sum_{m,n} b_{m,n}\,|\phi_m\rangle\langle\phi_n|\right)\left(\sum_{p,q} d_{q,p}\,|\phi_q\rangle\langle\phi_p|\right)|\psi\rangle \\ &= \left(\sum_{m,n}\sum_{p,q} b_{m,n}\,d_{q,p}\,|\phi_m\rangle\langle\phi_n|\phi_q\rangle\langle\phi_p|\right)|\psi\rangle \\ &= \left(\sum_{m,n}\sum_{p,q} b_{m,n}\,d_{q,p}\,|\phi_m\rangle\delta_{n,q}\langle\phi_p|\right)|\psi\rangle \\ &= \left(\sum_{m,n,p} b_{m,n}\,d_{n,p}\,|\phi_m\rangle\langle\phi_p|\right)|\psi\rangle \\ &= \left[\sum_{m,p}\left(\sum_{n} b_{m,n}\,d_{n,p}\right)|\phi_m\rangle\langle\phi_p|\right]|\psi\rangle \qquad (5.21)\end{aligned}$$

or, switching the order of operators,

$$\begin{aligned}
\widehat{D}\widehat{B}|\psi\rangle &= \left(\sum_{m,n} d_{m,n}\ |\phi_m\rangle\langle\phi_n|\right)\left(\sum_{q,p} b_{q,p}\ |\phi_q\rangle\langle\phi_p|\right)|\psi\rangle \\
&= \left(\sum_{p,q}\sum_{m,n} d_{m,n}\ b_{q,p}\ |\phi_m\rangle\langle\phi_n|\phi_q\rangle\langle\phi_p|\right)|\psi\rangle \\
&= \left(\sum_{p,q}\sum_{m,n} d_{m,n}\ b_{q,p}\ |\phi_m\rangle\delta_{n,q}\langle\phi_p|\right)|\psi\rangle \\
&= \left(\sum_{m,n,p} d_{m,n}\ b_{n,p}\ |\phi_m\rangle\langle\phi_p|\right)|\psi\rangle \\
&= \left[\sum_{m,p}\left(\sum_{n} d_{m,n}\ b_{n,p}\right)|\phi_m\rangle\langle\phi_p|\right]|\psi\rangle
\end{aligned} \tag{5.22}$$

The final result is thus, for any $|\psi\rangle$,

$$\widehat{B}\widehat{D}|\psi\rangle \neq \widehat{D}\widehat{B}|\psi\rangle \tag{5.23a}$$

because

$$\left(\sum_{n} b_{m,n}\ d_{n,p}\right)\ |\phi_m\rangle\langle\phi_p| \neq \left(\sum_{n} d_{m,n}\ b_{n,p}\right)|\phi_m\rangle\langle\phi_p| \tag{5.23b}$$

The order of operators thus obviously matters. We name *commutator* the operator resulting from the subtraction of two operator products:

$$[\widehat{B},\widehat{D}] \equiv \widehat{B}\widehat{D} - \widehat{D}\widehat{B} \tag{5.24}$$

✍ *Question 5.12:* **Properties of commutators (part I).** Consider four operators $\widehat{A},\widehat{B},\widehat{C},\widehat{D}$. Write $[\widehat{A}+\widehat{B},\widehat{C}+\widehat{D}]$ as a function of each commutator involving only a pair of operators.

Answer: Using the definition of commutators, it can be developed:

$$\begin{aligned}
[\widehat{A}+\widehat{B},\widehat{C}+\widehat{D}] &= (\widehat{A}+\widehat{B})(\widehat{C}+\widehat{D}) - (\widehat{C}+\widehat{D})(\widehat{A}+\widehat{B}) \\
&= \widehat{A}\widehat{C}+\widehat{A}\widehat{D}+\widehat{B}\widehat{C}+\widehat{B}\widehat{D}-\widehat{C}\widehat{A}-\widehat{D}\widehat{A}-\widehat{C}\widehat{B}-\widehat{D}\widehat{B}
\end{aligned}$$

Individual operators can be identified and gathered:

$$[\widehat{A}+\widehat{B},\widehat{C}+\widehat{D}] = [\widehat{A},\widehat{C}]+[\widehat{A},\widehat{D}]+[\widehat{B},\widehat{C}]+[\widehat{B},\widehat{D}]$$

✍ *Question 5.13:* **Properties of commutators (part II).** Show that if $\hat{A}$ and $\widehat{B}$ are any operators, we have $[\widehat{A}^2, \widehat{B}] = \widehat{A}[\widehat{A}, \widehat{B}] + [\widehat{A}, \widehat{B}]\widehat{A}$.

Answer: We only need to develop and insert the symmetric product $\widehat{A}\widehat{B}\widehat{A}$

$$[\widehat{A}^2, \widehat{B}] = \widehat{A}\widehat{A}\widehat{B} + (-\widehat{A}\widehat{B}\widehat{A} + \widehat{A}\widehat{B}\widehat{A}) - \widehat{B}\widehat{A}\widehat{A} = \widehat{A}[\widehat{A}, \widehat{B}] + [\widehat{A}, \widehat{B}]\widehat{A}$$

In a similar manner, it is an easy matter to derive the relation:

$$[\widehat{A}\widehat{B}, \widehat{C}\widehat{D}] = \widehat{A}([\widehat{B}, \widehat{C}]\widehat{D} + \widehat{C}[\widehat{B}, \widehat{D}]) + ([\widehat{A}, \widehat{C}]\widehat{D} + \widehat{C}[\widehat{A}, \widehat{D}])\widehat{B} \tag{5.25}$$

It is well known that the order in product matters for matrices. Operators can often have a matrix representation. Their matrix elements are inferred from their action $\widehat{B}$ on the eigenkets $|\phi_i\rangle$ of an operator $\widehat{A}$

$$\widehat{B}|\phi_i\rangle = \left(\sum_{m,n} b_{m,n}\ |\phi_m\rangle\langle\phi_n|\right)|\phi_i\rangle = \sum_{m,n} b_{m,n}\ |\phi_m\rangle\delta_{i,n} = \sum_m b_{m,i}\ |\phi_m\rangle$$

which is projected on other vectors of the same set $|\phi_j\rangle$:

$$\langle\phi_j|(\widehat{B}|\phi_i\rangle) = \sum_m b_{m,i}\ \langle\phi_j|\phi_m\rangle = b_{j,i}$$

Therefore, the matrix elements of an operator in the representation of the eigenspace of $\widehat{A}$ are found by computing each term:

$$\begin{pmatrix} \langle\phi_1|\widehat{B}|\phi_1\rangle & \langle\phi_1|\widehat{B}|\phi_2\rangle & \dots & \langle\phi_1|\widehat{B}|\phi_N\rangle \\ \langle\phi_2|\widehat{B}|\phi_1\rangle & \langle\phi_2|\widehat{B}|\phi_2\rangle & \dots & \langle\phi_2|\widehat{B}|\phi_N\rangle \\ \vdots & \vdots & \ddots & \vdots \\ \langle\phi_N|\widehat{B}|\phi_1\rangle & \langle\phi_N|\widehat{B}|\phi_2\rangle & \dots & \langle\phi_N|\widehat{B}|\phi_N\rangle \end{pmatrix} = \begin{pmatrix} b_{1,1} & b_{1,2} & \dots & b_{1,N} \\ b_{2,1} & b_{2,2} & \dots & b_{2,N} \\ \vdots & \vdots & \ddots & \vdots \\ b_{N,1} & b_{N,2} & \dots & b_{N,N} \end{pmatrix}$$

Of course, in its own eigenbasis, $\widehat{A}$ is diagonal:

$$\begin{pmatrix} a_1 & 0 & \dots & 0 \\ 0 & a_2 & \dots & 0 \\ \vdots & \vdots & \ddots & \vdots \\ 0 & 0 & \dots & a_N \end{pmatrix}$$

If $\widehat{A}$ and $\widehat{B}$ are observables, the commutator $[\widehat{A}, \widehat{B}]$ can be computed and, using (5.23), we readily find

$$[\widehat{A}, \widehat{B}] = \sum_{m \neq n} b_{m,n}(a_m - a_n)|\phi_m\rangle\langle\phi_n|$$

This commutator is zero if and only if each and every term of the double sum is zero:

- for the eigenvalues a_n that are not degenerate (i.e. there is no $a_m = a_n$ for $m \neq n$); $b_{m,n} = b_n\delta_{m,n}$. The operator $\widehat{B}$ is thus diagonal in the associated non-degenerate subspace and the sum on these terms reduces to $\sum_{n,\text{non-degen.}} b_n|\phi_n\rangle\langle\phi_n|$.
- for each degenerate eigenvalue, if some $|\phi_m\rangle$ are all the eigenkets of $\widehat{A}$ with the same eigenvalue, then any linear combination will also give an eigenket of $\widehat{A}$ with the same eigenvalue. This induces that, within the degeneracy subspace, it is possible to find linear combinations that are eigenkets of $\widehat{B}$ while, of course, still being eigenkets of $\widehat{A}$. This is called the diagonalization of $\widehat{B}$ in the degeneracy subspace of $\widehat{A}$:

$$\begin{pmatrix} b_{1,1} & 0 & \dots & 0 & 0 & 0 & \dots & 0 \\ 0 & b_{2,2} & \dots & 0 & 0 & 0 & \dots & 0 \\ \vdots & \vdots & \ddots & \vdots & \vdots & \vdots & \ddots & \vdots \\ 0 & 0 & \dots & b_{i,i} & b_{i,i+1} & 0 & \dots & 0 \\ 0 & 0 & \dots & b_{i+1,i} & b_{i+1,i+1} & 0 & \dots & 0 \\ \vdots & \vdots & \ddots & \vdots & \vdots & \vdots & \ddots & \vdots \\ 0 & 0 & \dots & 0 & 0 & 0 & \dots & b_{N,N} \end{pmatrix}$$

$$\Downarrow$$

$$\begin{pmatrix} b_{1,1} & 0 & \dots & 0 & 0 & 0 & \dots & 0 \\ 0 & b_{2,2} & \dots & 0 & 0 & 0 & \dots & 0 \\ \vdots & \vdots & \ddots & \vdots & \vdots & \vdots & \ddots & \vdots \\ 0 & 0 & \dots & b'_{i,i} & 0 & 0 & \dots & 0 \\ 0 & 0 & \dots & 0 & b'_{i+1,i+1} & 0 & \dots & 0 \\ \vdots & \vdots & \ddots & \vdots & \vdots & \vdots & \ddots & \vdots \\ 0 & 0 & \dots & 0 & 0 & 0 & \dots & b_{N,N} \end{pmatrix}$$

An important conclusion can thus be drawn: If two observables commute, they share a common eigenbasis set.

Question 5.14: The **Baker–Hausdorff formula** is of great use in many aspects of quantum mechanics. It is presented here as mere training to juggle with commutators and functions of operators. It will find its application in Volume 3 when the formalism of neutron scattering will be explained. The derivation is adapted from that which is proposed by Squires [83]. Consider two operators $\widehat{A}$ and $\widehat{B}$, the commutator of which is a non-zero scalar: $[\widehat{A}, \widehat{B}] = \gamma \in \mathbb{C}$. In other words, if $\widehat{A}$ and $\widehat{B}$ do not commute, but any operator commutes with their commutator, the following equality holds:

$$e^{\widehat{A}+\widehat{B}} = e^{\widehat{A}} e^{\widehat{B}} e^{-\gamma/2}$$

Obviously, the usual formula of exponential is retrieved when $\gamma = 0$. The fact that the operators do not commute necessitates the last exponential term as a correction factor.

1. Prove by recurrence that $\widehat{A}\widehat{B}^n - \widehat{B}^n\widehat{A} = n\gamma\widehat{B}^{n-1}$.
2. Let λ be any number. Use the equality of the previous question to show that $\widehat{A}e^{\lambda\widehat{B}} - e^{\lambda\widehat{B}}\widehat{A} = \gamma\lambda e^{\lambda\widehat{B}}$.
3. Consider a function $f(\lambda) = e^{\lambda\widehat{A}} e^{\lambda\widehat{B}} e^{-\lambda(\widehat{A}+\widehat{B})}$. Show that $\partial_\lambda f(\lambda) = \lambda\gamma f(\lambda)$.
4. Solve the differential equation found in the previous question and deduce Baker–Hausdorff's formula.

Answer:

1. The expression is obviously satisfied for $n = 1$. Now start from the nth-order equality and multiply by $\widehat{B}$ on the right:

$$(\widehat{A}\widehat{B}^n - \widehat{B}^n\widehat{A})\widehat{B} = n\gamma\widehat{B}^n$$

$$(\widehat{A}\widehat{B}^{n+1} - \widehat{B}^n\widehat{A}\widehat{B}) = n\gamma\widehat{B}^n$$

$$\widehat{A}\widehat{B}^{n+1} - \widehat{B}^n(\gamma - \widehat{B}\widehat{A}) = n\gamma\widehat{B}^n$$

$$\widehat{A}\widehat{B}^{n+1} + \widehat{B}^{n+1}\widehat{A} - \widehat{B}^n\gamma = n\gamma\widehat{B}^n$$

where the commutator was used on the penultimate line. The desired relation is thus demonstrated for order $n + 1$ and consequently at any order.

2. Use of (5.12) gives

$$\widehat{A}e^{\lambda\widehat{B}} - e^{\lambda\widehat{B}}\widehat{A} = \sum_{n=0} \frac{\lambda^n}{n!}(\widehat{A}\widehat{B}^n - \widehat{B}^n\widehat{A})$$

$$= \sum_{n=1} \frac{\lambda^n}{n!}(n\gamma\widehat{B}^{n-1}) = \gamma\lambda \sum_{n=1} \frac{\lambda^{n-1}}{(n-1)!}(\widehat{B}^{n-1})$$

$$= \gamma\lambda e^{\lambda\widehat{B}}$$

Answer: (continued)

3. The derivation is straightforward: $\partial_\lambda f(\lambda) = e^{\lambda\widehat{A}}(\widehat{A}e^{\lambda\widehat{B}} - e^{\lambda\widehat{B}}\widehat{A})e^{-\lambda(\widehat{A}+\widehat{B})} = e^{\lambda\widehat{A}}\gamma\lambda e^{\lambda\widehat{B}}e^{-\lambda(\widehat{A}+\widehat{B})} = \lambda\gamma f(\lambda)$ where use was made of the commutator found in the previous question.
4. The solution of the differential equation $\partial_\lambda f(\lambda) = \lambda\gamma f(\lambda)$ is $f(\lambda) = e^{\gamma\lambda^2/2}$ and valid for any value of λ. The scale factor is readily obtained from the $\lambda = 0$ case. Setting $\lambda = 1$ in equality $f(\lambda = 1) = e^{\gamma/2} = e^{\widehat{A}}e^{\widehat{B}}e^{-(\widehat{A}+\widehat{B})}$ immediately yields Baker–Hausdorff's formula.

Complete set of commuting observables (CSCO)

It has just been emphasized that if a set of observables $\widehat{A}$, $\widehat{B}$, $\widehat{C}$, $\widehat{D}, \ldots$ are such that each pair commutes, then they share a common eigenbasis set which can be orthogonal. In that common basis-set, it is then possible to write:

$$\begin{aligned}
\widehat{A}|\phi_i\rangle &= a_i|\phi_i\rangle \\
\widehat{B}|\phi_i\rangle &= b_i|\phi_i\rangle \\
\widehat{C}|\phi_i\rangle &= c_i|\phi_i\rangle \\
\widehat{D}|\phi_i\rangle &= d_i|\phi_i\rangle \\
&\vdots
\end{aligned} \tag{5.26}$$

If, for any common eigenket $|\phi_i\rangle$, the mere knowledge of its associated respective eigenvalues a_i, b_i, c_i, $d_i, \ldots$, suffices to *totally* and unambiguously define it, the set of observables is then said to form a complete set of commuting observables (CSCO). The associated eigenvalues are called *quantum numbers*. The eigenvalues of an operator are said to represent a *good quantum number* if they allow the state of the system to be determined. Hence, this is equivalent to saying that the operator belongs to the CSCO.

In practice, finding the appropriate CSCO boils down to obtaining unambiguous means of tagging eigenkets for a given problem. Frequently, the total energy operator, i.e. the Hamiltonian, will be chosen to be part of the CSCO so that the eigenkets will correspond to stationary states.

Consequently, all the observables included in the CSCO will describe conserved physical quantities. To use the above example, a common eigenket could thus be written as $|\phi_i\rangle = |a_i, b_i, c_i, d_i, \ldots\rangle$ where the only eigenvalues that are retained are those that are *necessary and sufficient* to fully describe the state.

If the Hamiltonian eigenvalues are not degenerate, then the sole indication of the energy is enough to define the state of a system. On the contrary, if they are degenerate, for example $|\phi_1\rangle$ and $|\phi_2\rangle$ share an energy value ε_{12}, it is possible to seek an observable, say $\widehat{A}$, that commutes with the Hamiltonian $\widehat{H}$ for which $|\phi_1\rangle$ and $|\phi_2\rangle$ correspond to two different eigenvalues a_1 and a_2. They will thus allow differentiation of the states with identical energy. These states can then be named respectively $|\varepsilon_{12}, a_1\rangle$ and $|\varepsilon_{12}, a_2\rangle$.

The infinite 2D symmetrical quantum well can be taken as a practical example. From Question *4.2* in Section 4.3, it can be readily seen that, if both sides have a width ℓ, then the eigenenergies for a particle of mass m are $\varepsilon = (\hbar\pi)^2(n_x^2 + n_y^2)/(2m\ell^2)$. The symmetry introduced by the two equivalent directions x and y creates degeneracy. As the Hamiltonian can be decomposed in two components $\widehat{H}_x$ and $\widehat{H}_y$, each of them acting on a unique direction, they obviously commute ($[\widehat{H}_x, \widehat{H}_y] = 0$) with respective eigenenergies $\varepsilon_x = n_x^2(\hbar\pi)^2/(2m\ell^2)$ and $\varepsilon_y = n_y^2(\hbar\pi)^2/(2m\ell^2)$. Since the pair of quantum numbers (n_x, n_y) suffices to thoroughly characterize a stationary quantum state, it can be denoted by $|n_x, n_y\rangle$. Any similar tagging of the kets would be acceptable, such as $|n_x^2, n_y\rangle$ or $|\varepsilon_x, \varepsilon_y\rangle$ or even $|\varepsilon, n_y\rangle$.

5.2.3. *Mean values and generalized indetermination*

Unless very specific conditions are met, it is generally not possible to predict the outcome of a particular measurement. However, quantum mechanics allows for the prediction of the mean result from a large number of identical experiments. The term "identical experiments" means the repetition of a measurement of the same physical quantity, represented by an observable $\widehat{A}$, on a system initially prepared in the *same* quantum state $|\psi\rangle$. We have seen that the result of each such experiment necessarily belongs to the spectrum of eigenvalues for the operator $\widehat{A}$ associated with the measured quantity. Each new measurement can be seen as a new "draw" among those possible values.

Obviously, this is a breakdown of classical causality according to which if nothing is changed to the causes (i.e. identical initial state and property), the effect (i.e. the result of the measurement) should always be the same.

In quantum mechanics, it is only possible to predict the probability law governing the frequency of possible results, not the result of a single measurement.[8] Here, each event, i.e. the outcome of a measurement of property A, must be seen as the realization of just one possible result according to a probability law given by $p(a_n) = p_n = |\langle\phi_n|\psi\rangle|^2$, where $|\phi_n\rangle$ is the eigenket[9] of $\widehat{A}$ associated with the eigenvalue a_n.

Consider a system in any arbitrary quantum state described by $|\psi\rangle$ that can be developed on the eigenbasis of an observable $\widehat{A}$, $\{|\phi_n\rangle\}$, according to $|\psi\rangle = \sum_n c_n|\phi_n\rangle$ with $c_n = \langle\phi_n|\psi\rangle$. If the eigenvalues of $\widehat{A}$ are $\{a_n\}$, the mean value of the associated quantity is denoted $\langle A\rangle_\psi$ and can be computed as

$$\begin{aligned}\langle A\rangle_\psi &= \sum_n p_n a_n = \sum_n |\langle\phi_n|\psi\rangle|^2 a_n = \sum_n \langle\psi|\phi_n\rangle\langle\phi_n|\psi\rangle a_n \\ &= \sum_n \langle\psi|a_n|\phi_n\rangle\langle\phi_n|\psi\rangle = \sum_n \langle\psi|\widehat{A}|\phi_n\rangle\langle\phi_n|\psi\rangle \\ &= \langle\psi|\widehat{A}\left(\sum_n |\phi_n\rangle\langle\phi_n|\right)|\psi\rangle = \langle\psi|\widehat{A}|\psi\rangle\end{aligned}$$

where, according to the fourth postulate, the probability of measuring a particular eigenvalue a_n was taken to be $p_n = c_n^* c_n = |\langle\phi_n|\psi\rangle|^2$ and the completeness of the eigenbasis-set was used.

Therefore, the mean value of a physical quantity represented by observable $\widehat{A}$ when the system is in state $|\psi\rangle$ is computed by

$$\langle A\rangle_\psi = \langle\psi|\widehat{A}|\psi\rangle \tag{5.27}$$

[8]The exception to such a statement is when the system is prepared, prior to the measurement, in an eigenstate of the observable.

[9]For clarity, the eigenvalues are supposed to be non-degenerate.

This is a very important result because, in many practical applications, the mean value is the only accessible quantity from experiment. This is generally the case when the macroscopic property of an ensemble of identical particles is measured, such as the energy, the magnetic or the electric dipole moment. If the particles are independent, the property is the mere sum of each particle contribution. For a large ensemble of N particles, all in the same state $|\psi\rangle$, the macroscopic result is thus $N\langle\psi|\widehat{A}|\psi\rangle$. This will be extensively used in Volume 2.

The dispersion of results around the mean value $\langle A\rangle_\psi$ can be estimated by the variance:

$$\sigma_\psi^2(A) = \langle A^2\rangle_\psi - \langle A\rangle_\psi^2 = \langle\psi|\widehat{A}^2|\psi\rangle - \langle\psi|\widehat{A}|\psi\rangle^2 \tag{5.28}$$

Its square root corresponds to the standard deviation[10] denoted by $\Delta_\psi(A)$. This is thus an estimate of the indetermination associated with the result of a measurement when the system is in state $|\psi\rangle$. This quantity only depends on the state of the system and the observable spectrum. Consequently, it is intrinsic to the system and not related in any way to the skill of the experimentalist. It can thus be seen as a lower bound to experimental uncertainty.

Question 5.15: What are the **mean value and standard deviation** when the system is prepared in an eigenstate of the property to be measured?

Answer: If the system is prepared in $|\phi_n\rangle$, an eigenstate of the operator $\widehat{A}$ associated with the physical quantity to be measured, the result can only be the eigenvalue a_n which corresponds to this eigenket. This is because the probabilities for each result $\{a_j\}$ are $p_j = |\langle\phi_j|\phi_n\rangle|^2 = \delta_{j,n}$. As the operator is Hermitian, the eigenbasis is orthogonal (and normalized) and all probabilities are zero except the one corresponding to a_n, which is one. The standard deviation is consequently zero.

[10]For a normal law, a probability of 68% is included in the interval $\pm\sigma$:

$$\sigma^2 = \int_{-\infty}^{+\infty} x^2\sqrt{\alpha/\pi}\exp(-\alpha x^2)dx$$

and $\int_{-\sigma}^{+\sigma}\sqrt{\alpha/\pi}\exp(-\alpha x^2)dx \approx 0.6827$.

Consider two observables $\widehat{A}$ and $\widehat{B}$ such that their commutator is

$$[\widehat{A}, \widehat{B}] = i\hbar\widehat{C}$$

By construction $\widehat{C}$ is Hermitian (see Question *5.16*). The system is in state $|\psi\rangle$ and a new operator can be constructed:

$$(\widehat{A} - \langle A\rangle_\psi) + i\lambda(\widehat{B} - \langle B\rangle_\psi)$$

where λ is a real scalar and we have used the notations $\langle A\rangle_\psi = \langle\psi|\widehat{A}|\psi\rangle$ and $\langle B\rangle_\psi = \langle\psi|\widehat{B}|\psi\rangle$. By definition of the modulus square

$$|[(\widehat{A} - \langle A\rangle_\psi) + i\lambda(\widehat{B} - \langle B\rangle_\psi)]\ |\psi\rangle|^2 \geqslant 0$$

Since the operators are Hermitian and λ is real, the mean value can be written

$$\langle\psi|(\widehat{A} - \langle A\rangle_\psi)^2 + \lambda^2(\widehat{B} - \langle B\rangle_\psi)^2 + i\lambda[\widehat{A}, \widehat{B}]|\psi\rangle \geqslant 0$$

or

$$\langle(\widehat{A} - \langle A\rangle_\psi)^2\rangle_\psi + \lambda^2\langle(\widehat{B} - \langle B\rangle_\psi)^2\rangle_\psi - \hbar\lambda\langle\widehat{C}\rangle_\psi \geqslant 0$$

This inequality is satisfied for any real λ only if

$$\hbar^2\langle\widehat{C}\rangle_\psi^2 - 4\langle(\widehat{A} - \langle A\rangle_\psi)^2\rangle_\psi\langle(\widehat{B} - \langle B\rangle_\psi)^2\rangle_\psi \leqslant 0$$

hence

$$\Delta_\psi(A)\ \Delta_\psi(B) \geq \frac{\hbar}{2}\langle C\rangle_\psi \tag{5.29}$$

This is a generalization of the indeterminacy (or uncertainty) principle that was encountered in Section 3.2.2. It bears an essential implication: it is not possible to determine simultaneously, with arbitrary precision, the values of two observables, the operators of which do not commute. The physical quantities are thus said to be *incompatible.*

We will refer to this result on a number of occasions as it represents one of the cornerstones of quantum mechanics: order matters.

✍ *Question 5.16:* **Hermicity of $\widehat{C}$.** Show that $\widehat{C}$, defined by $[\widehat{A}, \widehat{B}] = i\hbar\widehat{C}$, is Hermitian when $\widehat{A}$ and $\widehat{B}$ are Hermitian.

Answer: We simply need to compute

$$\begin{aligned}\widehat{C}^\dagger &= \frac{1}{-i\hbar}[\widehat{A}, \widehat{B}]^\dagger = \frac{1}{-i\hbar}((\widehat{A}\widehat{B})^\dagger - (\widehat{B}\widehat{A})^\dagger) = \frac{1}{-i\hbar}(\widehat{B}^\dagger\widehat{A}^\dagger - \widehat{A}^\dagger\widehat{B}^\dagger) \\ &= \frac{1}{i\hbar}(\widehat{A}^\dagger\widehat{B}^\dagger - \widehat{B}^\dagger\widehat{A}^\dagger) = \frac{1}{i\hbar}(\widehat{A}\widehat{B} - \widehat{B}\widehat{A}) = \widehat{C}\end{aligned} \tag{5.30}$$

where, in the last line, the hermicity of operators $\widehat{A}$ and $\widehat{B}$ was used.

The canonical commutator and position-momentum indeterminacy

Position and momentum are called the *canonical coordinates* (see Section 6.2). In one dimension, their commutator is also an operator and acts on any well-behaving test wavefunction $\psi(x)$ according to

$$\begin{aligned}[\widehat{x}, \widehat{p}_x]\psi(x) &= \widehat{x}\widehat{p}_x\psi(x) - \widehat{p}_x\widehat{x}\psi(x) \\ &= \widehat{x}\left(\frac{\hbar}{i}\frac{\partial}{\partial x}\psi(x)\right) - \frac{\hbar}{i}\frac{\partial}{\partial x}\left(\widehat{x}\psi(x)\right) \\ &= \widehat{x}\left(\frac{\hbar}{i}\frac{\partial}{\partial x}\psi(x)\right) - \frac{\hbar}{i}\frac{\partial}{\partial x}\left(x\psi(x)\right) \\ &= \widehat{x}\left(\frac{\hbar}{i}\frac{\partial}{\partial x}\psi(x)\right) - \frac{\hbar}{i}\psi(x) - \frac{\hbar}{i}x\frac{\partial}{\partial x}\psi(x) \\ &= i\hbar\psi(x)\end{aligned} \tag{5.31}$$

which is valid for any test wavefunction. Consequently, we obtain the *canonical commutator*:

$$[\widehat{x}, \widehat{p}_x] - i\hbar\widehat{\mathbb{1}}$$

Since the position and momentum observables do not commute, they are incompatible quantities and the general indeterminacy principle (5.29) applies so that, in any quantum state represented by $|\psi\rangle$

$$\Delta_\psi x \Delta_\psi p_x \geqslant \frac{\hbar}{2}$$

since $\langle\psi|\widehat{\mathbb{1}}|\psi\rangle = 1$. This shows that the indeterminacy principle, for which a heuristic justification was given in the previous chapter, is here more rigorously demonstrated in the light of its connections to the structure of the operators. It goes without saying that such a relation holds in every direction of space but no incompatibility is found between the momentum in one direction and the position in another because, for example $[\widehat{x}, \widehat{p}_y] = 0$.

Finally, consider

$$\begin{aligned}[\widehat{p}_x, \widehat{x}^2] &= \widehat{p}_x\ \widehat{x}^2 - \widehat{x}^2\widehat{p}_x = \widehat{p}_x\ \widehat{x}^2 - \widehat{x}\ \widehat{p}_x\ \widehat{x} + \widehat{x}\ \widehat{p}_x\ \widehat{x} - \widehat{x}^2\widehat{p}_x \\ &= [\widehat{p}_x, \widehat{x}]\ \widehat{x} + \widehat{x}\ [\widehat{p}_x, \widehat{x}] = -2i\hbar\widehat{x}\end{aligned}$$

It is then straightforward to show by recurrence that $[\widehat{p}_x, \widehat{x}^n] = -ni\hbar\widehat{x}^{n-1}$. As a consequence, if $f(\widehat{x})$ is an operator constructed from the sole position operator, it is written (as seen in (5.12)): $f(\widehat{x}) = \sum_n \frac{1}{n!}\frac{\partial^n f}{\partial x^n}\ \widehat{x}^n$, which implies

$$[\widehat{p}_x, f(\widehat{x})] = -i\hbar\sum_n \frac{n}{n!}\frac{\partial^n f}{\partial x^n}\ \widehat{x}^{n-1} = \frac{\hbar}{i}f'(\widehat{x}) \tag{5.32}$$

Question 5.17: **Commutators and the stability of the hydrogen atom.**

1. Let $\widehat{A}$ and $\widehat{B}$ be two Hermitian operators. Show that for any quantum state represented by $|\psi\rangle$

$$\langle\psi|i[\widehat{A}, \widehat{B}]|\psi\rangle = -2\,\mathrm{Im}\langle\widehat{A}\psi|\widehat{B}\psi\rangle$$

   where we made use of the notation according to which $|\widehat{A}\psi\rangle$ is the ket resulting from the action of $\widehat{A}$ on $|\psi\rangle$.

2. Show that

$$\sum_{j=1}^{3} i\left[\frac{1}{\widehat{r}}\widehat{p}_j\frac{1}{\widehat{r}}, \widehat{x}_j\right] = 3\hbar\frac{1}{\widehat{r}^2}$$

   where $\widehat{r} = \sqrt{\widehat{x}_1^2 + \widehat{x}_2^2 + \widehat{x}_3^2}$.

3. Show that

$$\left[\widehat{p}_j, \frac{1}{\widehat{r}}\right] = i\hbar\frac{\widehat{x}_j}{\widehat{r}^3}$$

4. From the preceding three relations, deduce

$$\hbar\langle\psi|\frac{1}{\widehat{r}^2}|\psi\rangle = -2\,\mathrm{Im}\sum_{j=1}^{3}\langle\psi|p_j\frac{x_j}{r^2}|\psi\rangle$$

Question 5.17: (continued)

5. Using the Cauchy–Schwarz inequality,

$$\left|\sum_{j=1}^{3}\langle\psi|p_j\frac{x_j}{r^2}|\psi\rangle\right|^2 \leqslant \langle\psi|p^2|\psi\rangle\langle\psi|\frac{1}{r^2}|\psi\rangle$$

show that for any quantum state representing an electron in a Coulomb potential, the mean energy is necessarily negative and finite. Deduce the stability of the nucleus–electron system.

Answer:

1. We first write the mean value for the commutator:

$$\langle\psi|i[\widehat{A},\widehat{B}]|\psi\rangle = i(\langle\psi|\widehat{A}\widehat{B}|\psi\rangle - \langle\psi|\widehat{B}\widehat{A}|\psi\rangle)$$

Then, both operators being Hermitian, they can act on the left (on the bra) or on the right (on the ket):

$$\langle\psi|i[\widehat{A},\widehat{B}]|\psi\rangle = i(\langle\widehat{A}\psi|\widehat{B}\psi\rangle - \langle\widehat{B}\psi|\widehat{A}\psi\rangle)$$

We note that $\langle\widehat{B}\psi|\widehat{A}\psi\rangle = (\langle\widehat{A}\psi|\widehat{B}\psi\rangle)^*$, which yields

$$\langle\psi|i[\widehat{A},\widehat{B}]|\psi\rangle = -2\,\mathrm{Im}\langle\widehat{A}\psi|\widehat{B}\psi\rangle$$

2. We start by noting that $[\widehat{x}_j, 1/\widehat{r}] = 0$. Hence, for each component j,

$$i\left[\frac{1}{\widehat{r}}\widehat{p}_j\frac{1}{\widehat{r}}, \widehat{x}_j\right] = i\left(\frac{1}{\widehat{r}}\widehat{p}_j\widehat{x}_j\frac{1}{\widehat{r}} - \frac{1}{\widehat{r}}\widehat{x}_j\widehat{p}_j\frac{1}{\widehat{r}}\right) = \hbar\frac{1}{\widehat{r}^2}$$

Then, gathering the three components

$$\sum_{j=1}^{3} i\left[\frac{1}{\widehat{r}}\widehat{p}_j\frac{1}{\widehat{r}}, \widehat{x}_j\right] = 3\hbar\frac{1}{\widehat{r}^2}$$

3. This demonstration is similar to the one relative to the canonical commutator. A test function $f(\boldsymbol{r})$ is first introduced and the action of the commutator on it gives

$$\left[\widehat{p}_j, \frac{1}{\widehat{r}}\right] f(\boldsymbol{r}) = \frac{\hbar}{i}\frac{\partial}{\partial x_j}\frac{1}{r}f(\boldsymbol{r}) - \frac{\hbar}{i}\frac{1}{r}\frac{\partial}{\partial x_j}f(\boldsymbol{r}) = i\hbar\frac{x_j}{r^3}f(\boldsymbol{r})$$

which brings the result.

4. Taking the mean value for state $|\psi\rangle$ of the relation found in question 2, we obtain

$$3\hbar\langle\psi|\frac{1}{\widehat{r}^2}|\psi\rangle = \sum_{j=1}^{3}\langle\psi|i\left[\frac{1}{\widehat{r}}\widehat{p}_j\frac{1}{\widehat{r}}, \widehat{x}_j\right]|\psi\rangle$$

We then use the relation of question 1

$$3\hbar\langle\psi|\frac{1}{\widehat{r}^2}|\psi\rangle = -2\,\mathrm{Im}\sum_{j=1}^{3}\langle\frac{1}{\widehat{r}}\widehat{p}_j\frac{1}{\widehat{r}}\psi|\widehat{x}_j\psi\rangle$$

Answer: (continued)
Finally, the commutator of question 3 allows writing

$$3\hbar\langle\psi|\frac{1}{\widehat{r}^2}|\psi\rangle = -2\,\mathrm{Im}\sum_{j=1}^{3}\langle\frac{1}{\widehat{r}^2}\widehat{p}_j\psi|\widehat{x}_j\psi\rangle - 2\,\mathrm{Im}\sum_{j=1}^{3}\langle\psi|i\hbar\frac{x_j}{r^4}\widehat{x}_j\psi\rangle$$

It is then possible to use $(\frac{1}{\widehat{r}^2}\widehat{p}_j)^\dagger = \widehat{p}_j\frac{1}{\widehat{r}^2}$

$$3\hbar\langle\psi|\frac{1}{\widehat{r}^2}|\psi\rangle = -2\,\mathrm{Im}\sum_{j=1}^{3}\langle\psi|\widehat{p}_j\frac{1}{\widehat{r}^2}\widehat{x}_j|\psi\rangle + 2\,\mathrm{Im}\langle\psi|i\hbar\frac{1}{r^2}|\psi\rangle$$

and finally

$$\hbar\langle\psi|\frac{1}{\widehat{r}^2}|\psi\rangle = -2\,\mathrm{Im}\sum_{j=1}^{3}\langle\psi|\widehat{p}_j\frac{1}{\widehat{r}^2}\widehat{x}_j|\psi\rangle$$

5. The Cauchy–Schwarz inequality imposes

$$\left|\hbar\langle\psi|\frac{1}{\widehat{r}^2}|\psi\rangle\right|^2 \leqslant 4\langle\psi|\widehat{p}^2|\psi\rangle\langle\psi|\frac{1}{\widehat{r}^2}|\psi\rangle$$

therefore,

$$\hbar^2\langle\psi|\frac{1}{\widehat{r}^2}|\psi\rangle \leqslant 4\langle\psi|\widehat{p}^2|\psi\rangle$$

Consequently, the electron mean energy in a coulombic potential created by a nucleus is bounded according to

$$\langle\psi|\widehat{H}|\psi\rangle = \langle\psi|\frac{\widehat{p}^2}{2m} - \frac{Ze^2}{4\pi\epsilon_0\widehat{r}}|\psi\rangle \geqslant \langle\psi|\hbar^2\frac{1}{8m\widehat{r}^2} - \frac{Ze^2}{4\pi\epsilon_0\widehat{r}}|\psi\rangle$$

The right-hand side term has an extremum for $r = \pi\epsilon_0\hbar^2/(mZe^2)$ which yields the existence of a lower bound to the total energy:

$$\langle\psi|\widehat{H}|\psi\rangle \geqslant -2m\left(\frac{Ze^2}{4\pi\epsilon_0\hbar}\right)^2$$

This result indicates that the electron cannot collapse on the nucleus because it would yield an infinite negative energy. Obviously, this is a consequence of the Heisenberg indetermination according to which, if the electron were to get infinitely close to the nucleus, its position would be so well defined that the momentum, hence the kinetic energy, standard deviation would become infinitely large.

5.3. An Application of Measurement Postulates to Quantum Cryptography

The postulates of quantum physics have been introduced to provide a set of consistent rules in order to explain physical phenomena resisting a classical description. Once the new rules were established, we realized that new properties were emerging and, consequently, a wide field of applications. At this stage, we already have the necessary tools to understand some basic aspects of what is known as *quantum cryptography*.

5.3.1. *The secret correspondence between Alice and Bob*

How to send a trunk containing secrets to a recipient while ensuring that (s)he will be the only one to have access to its content? The method is well known: the sender sends the trunk closed with a padlock. Upon receipt, the recipient adds his own padlock and returns everything to the sender. The original sender removes his initial padlock and returns the trunk a last time to the recipient, who obviously possesses the key of the remaining padlock. This method is effective on one condition: it is the intended recipient who placed his lock and returned the trunk. Clearly, it must be ensured that the trunk was not diverted.

The purpose of cryptography is to protect messages so that only the recipient can read them. All the finesse of a cryptography method lies in the way of transmitting the reading code to the recipient. This code is the *encryption key*, the key to open the padlock in our trunk analogy. We are not interested in the content of the message or even in cryptographic techniques but only in how to communicate an encryption key to a correspondent, ensuring that no spy could obtain enough knowledge to read the message.

Traditionally, the sender is named Alice (A). She will encrypt the code as a set of bits (i.e. a sequence of 0 and 1). Yet, she will ensure that an eavesdropper, named Eve (E), does not discretely intercept the code, then allowing her to read any further correspondence with an intended recipient named Bob (B). The correspondents will then use a method in which the quantum states of a system are interpreted in terms of 0 and 1. The method that we present here is an idealization of what currently exists and we adapt the principle as a direct illustration of the measurement postulates.

5.3.2. *A measurement that leaves its mark*

Let a measurable physical property of a quantum particle be represented by an observable, i.e. an Hermitian operator, $\widehat{P}_\phi$, the (orthogonal) eigenkets of which are $|\phi_0\rangle$ and $|\phi_1\rangle$ with the respective eigenvalues a_0 and a_1. According to the third postulate of quantum physics, any measurement of the observable represented by $\widehat{P}_\phi$ can only yield either a_0 or a_1. Immediately, after measuring the property, the particle is in the eigenstate associated with the eigenvalue that was obtained as result. Therefore after a measurement of $\widehat{P}_\phi$, the particle can only be in one of the two states $|\phi_0\rangle$ and $|\phi_1\rangle$. If we later happen to find that the particle is in a state other than one of these two, it would automatically indicate that a disturbance took place on the way.

Consider another property represented by operator $\widehat{P}_\chi$ with (orthogonal) eigenkets $|\chi_0\rangle$ and $|\chi_1\rangle$. If $[\widehat{P}_\phi, \widehat{P}_\chi] \neq 0$, then it is clear that the order in which a double measurement is carried out is essential. In particular, the system always ends up being in a specific state of the *last* measured property operator.

Take a simple example. The system is prepared in an eigenstate of $\widehat{P}_\phi$, say $|\phi_1\rangle$. If the property $\widehat{P}_\phi$ is measured, the outcome will necessarily be a_1. Now, let us suppose that the particle was claimed to be in a state $|\phi_1\rangle$. If a measurement of $\widehat{P}_\phi$ gives a_0 as a result, we would naturally conclude that an interim event modified the system. Since $\langle\phi_0|\phi_1\rangle = 0$, the probability of such an outcome was *a priori* zero. This disturbance that brought the system to a state that is no longer orthogonal to $|\phi_0\rangle$ can be created by an intermediate measurement that is not compatible with $\widehat{P}_\phi$, i.e. a property that does not have $|\phi_1\rangle$ among its eigenstates. This property thus corresponds to an operator which does not commute with $\widehat{P}_\phi$. It can be represented by $\widehat{P}_\chi$. The disturbance of a system initially in a quantum state, generated by the measurement of a property not accepting this state as an eigenstate, is the physical principle of quantum cryptography.

Question 5.18: **Successive measurements.** Consider two operators respectively represented by

$$\widehat{\sigma}_z = \begin{pmatrix} 1 & 0 \\ 0 & -1 \end{pmatrix} \quad \text{and} \quad \widehat{\sigma}_x = \begin{pmatrix} 0 & i \\ -i & 0 \end{pmatrix}$$

1. Can the measurement of the property represented by $\widehat{\sigma}_x$ take place indifferently before or after $\widehat{\sigma}_z$?
2. What are the possible outcomes of a measurement of $\widehat{\sigma}_x$?
3. If the outcome of $\widehat{\sigma}_x$ is 1, what is the probability that a measurement of $\widehat{\sigma}_z$ also gives 1? What is then the probability that a measurement of $\widehat{\sigma}_x$ will again give 1?

Answer:

1. Since these operators are Hermitian, they could legitimately represent some physical quantity (and they do, as explained in the next volume). However, their commutator

$$[\widehat{\sigma}_z, \widehat{\sigma}_x] = 2i \begin{pmatrix} 0 & 1 \\ 1 & 0 \end{pmatrix}$$

is not zero. The two properties are incompatible and the order in which the measurement is carried out does indeed matter.
2. The possible outcomes of a measurement of the property represented by $\widehat{\sigma}_x$ are its eigenvalues: $\{-1, 1\}$.
3. To compute probabilities, it is necessary to know what the normalized eigenvectors are. For $\widehat{\sigma}_x$, the components of the eigenvectors in $\widehat{\sigma}_z$ eigen-representation are $\frac{1}{\sqrt{2}}\begin{pmatrix} 1 \\ i \end{pmatrix}$ and $\frac{1}{\sqrt{2}}\begin{pmatrix} 1 \\ -i \end{pmatrix}$ for the eigenvalues -1 and 1, respectively. If the outcome of a measurement of $\widehat{\sigma}_x$ gives 1, the system's state ket becomes $\frac{1}{\sqrt{2}}\begin{pmatrix} 1 \\ -i \end{pmatrix}$. The eigenket corresponding to the $\widehat{\sigma}_z$ eigenvalue 1 is $\begin{pmatrix} 1 \\ 0 \end{pmatrix}$ and the probability is thus given by the modulus square of the scalar product:

$$\mathcal{P}_{(\sigma_z=1)} = \left| \frac{1}{\sqrt{2}}(1, 0) \begin{pmatrix} 1 \\ -i \end{pmatrix} \right|^2 = \frac{1}{2}$$

Once the measurement has given 1 for $\widehat{\sigma}_z$, the quantum state is projected to $\begin{pmatrix} 1 \\ 0 \end{pmatrix}$ and a new measurement of $\widehat{\sigma}_x$ could give 1 with probability:

$$\mathcal{P}_{(\sigma_x=1)} = \left| \frac{1}{\sqrt{2}}(1, i) \begin{pmatrix} 1 \\ 0 \end{pmatrix} \right|^2 = \frac{1}{2}$$

5.3.3. *Sharing a quantum key*

The goal is that Alice and Bob agree, from afar, on a secret key. Before Alice can send any message encrypted with this key, it is imperative to ensure that no spy could have intercepted it. The approach is detailed in a simple example in Table 5.1 and is schematically as follows:

1. $4N$ particles are presented successively. Each one is prepared at random in any state which does not match any of the eigenstates $\{|\phi_0\rangle, |\phi_1\rangle, |\chi_0\rangle, |\chi_1\rangle\}$.
2. On each particle, Alice makes a measurement of a property choosing at random $\widehat{P}_\phi$ or $\widehat{P}_\chi$. The $4N$ successive measurements give row "Prop. A" and their outcomes are represented by M_A in Table 5.1.
3. Following the measurement, each particle is in a state corresponding to an eigenvalue of the randomly chosen property. The row S_A in Table 5.1 provides an example.

Table 5.1. Chronology of detections and emissions for transmissions without spying. To preserve space, a sequence containing only 16 quantum states was used. Bit value C is the index of the eigenvalue that was measured. Initial n.r. means that the corresponding bits are not kept.

bits #	1	2	3	4	5	6	7	8
Prop. A	$\widehat{P}_\phi$	$\widehat{P}_\phi$	$\widehat{P}_\chi$	$\widehat{P}_\phi$	$\widehat{P}_\phi$	$\widehat{P}_\phi$	$\widehat{P}_\phi$	$\widehat{P}_\chi$
M_A	a_1	a_0	b_1	a_1	a_0	a_1	a_1	b_1
S_A	$\|\phi_1\rangle$	$\|\phi_0\rangle$	$\|\chi_1\rangle$	$\|\phi_1\rangle$	$\|\phi_0\rangle$	$\|\phi_1\rangle$	$\|\psi_1\rangle$	$\|\chi_1\rangle$
C	1	0	1	1	0	1	1	1
Prop. B	$\widehat{P}_\phi$	$\widehat{P}_\chi$	$\widehat{P}_\phi$	$\widehat{P}_\chi$	$\widehat{P}_\phi$	$\widehat{P}_\phi$	$\widehat{P}_\chi$	$\widehat{P}_\chi$
M_B	a_1	b_0	a_0	b_1	a_0	a_1	b_0	b_1
C_{A+B}	1	n.r.	n.r.	n.r.	0	1	n.r.	1

bits #	9	10	11	12	13	14	15	16
Prop. A	$\widehat{P}_\phi$	$\widehat{P}_\phi$	$\widehat{P}_\chi$	$\widehat{P}_\phi$	$\widehat{P}_\chi$	$\widehat{P}_\phi$	$\widehat{P}_\phi$	$\widehat{P}_\phi$
M_A	a_0	a_1	b_1	a_0	b_0	a_1	a_0	a_0
S_A	$\|\phi_0\rangle$	$\|\phi_1\rangle$	$\|\chi_1\rangle$	$\|\phi_0\rangle$	$\|\chi_0\rangle$	$\|\phi_1\rangle$	$\|\phi_0\rangle$	$\|\phi_0\rangle$
C	0	1	1	0	0	1	0	0
Prop. B	$\widehat{P}_\chi$	$\widehat{P}_\phi$	$\widehat{P}_\chi$	$\widehat{P}_\chi$	$\widehat{P}_\phi$	$\widehat{P}_\phi$	$\widehat{P}_\chi$	$\widehat{P}_\phi$
M_B	b_0	a_1	b_1	b_0	a_0	a_1	b_0	a_0
C_{A+B}	n.r.	1	1	n.r.	n.r.	1	n.r.	0

4. From the values derived from the measurements, it is possible to perform coding in a series of bits (row C). An example of correspondence is as follows: a_1 or b_1 gives a value of 1 and a_0 or b_0 gives a value of 0.
5. The particles pass into the same order on Bob's device. For each particle, Bob has no information as to what type of property (P_ϕ or P_χ) was measured by Alice. Therefore, he has no better choice than again performing randomly a series of measurements on one or the other of the properties (Prop. B).
6. Bob publicly communicates[11] to Alice the series of properties that he measured (Prop. B) without giving the outcomes (M_B).
7. In exchange, still publicly, Alice tells him which particles have undergone two consecutive measurements of the same property (comparison between Prop. B and Prop. A).
8. Since only Bob and Alice know the results of the measurements of these properties, then they are the only ones who know the series of bits which

[11] "Publicly" here means that piece of information has no reason to be secret. It is not necessary to give it extensive publicity.

are common. It will be called the "random set". This code is reported on row C_{A+B}.

5.3.4. *Spy, are you there?*

We see that the price of the draw of the properties measured by Bob is losing on average half of the information sent that will not be retained in the final key. Of the $4N$ particles used by Alice, only $2N$ (on average) will result in a relevant bit to constitute the common set.

The loss will actually be larger than this estimate. At this point we still do not know if this key is kept secret. For it to be so, we need to ensure that a spy could not make measurements on the particles before Bob would receive them. To conduct the check, he will once again have to sacrifice a part of the useful common set. This is where the quantum nature of particles comes in.

As Eve, the spy, is clever, she is aware of what properties can be measured. Like Alice and Bob, she knows that, for each particle, there is a choice between measuring $\widehat{P}_\phi$ and $\widehat{P}_\chi$. Just like Bob, she has one chance out of two of measuring the same property as Alice. However, she has no other option than to pick at random the properties to be measured since this has to be carried out before Bob receives the particles. It has to take place before Alice publicly communicates the list of the properties she has measured. Unfortunately, for the spy, it would be an unlikely chance that it coincides exactly with the sequence used by Bob.[12] It is even inevitable (when N is large) that some properties measured by the spy do not match. In this case, the measurement induces a change in the quantum state of the particle and the probability that Bob's and Alice's measurements of the common set coincide is no longer equal to 1. In order to detect an intermediate eavesdropper, it is therefore sufficient for Bob to take a subset of measurement results which should be identical to those of Alice. Suppose he chooses to sacrifice half (N) of the common set for a checking purpose.

[12]In fact, it would be enough to coincide with those Bob and Alice have in common. The probability is still low: $(1/2)^{(2N)}$.

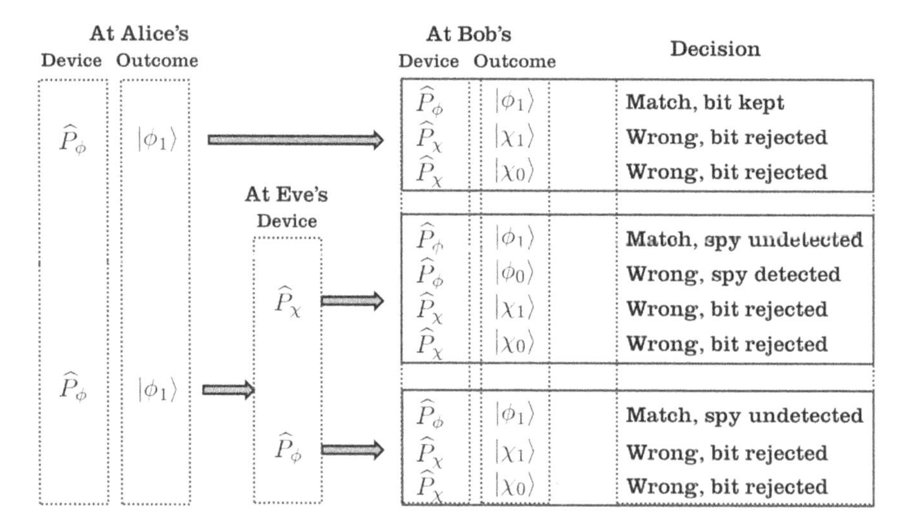

Fig. 5.6. The different scenarios paths of the same quantum state leaving Alice are listed. The gray arrows indicate the transmission from one protagonist to another. The cases with and without spying are envisaged. The decision to keep the bit depends on the adequacy of the measured properties. In all cases, the bit will not be retained if it belongs to the subset dedicated to the detection of a spy.

Bob publicly communicates his results for the subset to Alice. If no difference is observed, it is legitimate for the two correspondents to assume that the other bits that are retained, but not disclosed publicly, remained secret and may constitute a safe coding key. Otherwise, they would obviously have to start the process again, possibly with a larger number of particles to decrease the chance of the spy going unnoticed. This procedure is summarized in Fig. 5.6.

Remarks:

- The best way to ensure that no key bit was read is obviously to compare all of Alice's and Bob's results. But as the comparison can only be public, nothing would be secret.
- It is difficult to work on a sequence of single particles and no measurement is free of defects. The reality is therefore expected to differ somewhat from the ideal scheme that has just been described. In particular, experimental errors can lead to a difference between two measurement

results without an eavesdropper being involved. The limitation of such a "false positive" of eavesdropper detection is a major challenge of quantum cryptography.

- Both properties should have a selected spectrum in the same 2D space, so that the transcription in terms of bits is straightforward. Among the properties usually mentioned, photon polarization states are often quoted. In this case, for the measurements of the two incompatible properties, two analyzers whose axes' orientations are offset by an angle of $\pi/4$ can be chosen. We will see in the next volume that another property, which can be described in a 2D space, is the magnetic moment of the electron, proton, neutron, or some carbon isotopes.
- We have deliberately ignored a key fact. If the spy was able to copy each incident particle, then she could make as many measurements she wished. Such a procedure would allow her to wait for Bob to tell Alice what measurements were compatible, without changing the sequence of the transmitted quantum states. Her intervention would thus go unnoticed.

  Fortunately, a "no quantum cloning" theorem forbids quantum system states to be perfectly copied (see Question *5.19*). There is therefore an additional barrier to this highly reprehensible conduct which is that interfering in someone else's correspondence.
- This construction method of quantum encryption keys was proposed by Bennett and Brassard in 1984 hence the name "BB84" [7]. Many other schemes have been devised since then but the basic physical principles remain essentially the same. Tests of transmissions have been performed by means of optical fibres for distances over 300 km. In 2016, China launched a satellite with the aim of testing quantum cryptography (using the more sophisticated technology of entangled photons) over distances superior to a thousand kilometres.

Question 5.19: **Quantum no-cloning theorem.** In the same way that a photocopy of a document should be on a separate sheet of paper, reproducing the quantum state of a system S_1 requires a secondary system S_2 to allow for a support of the copied information. For the copy to be accurate and useful,

Question 5.19: (continued)
the procedure must be able to adapt to whatever the initial quantum state may be.

Consider a particle described in a Hilbert space with a basis set $\{|a_1\rangle, |a_2\rangle, \ldots, |a_N\rangle\}$. It is assumed that it is possible to build a machine to exactly copy this particle in any given state. So, if system S_1 is originally described by $|\psi\rangle$, after copying, there is a pair of systems $S_1 \cup S_2$ described by the state: $|\psi\rangle \otimes |\psi\rangle$.

1. Operator $\widehat{U}$ performs the quantum cloning and represents the copy machine, so that $\widehat{U}|\psi\rangle = |\psi\rangle \otimes |\psi\rangle$. Show that $\widehat{U}$ is a unitary operator.
2. Show that it is not possible to satisfy both a cloning of the states belonging to the state basis: $\widehat{U}|a_i\rangle = |a_i\rangle \otimes |a_i\rangle$, for any $i \leqslant N$ and the cloning of *any state* $|\psi\rangle = \sum_{i=1}^{N} \alpha_i |a_i\rangle$.

   Thus, this shows that such a machine able to clone systems in any quantum state cannot exist. Such a machine would only be capable of duplicating systems in specific states of the basis therefore quite useless for spying purposes.

Answer:

1. If, whatever the initial state, $\widehat{U}|\psi\rangle = |\psi\rangle \otimes |\psi\rangle$, the ket representing the final systems pair is required to be normalized:

$$(\langle\psi| \otimes \langle\psi|)\,(|\psi\rangle \otimes |\psi\rangle) = 1$$

   Therefore,

$$\langle\psi|\widehat{U}^\dagger \widehat{U}|\psi\rangle = 1$$

   for any $|\psi\rangle$. Hence, $\widehat{U}^\dagger \widehat{U} = \mathbb{1}$.
2. The operator that represents the hypothetical quantum cloning machine is unitary. It is thus linear (see Question *5.10*) and

$$\widehat{U}|\psi\rangle = \widehat{U} \sum_{i=1}^{N} \alpha_i |a_i\rangle = \sum_{i=1}^{N} \alpha_i \widehat{U}|a_i\rangle = \sum_{i=1}^{N} \alpha_i |a_i\rangle \otimes |a_i\rangle$$

   but also, as the tensor product is linear (see (5.3)),

$$\widehat{U}|\psi\rangle = |\psi\rangle \otimes |\psi\rangle = \sum_{i=1}^{N} \sum_{j=1}^{N} \alpha_i \alpha_j |a_i\rangle \otimes |a_j\rangle$$

   The only way to make these two expressions compatible is to consider quantum states such as $|\psi\rangle = |a_i\rangle$, i.e. which coincides with one of the states of the basis.

   A quantum cloning machine, if it existed, could only copy particles in a particular state.

5.4. Time Evolution of a State Ket

5.4.1. *General implications of the evolution postulate*

For a conservative system, the Hamiltonian is obtained from the total energy expression:

$$\widehat{H}(t) = \widehat{T} + \widehat{V}(\{\boldsymbol{r}\}, t)$$

where $\widehat{T}$ is the kinetic energy operator $\widehat{T} = \sum_j \frac{\widehat{p}_j^2}{2m_j}$ where $\widehat{\boldsymbol{p}}_j$ is the momentum operator associated with particle j with mass m_j. Operator $\widehat{V}(\{\boldsymbol{r}\}, t)$ represents the potential energy and, at every moment t, it only acts on the position variables $\boldsymbol{r}_j$ of the system.

If a particle is in a quantum state represented by wavefunction $\psi(\boldsymbol{r}, t)$, we need to solve

$$i\hbar\frac{d\psi(\boldsymbol{r}, t)}{dt} = -\frac{\hbar^2}{2m}\nabla_r^2\psi(\boldsymbol{r}, t) + \widehat{V}(\boldsymbol{r}, t)\psi(\boldsymbol{r}, t)$$

It thus generalizes what was found in the preceding chapter using plane wave properties.

As an example, if a state ket is expressed in the Hamiltonian eigenbasis $\{|\phi_n\rangle\}$, it is necessary to solve:

$$\widehat{H}|\phi_n\rangle = \varepsilon_n|\phi_n\rangle$$

The ε_n are the Hamiltonian eigenvalues (also called *eigenenergies*) and, consequently, are the possible outcomes of an energy measurement of the system. The temporal evolution of any quantum state will thus be given by

$$i\hbar\frac{d}{dt}|\psi(t)\rangle = i\hbar\frac{d}{dt}\sum_n c_n(t)|\phi_n\rangle = \widehat{H}\sum_n c_n(t)|\phi_n\rangle$$

The Hamiltonian and the time derivatives are both linear operators. The former only acts on kets and bras while the latter modifies time functions.

So we get

$$i\hbar \sum_n \frac{dc_n(t)}{dt} |\phi_n\rangle = \sum_n c_n(t)\ \widehat{H}|\phi_n\rangle$$
$$= \sum_n c_n(t)\ \varepsilon_n\ |\phi_n\rangle$$

Projecting onto an eigenstate of $\hat{H}$ and using orthogonality properties of the Hermitian operator eigenkets (as in Section 5.1.4), it is found that

$$\langle\phi_n| i\hbar \sum_m \frac{dc_m(t)}{dt} |\phi_m\rangle = \sum_m c_m(t)\ \varepsilon_m\ \langle\phi_n|\phi_m\rangle$$
$$i\hbar \sum_m \frac{dc_m(t)}{dt}\ \delta_{n,m} = \sum_m c_m(t)\ \varepsilon_m\ \delta_{n,m}$$
$$i\hbar \frac{dc_n(t)}{dt} = c_n(t)\ \varepsilon_n$$

Therefore $c_n(t) = c_n(0)e^{-i\varepsilon_n t/\hbar}$. Hence

$$\psi(t)\rangle = \sum_n c_n(0)\ e^{-i\varepsilon_n t/\hbar}|\phi_n\rangle \tag{5.33}$$

where $c_n(0)$ are determined from the initial conditions.

A key implication of this assumption is that, while quantum physics is probabilistic in its description of the system (and possible outcomes of a measurement), the time evolution of the wavefunction is fully determined (as far as the Schrödinger equation can be solved).

Question 5.20: **The spreading of a wave packet.** Let $\psi(x, t = 0) = \int C_0 e^{-\alpha(p-p_0)^2} e^{ipx/\hbar} dp$ be a Gaussian wave packet at $t = 0$ describing a particle of mass m.

1. What is the indetermination on the position of such a particle at $t = 0$?
2. Providing that the energy of the system can be expanded as

$$\varepsilon(p) \approx \varepsilon(p_0) + v_g(p - p_0) + g(p - p_0)^2$$

   where v_g and g are two real numbers, what is the wavefunction at $t > 0$?
3. Comment on the indetermination on the position at $t > 0$. What is the physical origin of such a behaviour?

Answer:

1. In order to compute any property using the wavefunction, the first thing is to have it normalized. Since $-\alpha(p^2 - ipx/(\alpha\hbar)) = -\alpha(p - ix/(2\alpha\hbar))^2 - x^2/(4\hbar^2\alpha)$, it can be applied to the Fourier transform of a Gaussian:

$$\psi(x, t=0) = \int_{-\infty}^{+\infty} C_0 e^{-\alpha(p-p_0)^2} e^{ipx/\hbar} dp = C_0 e^{ip_0 x/\hbar} \sqrt{\frac{\pi}{\alpha}} e^{-x^2/(4\alpha\hbar^2)}$$

and $\int_{-\infty}^{+\infty} |\psi(x, t=0)|^2 dx = 1$ gives $C_0 = \left(\frac{\alpha}{2\pi^3\hbar^2}\right)^{1/4}$. The mean position is 0, and the initial indetermination is

$$\Delta(x) = \left(\int_{-\infty}^{+\infty} x^2 |\psi(x, t=0)|^2 dx\right)^{1/2} = \hbar\sqrt{\alpha}$$

2. Using the time evolution, we get

$$\psi(x, t) = \int_{-\infty}^{+\infty} \left(\frac{\alpha}{2\pi^3\hbar^2}\right)^{1/4} e^{-\alpha(p-p_0)^2} e^{ipx/\hbar} e^{-i\varepsilon(p)t/\hbar} dp$$

With the proposed expansion, setting $p - p_0 = \hbar q$, it can be written:

$$\psi(x, t) = \left(\frac{\alpha}{2\pi^3\hbar^2}\right)^{1/4} e^{i(p_0 x - \varepsilon(p_0)t)/\hbar} \int_{-\infty}^{+\infty} e^{-(\alpha + igt)(\hbar q)^2} e^{iq(x - v_g t/\hbar)} \hbar dq$$

The result for the integral is somewhat similar to what we previously obtained with the change $x \to x - v_g t$ and $\alpha \to \alpha + igt/\hbar$:

$$\psi(x, t) = \frac{1}{(2\pi\hbar^2)^{1/4}} \frac{1}{(\alpha + igt)^{1/4}} e^{i(p_0 x - \varepsilon(p_0)t)/\hbar} e^{-(x - v_g t)^2/(4\hbar^2(\alpha + igt/\hbar))}$$

3. The maximum of the probability density translates in time with a speed given by the group velocity v_g. But another interesting point is that the width of the density distribution is now given by $\hbar(\alpha^2 + g^2t^2/\hbar^2)^{1/4}$. Obviously, the localization of the particle will never be better than it was at $t = 0$. The spreading of the wave packet originates from the different time evolution for each wave component in the initial expansion. One could see the initial configuration as a construction to optimize the interference condition. As time changes, wave phases evolve differently according to their respective $\varepsilon(p)$. Eventually, the phases become so different that the wave packet has totally dissolved: the particle becomes fully delocalized.

✍ *Question 5.21:* **Translation and evolution operators.** A displacement operator acts on a continuous variable, such as position, momentum or time, and changes its value. When this operator acts on the position, it is called a *translation operator*. When it changes the time value, it is named an *evolution operator*.

1. Show that the operator defined as $\widehat{T}(dx) = \mathbb{1} + dx\frac{i}{\hbar}\widehat{P}_x$ is the operator that changes the wavefunction $\psi(x)$ taken at position x into $\psi(x + dx)$, the value of the same wavefunction at a position shifted to $x + dx$.
2. How should the action of $\mathbb{1} - \frac{i}{\hbar}\widehat{X}dp$ be interpreted?

Question 5.21: (continued)

3. During the thirties, Stone [85] demonstrated that the previous results could be generalized to any displacement value when an operator $\widehat{T}(\ell) = e^{i\widehat{P}_x\ell/\hbar}$ was introduced. By using the definition of the function of operators (Section 5.2.2) check that $\widehat{T}(\ell)$ is indeed the operator that transforms $\psi(x)$ into $\psi(x+\ell)$.
4. To be convinced that we are dealing with a displacement operator, we can also consider $\widehat{U}(\Delta t) = e^{-i\widehat{H}\Delta t/\hbar}$ where $\widehat{H}$ is the system's Hamiltonian. Show that $\widehat{U}(\Delta t)$ is the time evolution operator which gives the state of the system at time $t + \Delta t$ from the knowledge of the wavefunction at t.
5. Are the displacement operators introduced above Hermitian?

Answer:

1. If a wavefunction is differentiable in x, in the first order in translation dx, it is found

$$\psi(x+dx) = \psi(x) + dx\frac{\partial}{\partial x}\psi(x)$$

The definition of the momentum operator along the x-axis gives

$$\psi(x+dx) = \psi(x) + dx\frac{i}{\hbar}[\widehat{P}_x\psi(x)]$$

It can then be verified that operator $\widehat{T}(dx)$ transforms $\psi(x)$ into $\psi(x+dx)$. Thus, it is an infinitesimal translation operator acting on the position space.

2. The action of $\mathbb{1} - dp\frac{i}{\hbar}\widehat{X}$ on a wavefunction in momentum representation gives

$$\begin{aligned}\widetilde{\psi}(p) - \langle p|dp\frac{i}{\hbar}\widehat{X}|\psi\rangle &= \widetilde{\psi}(p) - dp\frac{i}{\hbar}\int\langle p|x\rangle\langle x|\widehat{X}|\psi\rangle dx \\ &= \widetilde{\psi}(p) - dp\frac{i}{\hbar}\int\frac{e^{-ipx/\hbar}}{\sqrt{2\pi\hbar}}\langle x|\widehat{X}|\psi\rangle dx \\ &= \widetilde{\psi}(p) - dp\frac{i}{\hbar}\int x\frac{e^{-ipx/\hbar}}{\sqrt{2\pi\hbar}}\langle x|\psi\rangle dx \\ &= \widetilde{\psi}(p) + dp\frac{\partial}{\partial p}\int\frac{e^{-ipx/\hbar}}{\sqrt{2\pi\hbar}}\langle x|\psi\rangle dx \\ &= \widetilde{\psi}(p) + dp\frac{\partial}{\partial p}\widetilde{\psi}(p) = \widetilde{\psi}(p+dp) \qquad (5.34)\end{aligned}$$

where use was made of $\langle p|x\rangle = \frac{e^{-ipx/\hbar}}{\sqrt{2\pi\hbar}}$ as found in (5.19).

The operator, which can be denoted $\widehat{T}(dp)$, creates a dp translation in momentum space.

3. If $\widehat{T}(\ell)$ is applied to a wavefunction $\psi(x)$, making use of Section 5.2.2, it is indeed found that

$$\widehat{T}(\ell)\psi(x) = e^{i\widehat{P}_x\ell/\hbar}\psi(x) = \sum_n\frac{1}{n!}\left(\frac{i}{\hbar}\right)^n\widehat{P}_x^n\psi(x)\ell^n = \sum_n\frac{1}{n!}\frac{\partial^n}{\partial x^n}\psi(x)\ell^n = \psi(x+\ell)$$

Answer: (continued)

4. Let $|\psi(t)\rangle$ describe a quantum state for the system at time t. Application of $\widehat{U}(\Delta t)$ yields

$$\widehat{U}(\Delta t)|\psi(t)\rangle = e^{-i\widehat{H}\Delta t/\hbar}|\psi(t)\rangle = \sum_n \frac{1}{n!}\left(-\frac{i}{\hbar}\right)^n \widehat{H}^n|\psi(t)\rangle(\Delta t)^n$$

If $|\psi(t)\rangle$ is expressed in the eigenbasis of $\widehat{H}$, making use of the spectral decomposition $|\psi(t)\rangle = \sum_m c_m e^{-i\varepsilon_m t/\hbar}|\phi_m\rangle$, we get

$$\begin{aligned}\widehat{U}(\Delta t)|\psi(t)\rangle &= \sum_{n,m}\frac{1}{n!}c_m\left(-i\varepsilon_m/\hbar\right)^n|\phi_m\rangle e^{-i\varepsilon_m t/\hbar}(\Delta t)^n \\ &= \sum_m c_m e^{-i\varepsilon_m t/\hbar}\sum_n \frac{1}{n!}\left(-i\varepsilon_m\Delta t/\hbar\right)^n|\phi_m\rangle \\ &= \sum_m c_m e^{-i\varepsilon_m(t+\Delta t)/\hbar}|\phi_m\rangle = |\psi(t+\Delta t)\rangle \qquad (5.35)\end{aligned}$$

$\widehat{U}(\Delta t)$ is indeed the evolution operator for a system with Hamiltonian $\widehat{H}$.

5. Operators $\widehat{H}$, $\widehat{X}$ and $\widehat{P}_x$ are Hermitian. Consider the case of $\widehat{P}$. The question is whether $e^{i\widehat{P}x/\hbar}$ is also Hermitian. Its Hermitian conjugate (i.e. adjoint) is $e^{-i\widehat{P}^\dagger x/\hbar} = e^{i\widehat{P}(-x)/\hbar}$. It corresponds to a translation in the opposite direction. This is apparent if we write that for a unitary operator $\widehat{U}^\dagger = \widehat{U}^{-1}$. So, if $\widehat{U}$ operates as a translator, $\widehat{U}^\dagger$ automatically makes a translation of the same quantity in the opposite direction. Obviously the operator cannot be Hermitian unless $x = 0$. But it would then correspond to the identity.

5.4.2. *Application of a tunnelling dynamics to the MASER*

A two-state model for the flipping of ammonia

At this point, it is particularly interesting to consider the case of NH_3, the ammonia molecule, the geometry of which is sketched in Figure 5.7. It can roughly be described as a triangular pyramid with the three hydrogen atoms on the vertices at the basis and a nitrogen atom at the top. Such a molecule possesses a large number of degrees of freedom: nine electrons are moving around the four nuclei.[13] However, at room temperature, it turns out that only translation or rotation motions of the pyramid should be considered. The molecule is constructed from hydrogen atoms, which are usually eager to get rid of their electrons, and nitrogen, known to be an electrophile. Upon the formation of chemical bonds in the molecule, a

[13]Not to mention nuclear internal degrees of freedom.

significant electron transfer occurs between the atomic sites resulting in the creation of a 4.74×10^{-30} C.m. (≈ 1.42 Debye[14]) dipole moment. It is the rotation of this dipole moment (so of the pyramid itself) that can be observed spectroscopically, in much the same way the dipole moment of a water molecule can be excited in a microwave oven.

While full rotation of the molecule is not as easy as it seems in liquid phase, inversion of the dipole moment is observed because the nitrogen can find a shortcut by tunnelling through the hydrogen triangle[15] and relative masses. Since both conformations correspond to the exact same geometry of the molecule seen from a different perspective, they must be associated with the same energy. It is obviously an example of degeneracy as a result of (all things being equal) an atom placed in a double-well potential (Fig. 5.7).

This behaviour can conveniently be modelled by isolating these two degenerate states. When the nitrogen sits on the left of the hydrogen plane (the dipole moment points to the right), the state will be represented by $|\ell\rangle$. When it is on the right of the plane (the dipole moment points to the left), the state ket will be denoted by $|r\rangle$. In the very simplistic representation $\{|\ell\rangle, |r\rangle\}$, the dipole moment operator along z could roughly be represented by a matrix:

$$\widehat{D}_z = \begin{pmatrix} D & 0 \\ 0 & -D \end{pmatrix} \tag{5.36}$$

with $D \approx 1.42$ Debye. If the potential barrier between the two locations were much higher (or thicker) the tunnelling would be a lot less probable. Kets $|r\rangle$ and $|\ell\rangle$ would then correspond to the ground state of a particle in an (almost) infinite well. As a first approximation, their position representations would then be given by wavefunctions bearing a close resemblance to the sine (Fig. 5.7(b)). But here, because a tunnelling is possible, the sine function on the barrier side does not exactly go to zero and bears a long

[14] 1 Debye $= 3.34 \times 10^{-30}$ C.m. is a current unit for dipole moments.

[15] Nitrogen is significantly heavier than the hydrogen triplet. It would thus make more sense to consider the hydrogen triangle's displacement relative to the nitrogen atom. It would be even more accurate to talk about the motion of the reduced mass with respect to the centre of mass of the pyramid. However, physics is not affected by our choice of description.

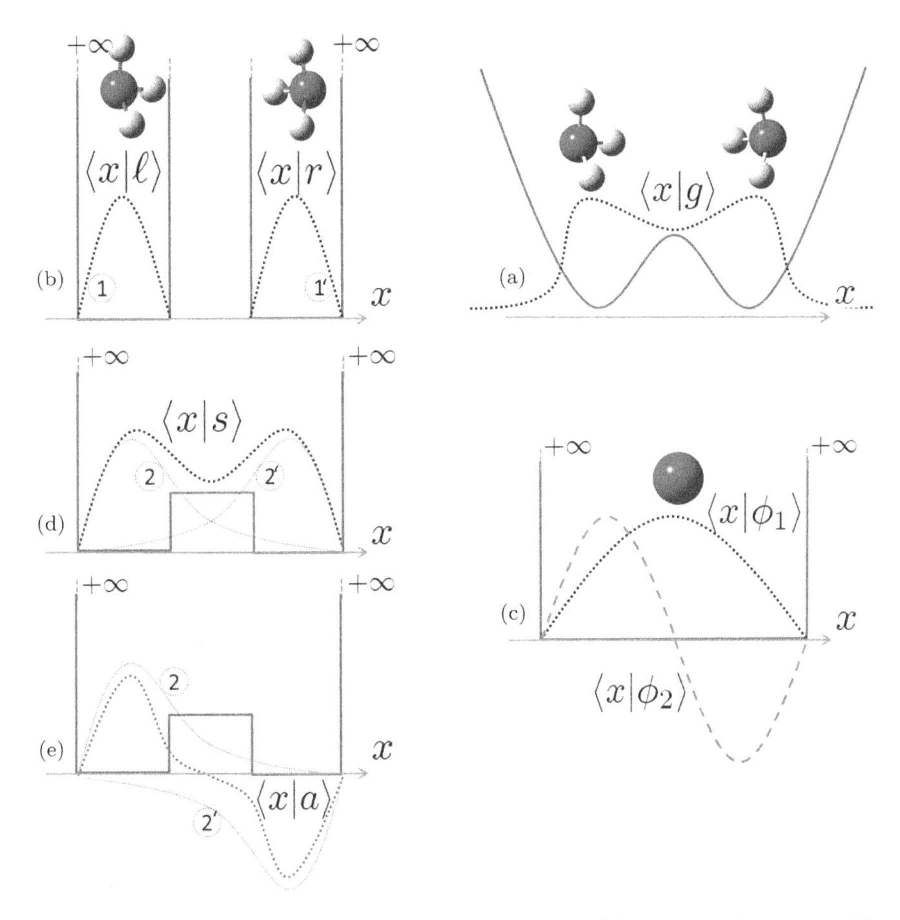

Fig. 5.7. Modelling the NH3 inversion in a 2-state space. (a) Inversion of nitrogen with respect to the hydrogen atoms triangle is performed by tunnelling through a potential barrier (solid line). Such a potential can be considered as the medium case between two extreme chemical configurations:

- On (b) both configurations could exist but the potential barrier would be too high to be tunnelled through (infinite stiffness of N–H bonds). The molecule would thus be stuck in either the $|\ell\rangle$ or $|r\rangle$ state.
- On (c) the nitrogen atom would be completely free to move with no repulsion from the hydrogen plane. The eigenstates would thus be those of the infinite quantum well, $|\phi_1\rangle$ and $|\phi_2\rangle$ (see Section 4.3).

The piecewise constant model can be used to infer the form of the ground state $|s\rangle$, symmetrical wavefunction (d) and the excited state $|a\rangle$, anti-symmetrical wavefunction (e) (see text).

tail that slips into the other well making a connection to the other sine-like function (Fig. 5.7(d) or (e)). The states represented by $|r\rangle$ and $|\ell\rangle$ are thus no longer eigenstates of the new Hamiltonian. As seen in Section 4.5, since the double well is symmetrical (even potential), the new wavefunction needs to be symmetrical or anti-symmetrical (4.11). This means that if we choose to limit our basis to the two eigenkets of the conformation operator, $|r\rangle$ and $|\ell\rangle$, the eigenstates of the symmetrical Hamiltonian would take the form

$$|s\rangle \propto |r\rangle + |\ell\rangle \quad \text{and} \quad |a\rangle \propto |r\rangle - |\ell\rangle$$

As usual, to be useful, these states need to be normalized. Denoting by $S = \langle \ell | r \rangle$ the *overlap integral*, we get

$$|s\rangle = \frac{1}{\sqrt{2(1+S)}}\left(|r\rangle + |\ell\rangle\right) \tag{5.37a}$$

$$|a\rangle = \frac{1}{\sqrt{2(1-S)}}\left(|r\rangle - |\ell\rangle\right) \tag{5.37b}$$

In the $\{|s\rangle, |a\rangle\}$ eigen-representation, the Hamiltonian is obviously diagonal and if we denote by ε_s and ε_a the respective eigenenergies, it takes the form

$$\widehat{H} = \begin{pmatrix} \varepsilon_s & 0 \\ 0 & \varepsilon_a \end{pmatrix} \tag{5.38}$$

As long as the potential remains symmetrical, as seen in Section 4.5, the eigen wavefunctions (which are the position space representations of $|s\rangle$ and $|a\rangle$) will still exist. We can track what happens to these functions as the barrier is progressively reduced in height. Both functions will modify to asymptotically coincide with the eigenfunctions of the flat infinite quantum well (see Section 4.3). Therefore $\langle x|s\rangle$ will eventually match the symmetrical ground state eigenfunction while $\langle x|a\rangle$ will correspond to the first anti-symmetrical excited state function. We can thereby infer, by continuity, that this order in energy should be preserved as long as the symmetry of the potential is preserved and the lowest eigenstate in our simple model is the symmetrical[16] combination $|s\rangle$: $\varepsilon_s < \varepsilon_a$.

[16]The proof requires the variational theorem and is given in Vol. 2.

The same observable can be written in the $\{|\ell\rangle, |r\rangle\}$ representation. The matrix elements are

$$\begin{aligned}
\langle\ell|\widehat{H}|\ell\rangle &= \left(\frac{1}{\sqrt{2}}(\sqrt{1+S}\langle s| - \sqrt{1-S}\langle a|)\right)\widehat{H} \\
&\quad\times\left(\frac{1}{\sqrt{2}}(\sqrt{1+S}|s\rangle - \sqrt{1-S}|a\rangle)\right) \\
&= \frac{\varepsilon_s+\varepsilon_a}{2} + S\frac{\varepsilon_s-\varepsilon_a}{2} \\
\langle r|\widehat{H}|r\rangle &= \frac{\varepsilon_s+\varepsilon_a}{2} + S\frac{\varepsilon_s-\varepsilon_a}{2} \\
\langle\ell|\widehat{H}|r\rangle &= \frac{\varepsilon_s-\varepsilon_a}{2} + S\frac{\varepsilon_s+\varepsilon_a}{2}
\end{aligned}$$

where space reversal symmetry was invoked to directly find $\langle r|\widehat{H}|r\rangle = \langle\ell|\widehat{H}|\ell\rangle$. Obviously, since the Hamiltonian is an observable, we have $\langle\ell|\widehat{H}|r\rangle = \langle r|\widehat{H}|\ell\rangle^*$. Denoting $\overline{\varepsilon} = \frac{\varepsilon_s+\varepsilon_a}{2}$ and $\Delta = \frac{\varepsilon_a-\varepsilon_s}{2} > 0$, we then obtain in the $\{|\ell\rangle, |r\rangle\}$ representation:

$$\widehat{H} = \begin{pmatrix} \overline{\varepsilon} - S\Delta & \overline{\varepsilon}S - \Delta \\ \overline{\varepsilon}S - \Delta & \overline{\varepsilon} - S\Delta \end{pmatrix} \tag{5.39}$$

This expression is better understood if we consider the case where the overlap S can be neglected. The Hamiltonian in the conformation representation is then

$$\widehat{H} = \begin{pmatrix} \overline{\varepsilon} & -\Delta \\ -\Delta & \overline{\varepsilon} \end{pmatrix} \tag{5.40}$$

The two conformations then have identical energies and would be eigenstates of the Hamiltonian if there was not this possibility for nitrogen to switch its location by tunnel effect. Such a jump between sites is then given by the off diagonal term, often quoted as *hopping integral* or *transfer integral*.

Since conformation states do not match the Hamiltonian eigenkets or, to put it another way, the dipole moment operator $\widehat{D}_z$ does not commute with $\widehat{H}$, neither $|\ell\rangle$ nor $|r\rangle$ are stationary states. Consequently, if the molecule is prepared at time $t = 0$ in state, say, $|\ell\rangle$, it will not remain in such a conformation and is likely to be detected in $|r\rangle$ later on. The initial state is chosen to be $|\psi(t=0)\rangle = |\ell\rangle$. Noticing that

$$|\ell\rangle = \frac{1}{\sqrt{2}}(\sqrt{1+S}|s\rangle - \sqrt{1-S}|a\rangle)$$

Schrödinger's evolution equation gives a state at time $t > 0$:

$$|\psi(t)\rangle = \frac{1}{\sqrt{2}}(\sqrt{1+S}e^{-i\varepsilon_s t/\hbar}|s\rangle - \sqrt{1-S}e^{-i\varepsilon_a t/\hbar}|a\rangle) \tag{5.41}$$

It is then possible to compute the probability of configuration switching as

$$\begin{aligned} P_{\ell\to r}(t) &= |\langle r|\psi(t)\rangle|^2 \\ &= \frac{1}{2}|(\sqrt{1+S}e^{-i\varepsilon_s t/\hbar}\langle r|s\rangle - \sqrt{1-S}e^{-i\varepsilon_a t/\hbar}\langle r|a\rangle)|^2 \\ &= \frac{1}{4}|(e^{-i\varepsilon_s t/\hbar} - e^{-i\varepsilon_a t/\hbar})|^2 = \frac{1}{4}|e^{-i\bar{\varepsilon}t/\hbar}(e^{+i\Delta t/\hbar} - e^{-i\Delta t/\hbar})|^2 \\ &= \sin^2(\Delta t/\hbar) \end{aligned}$$

where (5.37) was used. The switching between left and right conformations is thus found to occur every $\pi\hbar/(2\Delta)$. The hopping integral, which is characteristic of the nitrogen atom's ability to penetrate the potential barrier, now bears a particular significance: it directly yields the switching frequency $(\Delta/(\pi\hbar))$ and can easily be estimated from spectroscopy analysis.

In any conformation, whenever the molecule vibrates around its centre of mass, there is a change in its dipole moment. However, the probability densities associated with the symmetrical or anti-symmetrical wavefunction are symmetrical and therefore do not exhibit any net dipole moment. This is a consequence of the dipole moment operator not commuting with the Hamiltonian. Nevertheless, the molecule can couple with an electromagnetic field to absorb a photon corresponding to the splitting $\varepsilon_a - \varepsilon_s = 2\Delta$ and change its vibrational state. The splitting was first observed in 1932 in the 3 μm absorption line [26] from the NH_3 vibration spectrum then refined by observing a better adapted range in the far infrared spectrum [95]. Finally, the direct observation of $\varepsilon_a - \varepsilon_s = 2\Delta$ which falls in the microwave range [19] was published in 1934 (Fig. 5.8). The height of the potential barrier, using a more sophisticated model than our piece-wise constant picture, was hence estimated at 0.257 eV (2076 cm^{-1}) [71]. Recent value is now set to

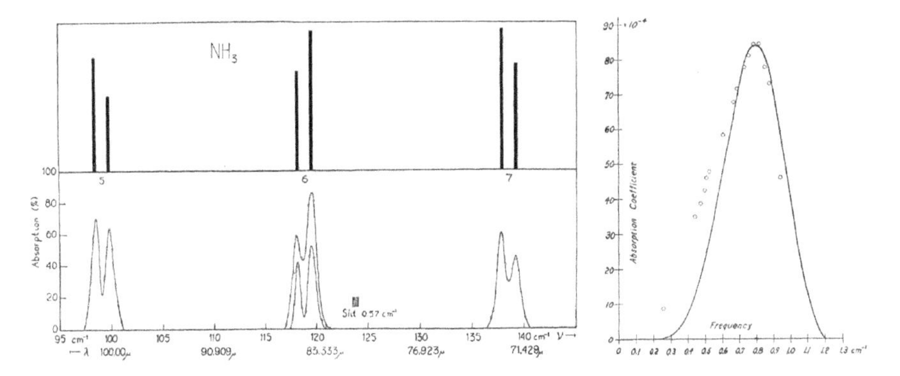

Fig. 5.8. Left: Far infrared absorption spectrum from [95]. Each doublet corresponds to a change in the vibration amplitude of the dipole moment. Improvement of instrumental resolution allowed Wright and Randall to observe the splitting and confirm their interpretation as arising from tunnel coupling of two conformations. Right: Microwave absorption between the split levels from [19]. Data values (dots) are corrected from asymmetry (solid line). Spectroscopy energy units are used: $1\,\mathrm{cm}^{-1} = 1.24 \times 10^{-4}$ eV $= 3 \times 10^4$ MHz. Note that the line splitting distance is twice the absorption line value in the radio spectrum. This is due to symmetry constraints in the quantum electromagnetic absorption process (known as *transition selection rules* — see Volume 3). Reprinted with permission from [19, 95]. Copyright © 1933 and 1934 by the American Physical Society.

0.22739 eV [82] while theory from *ab-initio* quantum computation gives an estimate of 0.219 eV [67].

Ammonia-based MASER

MASER is the acronym for Microwave Amplification by Stimulated Emission of Radiation. Such a device generates centimetre waves in much the same way a LASER generates light in shorter wavelengths range. The detailed study of the stimulated emission process requires a greater understanding of the interaction between light and matter (from a quantum aspect) than yet discussed in this volume. Suffice it to say that the basic principle relies on the fact that two possibilities for an excited molecule (or crystal or atom or electron) exist to relax from an excited state by emitting a photon. The photon can be emitted (almost) at any time and in any direction. This is called *spontaneous emission*. There is another process by which the de-excitation is somehow encouraged by the gregarious behaviour of photons: the greater the number of photons of a given sort (polarization and wave vector) already interacting with the excited species, the higher the probability that the relaxation will occur accompanied by the emission of a photon of the exact same sort. If, for example, all photons (of the correct energy) in the vicinity of the excited system are propagating towards direction $z > 0$ with a clockwise polarization state, the probability will be much higher that photons from the de-excitation process will be emitted with clockwise polarization in the z-direction. This is known as *stimulated emission*. The trick is then as follows: prepare the matter in a given excited state so that the relaxation will yield the required photon energy. The first emitted photons will be issued from a spontaneous process and in any direction. The system is then placed in a cavity with two mirrors facing each other, separated by a distance d. Only electromagnetic waves perpendicular to these mirrors with a wavelength $n\lambda = 2d$ will be allowed in a permanent regime (stationary waves) and any other wave will interfere with itself destructively. Therefore, there is a transition stage, dominated by spontaneous emissions, during which photons of proper orientation and wavelength are being selected. Photons which are spontaneously emitted in another direction than the narrow one defined by the pair of mirrors will just exit the cavity. Then the stimulated process, which has progressively increased, can take over with eventually most of the photons emitted from the medium bearing the exactly identical physical characteristics. A portion of the photons can then be extracted from the cavity by allowing one of the mirrors to be partially transparent.

MASER was the early form of LASER because it is in the microwave range that the stimulated emission probability is the highest (see Volume 3) and it is also in the range in which cavities are easier to fabricate. However, as we mentioned, such a device requires a cavity to sort useful photons and some excited matter to generate photons. The latter is called the *active medium*. At room temperature, it is more likely for the active medium to lay in low energy states than in the required excited state. It is then even more probable for it to absorb photons and become excited than the reverse. To obtain an efficient active medium, it is then essential to inverse

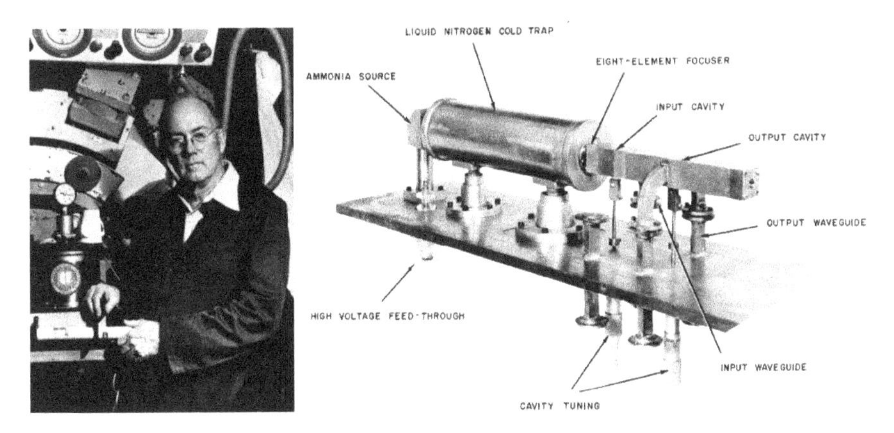

Fig. 5.9. Left: Charles Townes was awarded the Nobel Prize in 1964 for "Production of coherent radiation by atoms and molecules" which was the LASER technology starting point. Photo credits: by courtesy of Department of Physics, UC Berkeley. Right: Early version of the MASER (Philco Laser (1958), by courtesy of Smithsonian Institution Archives, Acc. 18-094, Science Service Records, E& MP 91.004.).

the population ratio: have a dominant number of centres in the targeted excited state. The process is quoted *population inversion.*

Ammonia was used in early MASER [45] (see Fig. 5.9) to take advantage of its emission line connecting the doublet states $|s\rangle$ and $|a\rangle$. As explained in the previous section, the photon energy obtained from this source is of the order of 10^{-4} eV which corresponds to a radiation frequency of 23.870 GHz.

Population inversion is facilitated by the existence of a different dipole moment for degenerate conformations. When the molecule is placed in an electric field $\boldsymbol{E}$ along the x-direction, the coupling changes the potential energy by $-\boldsymbol{D} \cdot \boldsymbol{E}$. In the $\{|\ell\rangle, |r\rangle\}$ representation, the Hamiltonian has a new form:

$$\widehat{H}_E = \widehat{H}_0 - \boldsymbol{E} \cdot \widehat{\boldsymbol{D}} = \begin{pmatrix} \bar{\varepsilon} - S\Delta & \bar{\varepsilon}S - \Delta \\ \bar{\varepsilon}S - \Delta & \bar{\varepsilon} - S\Delta \end{pmatrix} - E \begin{pmatrix} D & 0 \\ 0 & -D \end{pmatrix} \tag{5.42}$$

To simplify the remaining calculations and their interpretation, the overlap integral $S = \langle \ell | r \rangle$ is assumed to be negligible so that the new eigenenergies are given by

$$\varepsilon_{\pm} = \bar{\varepsilon} \pm \sqrt{D^2 E^2 + \Delta^2}$$

When the change in energy induced by the electric field $|\boldsymbol{D} \cdot \boldsymbol{E}|$ is weak compared to the hopping integral Δ, the eigenenergies become

$$\varepsilon_+ \approx \varepsilon_a + \frac{D^2E^2}{2\Delta}$$
$$\varepsilon_- \approx \varepsilon_s - \frac{D^2E^2}{2\Delta}$$

Now, when a molecule enters a region where there is a strong electric field gradient $\partial_x E$, i.e. a strong spatial variation of E along x, it is subject to a net force which depends on its energy state:

$$F_+ = -\partial_x \varepsilon_+ \approx -\frac{D^2E}{\Delta}\partial_x E$$
$$F_- = -\partial_x \varepsilon_- \approx +\frac{D^2E}{\Delta}\partial_x E$$

Molecules are thus deflected by the electric field gradient according to their initial energy and can be easily filtered. It becomes possible to remove low-energy state molecules and thereby reach an almost 100% excited molecular gas. Population inversion is then obtained and the active medium can now de-excite by stimulated emission, producing the MASER radiation. More details on the coupling with the electromagnetic field in the cavity will be given in Volume 3.

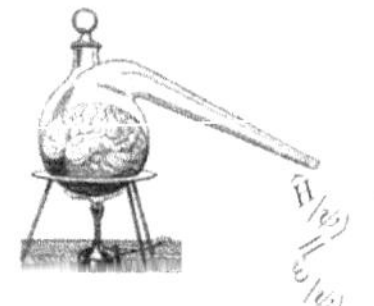

Chapter 5: Nuts & Bolts

- *Quantum states are represented by vectors named "**kets**", denoted by $|\psi\rangle$, defined in a **Hilbert space**.*
- ***Observable quantities** are represented by Hermitian operators. For a system in state $|\psi\rangle$, any (perfect) measurement of an observable property represented by operator $\widehat{A}$ can only give one of its eigenvalues $\{a_n\}$ corresponding to eigenkets $\{|\phi_n\rangle\}$ with respective **probabilities** $\{|\langle\phi_n|\psi\rangle|^2\}$. When the result is a_j, the system is systematically projected onto state $|\phi_j\rangle$.*

(Continued)

- *The set of eigenkets of an observable can be used as an orthogonal basis of a Hilbert space.* ***Wavefunctions*** $\psi(x)$ *are the coordinates of a particular quantum state* $|\psi\rangle$ *in the basis constructed from the continuous set of the eigenstates of the position operator:* $|\psi\rangle = \int \psi(x)|x\rangle dx$.
- ***Position*** *space and* ***momentum*** *space* ***representations*** *of the same quantum state are related:*

$$\psi(\boldsymbol{r}, t) = \frac{1}{(2\pi\hbar)^{3/2}} \int \widetilde{\psi}(\boldsymbol{p}, t) e^{i\boldsymbol{p}\cdot\boldsymbol{r}/\hbar} d^3p.$$

- *Hermitian operators which do not commute and such that* $[\widehat{A}, \widehat{B}] = i\hbar\widehat{C}$ *have a relation between their respective quantum mean standard deviations in state* $|\psi\rangle$*:*

$$\Delta_\psi(A)\Delta_\psi(B) \geqslant \frac{\hbar}{2}\langle C\rangle_\psi,$$

where $\langle C\rangle_\psi = \langle\psi|\widehat{C}|\psi\rangle$ *is the quantum expected value of observable* $\widehat{C}$ *while the system is in state* $|\psi\rangle$. *Application of this property to the position and momentum observables yields* ***Heisenberg's inequalities*** $\Delta_\psi(x)\Delta_\psi(p_x) \geqslant \frac{\hbar}{2}$ *because* $[\widehat{x}, \widehat{p}_x] = i\hbar\mathbb{1}$.

Part III

A Classical to Quantum World Fuzzy Border

Quantum physics is obviously a sophisticated intellectual construction. It is built upon a set of postulates which were carefully crafted by its founding fathers at the beginning of the twentieth century. By focusing our concerns on experiments that resisted, or even contradicted, physical models elaborated throughout more than 200 years* since the publication of Newton's Principia, we may have given the biased message that classical physics should be discarded. But a persistent mind would be legitimate in pointing out numerous discoveries and technical achievements still enjoyed today, such as a toast or a safe flight, and connecting them to classical physics accomplishments rather than to an outgrowth of the quantum theory.

To be valid, the quantum theory needs to provide an explanation for "old" *and* new experimental observations. Therefore, it should encompass classical physics as a limit case with a possibility of defining under which conditions the use of Newton's law remains pertinent. It should provide a criterion to gauge whether the mathematical artillery of quantum physics can be circumvented or not.

This part aims at defining a border that indicates the end of the classical world. A border beyond which classical ideas are no longer valid. To do so, we will first return, in a little more detail, to classical physics from the different perspective put forward in the mid-nineteenth century by Lagrange and Hamilton (Fig. III.1). It is from this formalism that the core of quantum physics was actually constructed and some key issues are better understood with some knowledge of this elegant formulation of classical physics. It is the aim of Chapter 6 to provide the minimal background of phase space mechanics to make a connection with the quantum formalism. It will also prove useful when, in Volume 2, we introduce elements of statistical physics.

Chapter 7 will then show how Schrödinger's equation relates to the classical picture. It will provide criteria that the system should comply with for the description in classical terms to be acceptable.

*Galileo Galilei's seminal writings "De mottu" (On Motion) and "Le Meccaniche" (Mechanics) respectively date back to 1590 and 1600.

(a) (b)

Fig. III.1 (a) Joseph Louis Lagrange (1736–1813) spent most of his life in Italy and Berlin before becoming director of the French Academy of Science. He became the first Professor of Mathematics at École Polytechnique and École Normale de l'An III (to become École Normale Supérieure) when they were created in 1794. (b) (Sir) William Rowan Hamilton (1805–1865) was appointed professor of astronomy at Trinity College Dublin even before graduating. Active and innovative in the field of optics and mathematics, he published relatively little, dedicating much of his time helping his colleagues. Photos credits: public domain.

6

Phase Space Classical Mechanics

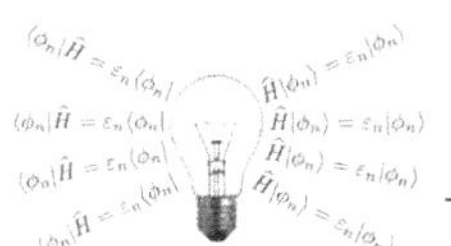

> *The content of this chapter will help you understand* ***the connections between quantum and classical physics, how symmetries lead to conservation laws,*** *and related topics.*

Lagrangian or Hamiltonian formalisms are the basis of *analytical mechanics*. This chapter provides essential keys to understand the physical grounds of this powerful branch of modern classical physics. We do not intend to be exhaustive nor always perfectly rigorous. The following sections aim to shine a new light on links between classical physics, as it was understood in the late nineteenth century, and common concepts of quantum physics.

For a more exhaustive understanding of the subject, it is a good idea to refer to [44].

6.1. Lagrangian and "Least Action Principle"

Let us consider a system $\mathcal{S}$ subjected to a combination of forces. We aim to establish its trajectory on the basis of an approach that is different, yet equivalent to that of Newton's law of motion.

Since there is a trajectory, the system $\mathcal{S}$ can be localized at each instant t in space by its coordinates[1] $\boldsymbol{x}(t)$ in an inertial frame. We then postulate that

- The state of the system, which interacts with its environment at every moment t, is totally conditioned by the triplet t, $\boldsymbol{x}$ and $\dot{\boldsymbol{x}} = d\boldsymbol{x}/dt$ by means of a function $\mathcal{L}(t, \boldsymbol{x}, \dot{\boldsymbol{x}})$, named "Lagrangian" of the system, after J. L. Lagrange (Fig. III.1(a)) who was the promoter of this alternative formulation of mechanics.
- The path taken by the system to go from $\boldsymbol{x}_1$ at moment t_1 to $\boldsymbol{x}_2$ at moment t_2 is the one that minimizes the quantity $\mathcal{A}$, named *action*, and defined by

$$\mathcal{A} = \int_{t_1}^{t_2} \mathcal{L}(t, \boldsymbol{x}, \dot{\boldsymbol{x}}) dt \tag{6.1}$$

The basis of analytical mechanics can be summarized in these two postulates. The importance of having an accurate expression of the Lagrangian for the system is then obvious. Its knowledge can completely determine the system state at any time between t_1 and t_2.

6.1.1. *Lagrange's equations*

Let $\mathcal{L}(t, \boldsymbol{x}, \dot{\boldsymbol{x}})$ be the system Lagrangian. We note $\boldsymbol{x}(t)$ the actual path taken by the system and $\boldsymbol{x}'(t) = \boldsymbol{x}(t) + \delta\boldsymbol{x}(t)$ another path, which is slightly wrong and differs from the actual path by the infinitely small amount $\delta\boldsymbol{x}(t)$ (Fig. 6.1).

Based on the assumptions outlined above, the action corresponding to the true trajectory is minimal. Any infinitesimal deviation from this should have no first-order effect on the action. By using a development near the correct trajectory, we thus find

$$\mathcal{L}(t, \boldsymbol{x} + \delta\boldsymbol{x}, \dot{\boldsymbol{x}} + \delta\dot{\boldsymbol{x}}) = \mathcal{L}(t, \boldsymbol{x}, \dot{\boldsymbol{x}}) + \vec{\nabla}_{\boldsymbol{x}}\mathcal{L} \cdot \delta\boldsymbol{x} + \vec{\nabla}_{\dot{\boldsymbol{x}}}\mathcal{L} \cdot \delta\dot{\boldsymbol{x}} \tag{6.2}$$

[1] If the system consists of a set of N identical (for simplicity) particles, $\boldsymbol{x}(t)$ encapsulates all coordinates of all particles. It is then a vector in a $3N$-dimensional space.

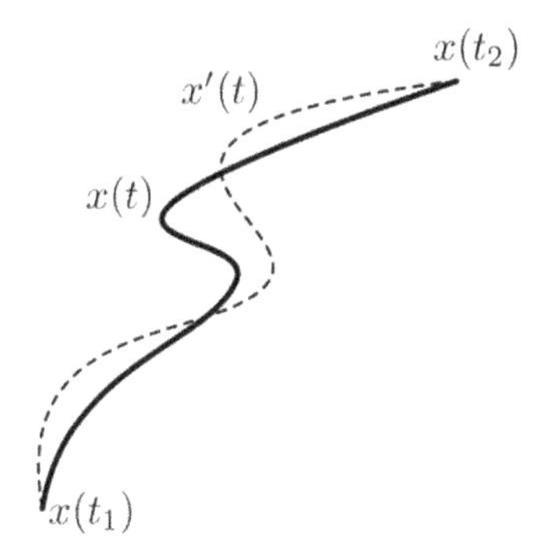

Fig. 6.1. Two possible trajectories for system $\mathcal{S}$ with the "off track" path $\boldsymbol{x}'(t) = \boldsymbol{x}(t) + \delta\boldsymbol{x}(t)$. The solid line represents the true trajectory $\boldsymbol{x}(t)$, the one that minimizes the action.

where the vector differential operators $\vec{\nabla}_{\boldsymbol{x}}$ and $\vec{\nabla}_{\dot{\boldsymbol{x}}}$ calculate gradients with respect to position and velocity coordinates. Near the true trajectory, the first-order change in action needs to be zero:

$$\delta\mathcal{A} = \int_{t_1}^{t_2} (\vec{\nabla}_{\boldsymbol{x}}\mathcal{L} \cdot \delta\boldsymbol{x} + \vec{\nabla}_{\dot{\boldsymbol{x}}}\mathcal{L} \cdot \delta\dot{\boldsymbol{x}})dt = 0 \tag{6.3}$$

We note that the second term can be integrated by parts, so that

$$\int_{t_1}^{t_2} \vec{\nabla}_{\dot{\boldsymbol{x}}}\mathcal{L} \cdot \frac{d\delta\boldsymbol{x}}{dt}dt = [\vec{\nabla}_{\dot{\boldsymbol{x}}}\mathcal{L} \cdot \delta\boldsymbol{x}]_{t_1}^{t_2} - \int_{t_1}^{t_2} \frac{d}{dt}(\vec{\nabla}_{\dot{\boldsymbol{x}}}\mathcal{L}) \cdot \delta\boldsymbol{x}dt \tag{6.4}$$

By assumption, all trajectories agree on the starting and ending points, $\delta\boldsymbol{x}(t_1) = \delta\boldsymbol{x}(t_2) = 0$. Only the integral term remains and we are left with

$$\delta\mathcal{A} = \int_{t_1}^{t_2} \left(\vec{\nabla}_{\boldsymbol{x}}\mathcal{L} - \frac{d}{dt}\vec{\nabla}_{\dot{\boldsymbol{x}}}\mathcal{L}\right) \cdot \delta\boldsymbol{x}\, dt = 0 \tag{6.5}$$

The latter expression should be satisfied for any infinitesimal deviation $\delta\boldsymbol{x}(t)$ with respect to the path of least action. This is only possible if, at every moment t,

$$\vec{\nabla}_{\boldsymbol{x}}\mathcal{L} - \frac{d}{dt}\vec{\nabla}_{\dot{\boldsymbol{x}}}\mathcal{L} = \vec{0} \tag{6.6}$$

This vector equality can be decomposed into several scalar equalities, called *Lagrange's equations*, one for each system coordinate.[2] We then have

$$\frac{\partial \mathcal{L}}{\partial x_{i,n}} = \frac{d}{dt}\frac{\partial \mathcal{L}}{\partial \dot{x}_{i,n}}$$

where i stands for one of the coordinates of the nth particle in the system. For N particles in three dimensions, knowledge of the system's behaviour requires a set of $3N$ Lagrange equations to be solved.

Lagrange equations provide an alternative formulation of classical physics. They obviously describe the same physical reality but from a different perspective. The transition from one description to the other is done by means of an appropriate Lagrangian. A wise choice for $\mathcal{L}$ will allow us to make the connection and, for example, find Newton's second law of motion.[3] First, let us study what symmetry considerations can bring in terms of conservation laws.

6.2. From Lagrange to Hamilton

We assume that $\mathcal{S}$ is isolated. Without outside forces, the system's Lagrangian cannot explicitly depend on t because no clock signal can reach it. Therefore, the time axis is homogeneous for $\mathcal{S}$ and $\partial\mathcal{L}/\partial t = 0$. Variations of the system properties over time can only originate from the changes in its position and velocity when t evolves

$$\frac{d\mathcal{L}}{dt} = \vec{\nabla}_{\boldsymbol{x}}\mathcal{L}\cdot\frac{d\boldsymbol{x}}{dt} + \vec{\nabla}_{\dot{\boldsymbol{x}}}\mathcal{L}\cdot\frac{d\dot{\boldsymbol{x}}}{dt} \tag{6.7}$$

$$= \frac{d}{dt}(\vec{\nabla}_{\dot{\boldsymbol{x}}}\mathcal{L})\cdot\frac{d\boldsymbol{x}}{dt} + \vec{\nabla}_{\dot{\boldsymbol{x}}}\mathcal{L}\cdot\frac{d\dot{\boldsymbol{x}}}{dt} \tag{6.8}$$

$$= \frac{d}{dt}[\vec{\nabla}_{\dot{\boldsymbol{x}}}\mathcal{L}\cdot\dot{\boldsymbol{x}}] \tag{6.9}$$

where Lagrange's equations (6.6) have been used. We thus find

$$\frac{d}{dt}[\vec{\nabla}_{\dot{\boldsymbol{x}}}\mathcal{L}\cdot\dot{\boldsymbol{x}} - \mathcal{L}] = 0 \tag{6.10}$$

---

[2]And each particle in the system.

[3]Observe that we have introduced neither the notion of force nor that of energy.

The end result is of utmost importance: we have found a quantity that is conserved because the system is isolated. We will call[4] it *Hamilton's function*:

$$\mathcal{H} \equiv \overrightarrow{\nabla}_{\dot{x}}\mathcal{L} \cdot \dot{\boldsymbol{x}} - \mathcal{L} \tag{6.11}$$

From its definition, it appears that this function is defined up to a time constant term. This quantity is thus analogous to Newton's total energy. This similarity, and the necessity to build connections to our familiar view of mechanics, encourages us to express the Lagrangian with units of energy. However, note that at this stage we still do not know the explicit form of the Lagrangian.

Let us suppose now that space is homogeneous. This means that any rigid body translation leaves the system unchanged. This property must be reflected in the Lagrangian. When we apply the transformation $\boldsymbol{x} \to \boldsymbol{x} + \delta\boldsymbol{\ell}$, where $\delta\boldsymbol{\ell}$ is any small displacement, which is identical for any constituent of the system $\mathcal{S}$, the Lagrangian's variation is then zero and

$$\delta\mathcal{L} = \overrightarrow{\nabla}_{x}\mathcal{L} \cdot \delta\boldsymbol{\ell} = \frac{d}{dt}(\overrightarrow{\nabla}_{\dot{x}}\mathcal{L}) \cdot \delta\boldsymbol{\ell} = 0 \quad \text{for any } \delta\boldsymbol{\ell} \tag{6.12}$$

where we once again used the Lagrange equations (6.6). Thus, as a result of the homogeneity of space, we have demonstrated another conservation law which can therefore be associated with momentum:

$$\boldsymbol{p} \equiv \overrightarrow{\nabla}_{\dot{x}}\mathcal{L} \tag{6.13}$$

and Hamilton's function can be written as

$$\mathcal{H} = \boldsymbol{p} \cdot \dot{\boldsymbol{x}} - \mathcal{L} \tag{6.14}$$

Vector $\boldsymbol{p}$ is said to be the *conjugate momentum* of position vector $\boldsymbol{x}$. The latter evocation can help us interpret the Lagrange equations in terms of classical mechanics. It leads to

$$\frac{d\boldsymbol{p}}{dt} = \frac{d}{dt}(\overrightarrow{\nabla}_{\dot{x}}\mathcal{L}) = \vec{\nabla}_{x}\mathcal{L} \tag{6.15}$$

[4]This function is also called *Hamiltonian*. We will keep this term for the *total energy* operator used in quantum physics.

This expression allows another step towards Newton's law of motion. Suppose the forces to which the system $\mathcal{S}$ is subjected are derivable from a scalar potential $\boldsymbol{F} = -\vec{\nabla}_{\boldsymbol{x}} V(\boldsymbol{x})$. The similarity between Newton's law of motion

$$\frac{d}{dt} m\dot{\boldsymbol{x}} = -\vec{\nabla}_{\boldsymbol{x}} V(\boldsymbol{x})$$

and (6.15) encourages us to set

$$\mathcal{L} = \frac{1}{2} m\dot{\boldsymbol{x}}^2 - V(\boldsymbol{x}) \tag{6.16}$$

which yields

$$\boldsymbol{p} = m\dot{\boldsymbol{x}} \tag{6.17}$$

✍ *Question 6.1:* **A mechanical model of vibrations on a polymer chain.** The purpose of this question is to make use of Lagrange's equation to determine the equation of motion for the vibration of a polymer chain. The atoms in the chain have two types of bond (see Fig. 6.2): the short bond is characterized by an effective stiffness[a] k_s and the long bond by stiffness k_ℓ. Therefore, two types of atom need to be considered: those with the short bond on the left and the long bond on the right, the displacement of which is noted y_n, and those with the long bond on the left and the short bond on the right, with displacement x_n. The equilibrium distance between two nearest identical atoms (i.e. with the same environment) is a, while the two atoms connected by a short bond are separated by d (see Fig. 6.2). Therefore for the nth pair of atoms connected by a short bond, the atom on the left is positioned at na and the atom on the right is at $na + d$. Both types of stiffness obviously only account for the nearest neighbours interaction. We assume an additional interaction between an atom and its second closest neighbours with an associated stiffness k_2. Each atom type is described by a displacement wave in a steady state such as $x_n = Xe^{i(qna-\omega(q)t)}$ or $y_n = Ye^{i(q(na+d)-\omega(q)t)}$. Find the dispersion relation, i.e. the possible angular frequencies as a function of the wave vector q. For convenience, a periodic limit condition can be assumed so that for a N-atom chain, the displacement of atom N is the same as that of atom 1.

[a]The relation between an interatomic potential and its effective stiffness will be explained in Volume 2.

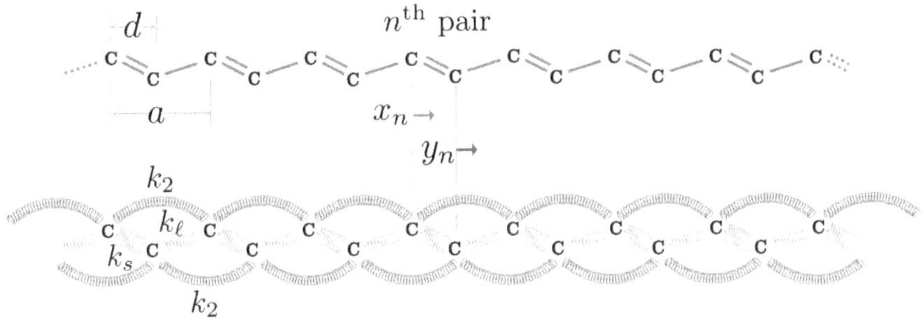

Fig. 6.2. A mechanistic model with first- and second-neighbour interaction to represent the vibration of a quasi-infinite polymer chain.

Answer: Hooke's law for a given atom n positioned at na gives the potential energy $V_1(n) = \frac{1}{2}k_\ell(x_n - y_{n-1})^2 + \frac{1}{2}k_s(x_n - y_n)^2$ for the nearest neighbours interaction. The second closest neighbours interaction has the form: $V_2(n) = \frac{1}{2}k_2(x_n - x_{n+1})^2 + \frac{1}{2}k_2(y_n - y_{n+1})^2$. For the global chain, we thus have the Lagrangian: $\mathcal{L} = \sum_n \frac{1}{2}m(\dot{x}_n^2 + \dot{y}_n^2) - \sum_n V_1(n) - \sum_n V_2(n)$ and the equations of motion for each atom type of the nth pair are:

$$\frac{d}{dt}\frac{\partial}{\partial \dot{x}_n}\mathcal{L} = \frac{\partial}{\partial x_n}\mathcal{L}$$

$$m\ddot{x}_n = -k_s(x_n - y_n) - k_\ell(x_n - y_{n-1}) - k_2(2x_n - x_{n-1} - x_{n+1})$$

$$\frac{d}{dt}\frac{\partial}{\partial \dot{y}_n}\mathcal{L} = \frac{\partial}{\partial y_n}\mathcal{L}$$

$$m\ddot{y}_n = -k_s(y_n - x_n) - k_\ell(y_n - x_{n+1}) - k_2(2y_n - y_{n-1} - y_{n+1})$$

Because of translational symmetry, all other equations have the exact same form. In the steady state, the displacements are given by $x_n = Xe^{i(qna-\omega(q)t)}$ and $y_n = Ye^{iqd}e^{i(qna-\omega(q)t)}$. A solution to the system of linear coupled equations exists only if, for a given wave vector q, the displacement waves have one of the two possible frequencies:

$$\omega_\pm^2(q) = \Omega_s^2 + \Omega_\ell^2 + 4\Omega_2^2\sin^2\left(\frac{qa}{2}\right) \pm \sqrt{\Omega_s^4 + \Omega_\ell^4 + 2\Omega_s^2\Omega_\ell^2\cos(qa)}$$

where $\Omega_2^2 = k_2/m$, $\Omega_\ell^2 = k_\ell/m$ and $\Omega_s^2 = k_s/m$ are the resonant angular frequencies corresponding to stiffness k_2, k_ℓ and k_s respectively.

Expression (6.17) does not cause any surprise, as it is the usual definition of momentum. Newton's second law of motion states that $m\dot{\boldsymbol{x}}$ is a constant quantity if no net force is acting on the system. Everything then fits perfectly well.

However, before we rejoice too quickly, let us consider what the Lagrangian should be for a charged particle subject in addition to an electromagnetic field. Newton's law gives

$$\frac{d}{dt}m\dot{\boldsymbol{x}} = q\left[\boldsymbol{E} + \dot{\boldsymbol{x}} \times \boldsymbol{B}\right] - \vec{\nabla}_{\boldsymbol{x}}V(\boldsymbol{x}) \tag{6.18}$$

where $\boldsymbol{E}$ and $\boldsymbol{B}$ are the electric and magnetic fields respectively. Expression (6.18) can be rewritten introducing the scalar and vector potentials, Φ and $\boldsymbol{A}$, respectively[5]:

$$\frac{d}{dt}m\dot{\boldsymbol{x}} = q\left[-\frac{\partial \boldsymbol{A}}{\partial t} - \vec{\nabla}_{\boldsymbol{x}}\Phi + \dot{\boldsymbol{x}} \times (\vec{\nabla}_{\boldsymbol{x}} \times \boldsymbol{A})\right] - \vec{\nabla}_{\boldsymbol{x}}V(\boldsymbol{x}) \tag{6.19}$$

Finding the Lagrangian for a charged particle in the presence of these potentials is made easier if we gather the time derivatives:

$$\frac{d}{dt}m\dot{\boldsymbol{x}} + q\frac{\partial \boldsymbol{A}}{\partial t} = -\vec{\nabla}_{\boldsymbol{x}}(V + q\Phi) + q\dot{\boldsymbol{x}} \times (\vec{\nabla}_{\boldsymbol{x}} \times \boldsymbol{A}) \tag{6.20}$$

Using the equality $\dot{\boldsymbol{x}} \times (\vec{\nabla}_{\boldsymbol{x}} \times \boldsymbol{A}) = \vec{\nabla}_{\boldsymbol{x}}(\dot{\boldsymbol{x}} \cdot \boldsymbol{A}) - (\dot{\boldsymbol{x}} \cdot \vec{\nabla}_{\boldsymbol{x}})\boldsymbol{A}$, it is found that

$$\frac{d}{dt}\left[m\dot{\boldsymbol{x}} + q\boldsymbol{A}\right] - q(\dot{\boldsymbol{x}} \cdot \vec{\nabla}_{\boldsymbol{x}})\boldsymbol{A} = -\vec{\nabla}_{\boldsymbol{x}}(V + q\Phi - q\dot{\boldsymbol{x}} \cdot \boldsymbol{A}) - q(\dot{\boldsymbol{x}} \cdot \vec{\nabla}_{\boldsymbol{x}})\boldsymbol{A} \tag{6.21}$$

where, on the left-hand side of the equality, the transition from a partial derivative with respect to time to a total derivative induced a correction term. We then obtain

$$\vec{\nabla}_{\boldsymbol{x}}\left[-V(\boldsymbol{x}) - q\Phi(\boldsymbol{x}, t) + q\dot{\boldsymbol{x}} \cdot \boldsymbol{A}(\boldsymbol{x}, t)\right] - \frac{d}{dt}\left[m\dot{\boldsymbol{x}} + q\boldsymbol{A}(\boldsymbol{x}, t)\right] = 0 \tag{6.22}$$

In order for Lagrange's equations (6.6) to yield such an equality, we see that it is possible to propose the following form of Lagrangian:

$$\mathcal{L} = \frac{m}{2}\dot{\boldsymbol{x}}^2 + q\dot{\boldsymbol{x}} \cdot \boldsymbol{A}(\boldsymbol{x}, t) - q\Phi(\boldsymbol{x}, t) - V(\boldsymbol{x}, t) \tag{6.23}$$

which is valid for a charged particle in an electromagnetic field $(\boldsymbol{A}, \Phi)$ and a potential V. Consequently, the momentum becomes

$$\boldsymbol{p} = \vec{\nabla}_{\dot{\boldsymbol{x}}}\mathcal{L} = m\dot{\boldsymbol{x}} + q\boldsymbol{A}(\boldsymbol{x}, t) \tag{6.24}$$

[5]Remember from Section 3.1.1 that $\boldsymbol{E} = -\frac{\partial \boldsymbol{A}}{\partial t} - \vec{\nabla}_{\boldsymbol{x}}\Phi$ and $\boldsymbol{B} = \vec{\nabla}_{\boldsymbol{x}} \times \boldsymbol{A}$.

It then transpires that $\boldsymbol{p}$, the conjugate momentum of $\boldsymbol{x}$, does not match the usual $m\dot{\boldsymbol{x}}$. The latter conventional expression holds only if the forces are derived from a potential, that is to say when the Lagrangian can be written as a mere difference between the kinetic energy and potential energy (Eq. (6.16)).

The Lagrangian formulation of mechanics is useful to establish the motion equations of a complex system of interacting particles.

There is also a Hamiltonian formulation of a system dynamics, i.e. an alternate presentation proposed by W. R. Hamilton (Fig. III.1(b)), which resorts to Hamilton's function and its dynamics.

Returning, in the most general case, to the expression of the Hamilton function. Note that any variation around the true trajectory (given by a change $d\boldsymbol{x}$ and $d\dot{\boldsymbol{x}}$) necessarily results in a change of $\mathcal{H}$ according to

$$\begin{aligned} d\mathcal{H} &= d\boldsymbol{p} \cdot \dot{\boldsymbol{x}} + \boldsymbol{p} \cdot d\dot{\boldsymbol{x}} - d\mathcal{L} \\ &= d\boldsymbol{p} \cdot \dot{\boldsymbol{x}} + \boldsymbol{p} \cdot d\dot{\boldsymbol{x}} - \vec{\nabla}_{\boldsymbol{x}}\mathcal{L} \cdot d\boldsymbol{x} - \vec{\nabla}_{\dot{\boldsymbol{x}}}\mathcal{L} \cdot d\dot{\boldsymbol{x}} \\ &= d\boldsymbol{p} \cdot \dot{\boldsymbol{x}} - \dot{\boldsymbol{p}} \cdot d\boldsymbol{x} \end{aligned} \tag{6.25}$$

where we have used the definition of the momentum (6.14) and the Lagrange equations (valid only on the true trajectory that minimizes the action). Thus, it is found that the variation of the Hamilton function around a real trajectory depends only on the position and momentum variables. Thus:

$$\dot{\boldsymbol{x}} = \vec{\nabla}_{\boldsymbol{p}}\mathcal{H} \tag{6.26a}$$

$$\dot{\boldsymbol{p}} = -\vec{\nabla}_{\boldsymbol{x}}\mathcal{H} \tag{6.26b}$$

These $6N$ equations are known as *Hamilton's equations*. They are the equivalent of Lagrange's equations when a *phase space* representation is adopted, i.e. the space generated by the two vectors $\boldsymbol{x}$ and $\boldsymbol{p}$. It is easy to verify that (6.26b) allows to retrieve Newton's second law of motion when $\mathcal{H} = p^2/2m + V(\boldsymbol{x})$, hence the mere sum of kinetic and potential energies.

Question 6.2: **Gauges for a particle in a static magnetic field.** As briefly reminded in this section, the magnetic field is obtained from the vector potential. However, while $\boldsymbol{B}$ is clearly an observable, $\boldsymbol{A}$ is not defined in a unique manner. Obviously, this comes from $\vec{\nabla} \times \vec{\nabla} f(\boldsymbol{r}) = 0$ and such a degree of freedom is known as the *gauge choice*. It does not modify the end result for any (differentiable) function $f(\boldsymbol{r})$:

$$\boldsymbol{B} = \vec{\nabla} \times (\boldsymbol{A}(\boldsymbol{r}) + \vec{\nabla} f(\boldsymbol{r})).$$

Depending on the symmetry of the problem, different choices of $f(\boldsymbol{r})$ will be convenient. For example, the symmetric gauge which sets $\boldsymbol{A} = \boldsymbol{B} \times \boldsymbol{r}/2$ is widely used in problems with cylindrical symmetry.

Here, we consider a particle with charge q in a static and uniform magnetic field B oriented along $\boldsymbol{e}_z$.

1. Show that the symmetric gauge gives the correct magnetic field.
2. Show that, with the appropriate gauge choice, i.e. the right condition on $f(\boldsymbol{r})$, Hamilton's function can be written $\mathcal{H} = \frac{1}{2m}(\boldsymbol{p} + qxB\boldsymbol{e}_y)^2$. Such a choice is known as the "Landau gauge".
3. Show that Hamilton's equations yield Newton's second law of motion for a charged particle in a magnetic field.

Answer:

1. If we introduce the electromagnetic Lagrangian into Hamilton's function (6.14), we obtain: $\mathcal{H} = \boldsymbol{p} \cdot \dot{\boldsymbol{r}} - (\frac{m}{2}\dot{\boldsymbol{r}}^2 + q\dot{\boldsymbol{r}} \cdot \boldsymbol{A}(\boldsymbol{r}))$.

   Then, using (6.26a), $\dot{\boldsymbol{r}} = \frac{1}{m}(\boldsymbol{p} - q\boldsymbol{A}(\boldsymbol{r}))$, we obtain $\mathcal{H} = \frac{1}{2m}(\boldsymbol{p} - q\boldsymbol{A}(\boldsymbol{r}))^2$ where only a static magnetic field was assumed. For a constant magnetic field pointing towards $\boldsymbol{e}_z$, $\boldsymbol{B} = \vec{\nabla} \times (\boldsymbol{A}(\boldsymbol{r}) + \vec{\nabla} f(\boldsymbol{x}))$ becomes

$$B = \frac{\partial}{\partial x}(A_y + \partial_y f) - \frac{\partial}{\partial y}(A_x + \partial_x f)$$

   For the symmetric gauge, $\boldsymbol{A}(\boldsymbol{r}) = -yB/2\boldsymbol{e}_x + xB/2\boldsymbol{e}_y$ and the action of the curl operator readily yields $\vec{\nabla} \times \boldsymbol{A} = B\boldsymbol{e}_z$. This confirms that the symmetric gauge is a legitimate choice with this uniform field.
2. The choice of function f is arbitrary. It can be taken such that $A_x + \partial_x f = 0$ and $\partial_y f = 0$. This is equivalent to setting $\boldsymbol{A} = -xB\ \boldsymbol{e}_y$ known as the Landau gauge. Hence,

$$\mathcal{H} = \frac{1}{2m}(\boldsymbol{p} + qxB\boldsymbol{e}_y)^2$$

3. Hamilton's equations yield

$$\begin{cases} \dot{p_x} = -\dfrac{\partial}{\partial x}\mathcal{H} = -\dfrac{qB}{m}(p_y + qxB) = -qB\dot{y} \\ \dot{p_y} = -\dfrac{\partial}{\partial y}\mathcal{H} = 0 \end{cases}$$

Answer: (continued)

According to the definition of the momentum, we obtain

$$\begin{cases} m\ddot{x} = -qB\dot{y} \\ m\ddot{y} = qB\dot{x} \end{cases}$$

6.3. Constrained Trajectories

The least action principle is valid for any generalized coordinate system. A proper choice is usually motivated by the symmetry of the problem or the constraints imposed on the motion. A trivial example is that of a rigid pendulum for which the oscillating mass is attached to the center of rotation by a rigid rod instead of a string. There is thus no reason for using cartesian coordinates since the distance to the origin cannot change and the θ and φ angles suffice to determine the motion.

6.3.1. *From holonomic constraint...*

There are cases for which no easy coordinate system is adapted to the constraint. So a more systematic strategy is required. Holonomic constraints are those that can be expressed in the form $f(\boldsymbol{x}) = 0$. In its motion, the system can even be subjected to several of such independent constraints but, for simplicity, a unique constraint will be considered here. It will also be assumed that the constraint is limited to the positions. The existence of $f(\boldsymbol{x}) = 0$ implies that the coordinates are no longer independent, i.e. the variation of one of them, say x_i, can be written as a function of the changes experienced by the others:

$$dx_i = \sum_{j \neq i} \left[\frac{\partial x_i}{\partial x_j}\right]_{j'} dx_j \tag{6.27}$$

where the prime on j means that all coordinates but the jth are kept fixed. These partial derivatives are obviously a function of the constraint and are found by using $df(\boldsymbol{x}) = 0$ so that

$$df(\boldsymbol{x}) = \vec{\nabla} f \cdot d\boldsymbol{x} = \left[\frac{\partial f}{\partial x_i}\right]_{i'} dx_i + \sum_{j \neq i} \left[\frac{\partial f}{\partial x_j}\right]_{j'} dx_j = 0$$

and using (6.27), for any variation of the remaining independent coordinates:

$$\sum_{j\neq i}\left[\frac{\partial f}{\partial x_i}\right]_{i'}\left[\frac{\partial x_i}{\partial x_j}\right]_{j'} dx_j + \sum_{j\neq i}\left[\frac{\partial f}{\partial x_j}\right]_{j'} dx_j = 0$$

This equality holds for any set of dx_j if

$$\left[\frac{\partial f}{\partial x_i}\right]_{i'}\left[\frac{\partial x_i}{\partial x_j}\right]_{j'} + \left[\frac{\partial f}{\partial x_j}\right]_{j'} = 0$$

which gives the partial derivatives[6]

$$\left[\frac{\partial x_i}{\partial x_j}\right]_{j'} = -\frac{\left[\frac{\partial f}{\partial x_j}\right]_{j'}}{\left[\frac{\partial f}{\partial x_i}\right]_{i'}} \tag{6.28}$$

These relations can now be employed to introduce the constraint into the least action principle and thereby into the equations of motion derived from Lagrange's equations. Rewriting (6.5) with the constraint yields

$$\delta\mathcal{A} = \int_{t_1}^{t_2}\sum_{j\neq i}\left\{\frac{\partial\mathcal{L}}{\partial x_j} - \frac{d}{dt}\frac{\partial\mathcal{L}}{\partial\dot{x}_j} - \frac{\left[\frac{\partial f}{\partial x_j}\right]_{j'}}{\left[\frac{\partial f}{\partial x_i}\right]_{i'}}\left(\frac{\partial\mathcal{L}}{\partial x_i} - \frac{d}{dt}\frac{\partial\mathcal{L}}{\partial\dot{x}_i}\right)\right\}\delta x_j\,dt = 0 \tag{6.29}$$

which can be reformulated according to

$$\delta\mathcal{A} = \int_{t_1}^{t_2}\sum_{j\neq i}\left\{\frac{\partial(\mathcal{L}-\lambda f)}{\partial x_j} - \frac{d}{dt}\frac{\partial\mathcal{L}}{\partial\dot{x}_j}\right\}\delta x_j\,dt = 0 \tag{6.30}$$

where $\lambda = (\frac{\partial\mathcal{L}}{\partial x_i} - \frac{d}{dt}\frac{\partial\mathcal{L}}{\partial\dot{x}_i})/[\frac{\partial f}{\partial x_i}]_{i'}$. In this case, if f depends only on the positions, the Lagrange equations are written as

$$\frac{d\boldsymbol{p}}{dt} = \frac{d}{dt}(\vec{\nabla}_{\dot{\boldsymbol{x}}}\mathcal{L}) = \vec{\nabla}_{\boldsymbol{x}}(\mathcal{L}-\lambda f), \tag{6.31}$$

so that the constraint turns out to act on the system as an additional force $-\lambda\vec{\nabla}_{\boldsymbol{x}}(f)$ which imposes to the system to follow a different trajectory than the one which strictly corresponds to the least action.

[6]This relation is frequently used in thermodynamics and statistical physics. Volume 2 will be no exception.

If $\lambda = 0$, this means that no weight is given to the constraint, i.e. the strict least action principle is satisfied.

6.3.2. *... to Lagrange multipliers*

What has just been derived is of utmost importance for any kind of minimization process, not just the action, and is known as the "Lagrange multiplier" technique.

Let $\mathcal{L}(\boldsymbol{x})$ be a quantity to be minimized. This means that a set of values $\boldsymbol{x}_m$ for the parameters $\boldsymbol{x}$ is searched so that $\mathcal{L}(\boldsymbol{x}_m)$ is minimal. If the parameters are constrained with a relation that can be expressed by $f(\boldsymbol{x}) = 0$, the parameters are no longer independent. The general strategy would then be to express one parameter as a function of the others $x_i = g(x_{i'})$ and replace it into $\mathcal{L}(\boldsymbol{x})$ while solving the set of equations $\partial\mathcal{L}/\partial x_{i'} = 0$. This procedure is not always possible because $f(\boldsymbol{x})$ may have a complex expression (or even be given in a numerical form). Instead the Lagrange multiplier technique proposes to directly treat all parameters on an equal footing by minimizing $\mathcal{L}(\boldsymbol{x}) - \lambda f(\boldsymbol{x})$ with respect to $\boldsymbol{x}$ and λ by solving

$$\vec{\nabla}_{\boldsymbol{x}}(\mathcal{L} - \lambda f) = 0 \tag{6.32}$$

$$\frac{\partial}{\partial\lambda}(\mathcal{L} - \lambda f) = 0 \tag{6.33}$$

Parameter λ is the "Lagrange multiplier" and does not even always need to be determined. Obviously, the second equation automatically expresses the constraint.

6.4. From Hamilton to Hamilton–Jacobi

In this section, we propose to return to the action by linking it to the Hamilton function. As it will be seen in the next chapter, these two quantities play a central role in quantum physics and, in particular, its connection with classical physics.

By the definition of the action, we know that it is connected to the Lagrangian:

$$\mathcal{L} = \frac{d\mathcal{A}}{dt} \tag{6.34}$$

Furthermore, the temporal variation of the action may be generated by varying the time over which it is calculated but also by a change in positions:

$$\frac{d\mathcal{A}}{dt} = \frac{\partial \mathcal{A}}{\partial t} + \dot{\boldsymbol{x}} \cdot \overrightarrow{\nabla}_{\boldsymbol{x}} \mathcal{A} \tag{6.35}$$

The last term is found by considering the variation of the action when the ending point of a true trajectory is slightly modified arbitrarily. The calculation is in all respects similar to that conducted in (6.4) and we get

$$\begin{aligned}\delta\mathcal{A} &= \int_{t_1}^{t_2} \overrightarrow{\nabla}_{\dot{\boldsymbol{x}}}\mathcal{L} \cdot \frac{d\delta\boldsymbol{x}}{dt} dt \\ &= \overrightarrow{\nabla}_{\dot{\boldsymbol{x}}}\mathcal{L} \cdot \delta\boldsymbol{x}(t_2) - \int_{t_1}^{t_2} \frac{d}{dt}(\overrightarrow{\nabla}_{\dot{\boldsymbol{x}}}\mathcal{L}) \cdot \delta\boldsymbol{x} dt = \overrightarrow{\nabla}_{\dot{\boldsymbol{x}}}\mathcal{L} \cdot \delta\boldsymbol{x}(t_2)\end{aligned} \tag{6.36}$$

where only the first term exists because the real path is chosen, thus minimizes the action for a given destination. So,

$$\overrightarrow{\nabla}_{\boldsymbol{x}}\mathcal{A} = \overrightarrow{\nabla}_{\dot{\boldsymbol{x}}}\mathcal{L} = \boldsymbol{p} \tag{6.37}$$

Equation (6.35) can thus be rewritten by substituting (6.37) and (6.34):

$$\frac{\partial \mathcal{A}}{\partial t} = \mathcal{L} - \dot{\boldsymbol{x}} \cdot \boldsymbol{p} = -\mathcal{H} \tag{6.38}$$

This relation between the action and Hamilton's function is named the *Hamilton–Jacobi equation.* In the case of a system subjected to a potential $V(\boldsymbol{x})$, it is thus found that

$$\frac{\partial \mathcal{A}}{\partial t} = -\frac{p^2}{2m} - V(\boldsymbol{x}) = -\frac{|\overrightarrow{\nabla}_{\boldsymbol{x}}\mathcal{A}|^2}{2m} - V(\boldsymbol{x}) \tag{6.39}$$

Hamilton's function is as equally essential a quantity as the Lagrangian of a particle. It converges to the total energy on a true trajectory and yields the fundamental principle of mechanics (through the Hamilton equations). The relevant quantities for characterizing the system under study is then the position and momentum. This explains why the position, momentum and $\mathcal{H}$ are also found at the very heart of a quantum description.

To end this brief overview of analytical mechanics as the underlying structure of modern physics, we will establish a relationship to predict the temporal behaviour of any property of the system described by a function of Hamilton $\mathcal{H}$.

For simplicity, consider a property P described by a scalar which varies with time, position and momentum of its particles: $P(t, \boldsymbol{x}, \boldsymbol{p})$. If the system is not stationary, the temporal variation of the property is written as

$$\frac{dP(t, \boldsymbol{x}, \boldsymbol{p})}{dt} = \frac{\partial P}{\partial t} + \vec{\nabla}_{\boldsymbol{x}} P \cdot \dot{\boldsymbol{x}} + \vec{\nabla}_{\boldsymbol{p}} P \cdot \dot{\boldsymbol{p}}$$

The true trajectory for the system's motion is such that it satisfies Hamilton's equations (6.26) and

$$\frac{dP(t, \boldsymbol{x}, \boldsymbol{p})}{dt} = \frac{\partial P}{\partial t} + \vec{\nabla}_{\boldsymbol{x}} P \cdot \vec{\nabla}_{\boldsymbol{p}} \mathcal{H} - \vec{\nabla}_{\boldsymbol{p}} P \cdot \vec{\nabla}_{\boldsymbol{x}} \mathcal{H}$$

Poisson's brackets are then introduced as

$$\{A, B\}_i = \frac{\partial A}{\partial x_i}\frac{\partial B}{\partial p_i} - \frac{\partial B}{\partial x_i}\frac{\partial A}{\partial p_i} \tag{6.40}$$

for the ith coordinates. Thus, when a property does not explicitly depend on time,

$$\frac{dP(\boldsymbol{x}, \boldsymbol{p})}{dt} = \sum_{i=1}^{3N} \left(\frac{\partial P}{\partial x_i}\frac{\partial \mathcal{H}}{\partial p_i} - \frac{\partial \mathcal{H}}{\partial x_i}\frac{\partial P}{\partial p_i} \right) = \sum_{i=1}^{3N} \{P, \mathcal{H}\}_i \tag{6.41}$$

Such a property is constant in time if the sum of its Poisson brackets is identically zero.

Question 6.3: **Remarkable Poisson brackets.** Compute the following Poisson brackets: $\{x, p\}$, $\{p, f(x)\}$, $\{p, \mathcal{H}\}$ and $\{x, \mathcal{H}\}$, where $\mathcal{H}$ is the Hamilton function for a system subjected to forces deriving from a potential and $f(x)$ is any derivable function. Conclude that this approach is consistent with traditional classical physics.

Answer:

$$\{x, p\} = \frac{\partial x}{\partial x}\frac{\partial p}{\partial p} - \frac{\partial p}{\partial x}\frac{\partial x}{\partial p} = 1$$

$$\{p, f(x)\} = -\frac{\partial f(x)}{\partial x}$$

This result should be compared to commutators of operators and, more specifically, the momentum operator:

$$\{p, \mathcal{H}\} = -\frac{\partial \mathcal{H}}{\partial x} = F(x)$$

Answer: (continued)
where $F(x) = -\frac{\partial V}{\partial x}$ is the force that derives from the potential (conservative system). Therefore, we can just as well write

$$\{p, \mathcal{H}\} = \frac{\partial p}{\partial t}$$

$$\{x, \mathcal{H}\} = \frac{\partial \mathcal{H}}{\partial p} = \frac{p}{m}$$

which is indeed the velocity for a conservative system.

6.5. Reconnecting to Quantum Physics

As it will be shown in the next chapter, classical physics should be seen as a particular case of quantum physics. It will be explained how, and under which conditions, the equations of Hamilton or Newton can be recovered from those implied by quantum formalism. Consequently, the difficulty to proceed the other way around, i.e. starting from a classical description to end up with a quantum picture, is not trivial by any means. There is no systematic recipe for this but, in many cases, the "quantum condition" can be considered as a good starting point. The similarities are better shown if the "Heisenberg picture" is used. In this representation of quantum physics, kets are time independent while observables vary with time. Therefore, this is the opposite of Schrödinger's picture we have been using so far. This new point of view is interesting because it is closer to the classical treatment where there is no state but only variables which are related to measurable quantities. The observable at time t is obtained from its form at $t = 0$ by means of the evolution operator as seen in Question *5.21*:

$$\widehat{G}(t) = e^{i\widehat{H}t/\hbar}\widehat{G}(0)e^{-i\widehat{H}t/\hbar}$$

Taking the time derivative of this expression shows that the observable must satisfy an evolution equation so that

$$\begin{aligned}\frac{d}{dt}\widehat{G}(t) &= \frac{i}{\hbar}(\widehat{H}e^{i\widehat{H}t/\hbar}\widehat{G}(0)e^{-i\widehat{H}t/\hbar} - e^{i\widehat{H}t/\hbar}\widehat{G}(0)\widehat{H}e^{-i\widehat{H}t/\hbar}) \\ &= \frac{i}{\hbar}(\widehat{H}\widehat{G}(t) - \widehat{G}(t)\widehat{H}) \\ &= \frac{i}{\hbar}[\widehat{H}, \widehat{G}(t)] \end{aligned} \tag{6.42}$$

This equation was derived by Dirac, who named it "Heisenberg's equation of motion". It bears such a close resemblance to (6.41) that it cannot be a mere coincidence. Obviously, if the observable commutes with the Hamiltonian, it must correspond to a conserved quantity, i.e. its quantum mean values are constant. Moreover, if the observable $\widehat{G}$ is one of the canonical variables of the problem (as in Question *6.3*), for example x or p, both expressions are equivalent providing that the transformations $x \to \widehat{x}$ and $p \to \widehat{p}$ are accompanied by the commutation condition $[\widehat{x}, \widehat{p}] = i\hbar\mathbb{1}$. Sometimes $\frac{i}{\hbar}[\widehat{H}, \widehat{G}(t)]$ is referred to as the "Quantum Poisson Bracket". Note that this last quantity is more general than its classical counterpart because it can be applied to observables which do not have any classical representation. In particular, it is the case of the spin observable which will be introduced in Volume 2.

Heisenberg's equation gives the time derivative of an observable which does not commute with the Hamiltonian $\widehat{H}$. The eigenvalues of this observable will thus change over time and so can be used as a time indicator. Reversing (6.42) to $[\widehat{H}, \widehat{G}(t)] = -i\hbar\frac{d}{dt}\widehat{G}(t)$ and using the generalized Heisenberg relation (5.29), it is found that in a given quantum state $|\psi\rangle$, the following relation holds:

$$\frac{\Delta_\psi G}{\langle \dot{G} \rangle_\psi} \Delta_\psi E \geq \frac{\hbar}{2},$$

assuming that the property was chosen so that its mean quantum value is not zero in the given state. Obviously, the fraction on the left is a time-like quantity. It is derived from the variations (in the quantum state) of the values taken by observable $\widehat{G}(t)$, which thereby acts as a clock internal to the system. The time intervals, given by this internal clock, are noted $\Delta_\psi t_G$ and represent how long it takes for the system to have its property G change by an amount in the order of its standard deviation, $\Delta_\psi G$. It is thus possible to establish an energy-time Heisenberg indetermination relation as $\Delta_\psi \varepsilon \Delta_\psi t_G \geq \hbar/2$. Attention must be paid however to the fact that the time, to which this relation is referring, is not "the" time, which is still considered as an outside parameter, not related to any fixed "time operator" but a quantity derived from the variation of an arbitrarily chosen property. As a matter of fact, there has been a long-lasting debate[7] over the necessity to establish (in doing so, clarifying the meaning of) the time-energy inequality on the basis of a commutator between the Hamiltonian and a time operator. Interested readers are encouraged to explore further using, for example, [16] which presents a detailed overview of different approaches.

Chapter 6: Nuts & Bolts

- ***Analytical mechanics*** *is an equivalent formulation of classical physics to explain the dynamics of systems. It states the existence of a "**Lagrangian**" $\mathcal{L}(t, \boldsymbol{x}, \dot{\boldsymbol{x}})$, a function of time t, position $\boldsymbol{x}$ and speed $\dot{\boldsymbol{x}}$, the integral of which, named "action" along the actual trajectory, is expected to be minimal.*
- *The equations which rule the dynamics of the system are **Lagrange equations** $\vec{\nabla}_{\boldsymbol{x}}\mathcal{L} = \frac{d}{dt}\vec{\nabla}_{\dot{\boldsymbol{x}}}\mathcal{L}$ (in a 3N-dimensional space for an N-particle system).*
- *Conservation with respect to time invariance implies that the **Hamilton function**, i.e. the energy function, $\mathcal{H} \equiv \vec{\nabla}_{\dot{\boldsymbol{x}}}\mathcal{L}\cdot\dot{\boldsymbol{x}} - \mathcal{L}$ is constant. Similarly, homogeneity of space implies that the momentum $\boldsymbol{p} \equiv \vec{\nabla}_{\dot{\boldsymbol{x}}}\mathcal{L}$ is a constant quantity. If an electromagnetic field is acting on a charged system, its momentum thus becomes $\boldsymbol{p} = m\dot{\boldsymbol{x}} + q\boldsymbol{A}(\boldsymbol{x}, t)$ where $\boldsymbol{A}(\boldsymbol{x}, t)$ is the vector potential.*

[7] After it was (of course) initiated by Heisenberg and Bohr, the debate basically started in 1945, with Mandelstam and Tamm [70], and is still continuing today.

(Continued)

- *The* ***phase space*** *is the classical space where the axes correspond to positions and momenta in each direction. Dynamics in phase space is governed by* ***Hamilton's equations*** $\dot{\boldsymbol{x}} = \vec{\nabla}_{\boldsymbol{p}}\mathcal{H}$ *and* $\dot{\boldsymbol{p}} = -\vec{\nabla}_{\boldsymbol{x}}\mathcal{H}$. *The latter is the exact equivalent of Newton's second law when the forces are derived from a potential.*
- *For a given* $\mathcal{H}$, *the time variation of any property* P *is computed from a "****Poisson bracket****" by* $\frac{dP(\boldsymbol{x},\boldsymbol{p})}{dt} = \sum_{i=1}^{3N} \{P, \mathcal{H}\}_i$ *which provides a connection to "Heisenberg's quantum equation of motion".*

7

Quantum Criteria (Who Needs Quantum Physics?)

The content of this chapter will help you understand ***why it is often fine to use classical physics, how to predict the necessity of a quantum description****, and related topics.*

While this first volume has only scratched the surface of quantum physics and relevant applications, it has not been possible to hide the complexity of its practice. Students tend to be overwhelmed by the burden of mathematical concepts required even to explain what appears to be elementary problems when thought of in terms of classical physics. Volume 2 will delve into more realistic problems, dealing with atoms, molecules and solids. It is certainly legitimate to question the necessity of a systematic recourse to the quantum formalism. Therefore, the goal of this chapter is to identify when quantum physics cannot be avoided and under which conditions classical mechanics provides a pertinent snapshot.

Classical physics only requires a limited number of postulates (i.e. the existence of a Lagrangian and the least action principle, for example). Given the number of rules to account for quantum behaviour, it is expected that the quantum theory represents a more general framework than classical physics to provide an explanation of the physical world. However, rigorous proof of this belief turns out to be a rather complicated matter. As it will become apparent throughout the sections of this chapter, the reduction of QM to Newton's mechanics does not simply boil down to a single recipe such as setting $\hbar \to 0$ but necessitates a more pragmatic approach to answer the following question: to what extent are quantum effects still observable and pertinent components of measured properties?

It would be over-simplistic to draw the quantum borderline as a separation between the macroscopic and microscopic worlds. Quantum consequences can manifest themselves in many aspects of our macroscopic observations: nuclear radiation is a natural example and we know that superconductivity or superfluidity are large-scale effects of pure quantum properties. One mole of sodium atoms still radiates a discrete spectrum of light. DNA's cohesion, crystals' shapes, conduction properties of solids and even the stability of white dwarfs, all result from laws relying on quantum physics only.

7.1. Ehrenfest's Theorem

As previously stated, measurable physical quantities are represented by observables, that is to say, Hermitian operators whose eigenvectors form a complete basis of the state space.

When the system is prepared in a state $|\psi(t)\rangle$ at time t, the expectation value for an observable $\widehat{A}$ is given by $\langle\psi(t)|\widehat{A}|\psi(t)\rangle$. It should be reminded that the measurement of a physical quantity seldom gives its expected value since, according to the measurement postulate (5.1.3), the result can only be one of the eigenvalues of the operator $\widehat{A}$. However, since the expectation value depends on the quantum state considered, if the state is time dependent, it will also be the case for $\langle\psi(t)|\widehat{A}|\psi(t)\rangle$. The question then arises of the temporal evolution of such a quantity. Consider the time derivative:

$$\frac{d}{dt}(\langle\psi(t)|\widehat{A}|\psi(t)\rangle) = \left(\frac{d}{dt}\langle\psi(t)|\right)\widehat{A}|\psi(t)\rangle + \langle\psi(t)|\left(\frac{\partial}{\partial t}\widehat{A}\right)|\psi(t)\rangle + \langle\psi(t)|\widehat{A}\left(\frac{d}{dt}|\psi(t)\rangle\right)$$

If the system is described by a Hamiltonian $\widehat{H}$, Schrödinger's equation given in Section 5.1.6 and its adjoint formulation: $-i\hbar\partial_t(\langle\psi(t)|) = \langle\psi(t)|\widehat{H}$ can be used to determine the state evolution:

$$\begin{aligned}\frac{d}{dt}(\langle\psi(t)|\widehat{A}|\psi(t)\rangle) &= \langle\psi(t)|\left(\frac{\partial}{\partial t}\widehat{A}\right)|\psi(t)\rangle \\ &\quad + \langle\psi(t)|\left(\frac{-\widehat{H}}{i\hbar}\right)\widehat{A}|\psi(t)\rangle + \langle\psi(t)|\widehat{A}\left(\frac{\widehat{H}}{i\hbar}\right)|\psi(t)\rangle \\ &= \langle\psi(t)|\left(\frac{\partial}{\partial t}\widehat{A} + \frac{1}{i\hbar}[\widehat{A},\widehat{H}]\right)|\psi(t)\rangle \end{aligned} \tag{7.1}$$

which gives the following important result:

$$\frac{d}{dt}(\langle\psi(t)|\widehat{A}|\psi(t)\rangle) = \langle\psi(t)|\frac{\partial}{\partial t}\widehat{A}|\psi(t)\rangle + \langle\psi(t)|\frac{1}{i\hbar}[\widehat{A},\widehat{H}]|\psi(t)\rangle \tag{7.2}$$

This is known as *Ehrenfest's theorem* as it was proposed by P. Erhenfest (Fig. 7.1) in 1927 [30]. If the observable does not depend explicitly on time, then the first term on the right-hand side of the equality vanishes and

$$\frac{d}{dt}(\langle\psi(t)|\widehat{A}|\psi(t)\rangle) = \langle\psi(t)|\frac{1}{i\hbar}[\widehat{A},\widehat{H}]|\psi(t)\rangle \tag{7.3}$$

As an example of such an observable, consider the momentum. For a system subjected to a potential that only depends on the position (not the time), according to Ehrenfest's theorem, the derivative of the expected value of the momentum along one particular axis, in a given quantum state $|\psi(t)\rangle$, is

$$\begin{aligned}\frac{d}{dt}(\langle\psi(t)|\widehat{p}_x|\psi(t)\rangle) &= \langle\psi(t)|\frac{1}{i\hbar}[\widehat{p}_x,\widehat{H}]|\psi(t)\rangle \\ &= \langle\psi(t)|\frac{1}{i\hbar}[\widehat{p}_x,\widehat{V}(\boldsymbol{r})]|\psi(t)\rangle \\ &= \langle\psi(t)| - \partial_x\widehat{V}(\boldsymbol{r})|\psi(t)\rangle \end{aligned} \tag{7.4}$$

Fig. 7.1. Paul Ehrenfest (1880–1933). Ehrenfest's teachers were Boltzmann (in Vienna), Hilbert and Klein (in Göttingen). Lorentz (in Leiden) was his PhD advisor. Under the condition that he would read Tolstoy and never smoke tobacco, he married Tatyana Alexeyevna Afanasyeva, also a physicist, with whom he had four children (Vassili was afflicted with Down's syndrome) and collaborated in particular on the classic book *The Conceptual Foundations of the Statistical Approach in Mechanics* [31]. Erhenfest was close to Einstein and they played sonatas (piano and violin) on every occasion. Einstein wrote "he was not only the best teacher in our profession, I have ever known; he was also passionately preoccupied with the development and destiny of men, especially his students. Unfortunately, the accolades of his students and colleagues were not enough to overcome his deep-rooted sense of inferiority and insecurity". Indeed, Amsterdam was the scene of a tragic conclusion when Paul Ehrenfest took Vassili's and his own life. Credits: reprinted with kind permission of Austrian Central Library for Physics.

where $[\widehat{p}_x, \widehat{p}^2] = 0$ and (5.32) were used. The exact same result can readily be derived for any direction so that a compact way to write this result is

$$\frac{d}{dt}\left(\langle\psi(t)|\widehat{\boldsymbol{p}}|\psi(t)\rangle\right) = \langle\psi(t)| - \vec{\nabla}_r\widehat{V}(\boldsymbol{r})|\psi(t)\rangle$$

This equality is thus the quantum equivalent of Hamilton's equation of classical physics (6.26b). It is possible to define the gradient of the potential operator as the force operator: $\widehat{F}(\boldsymbol{r}) = -\vec{\nabla}\widehat{V}(\boldsymbol{r})$ so that

$$\frac{d}{dt}\left(\langle\psi(t)|\widehat{p}|\psi(t)\rangle\right) = \langle\psi(t)|\widehat{F}(\boldsymbol{r})|\psi(t)\rangle \tag{7.5}$$

Obviously, this expression takes us back to classical mechanics in its similarity to Newton's second law of motion. Does this result imply that classical physics is not dead and its validity persists for averaged quantities, when a large number of measurements are performed on a series of identical systems, prepared in identical quantum states? Note that (7.5) still contains a quantum subtlety. The time derivative of the momentum expected value is *not* equal to the force at the expected position. What is involved here is the expected value for the force. Thus, there is still no sign of a trajectory in the classical sense. This subtlety diminishes as the force's expectation value and the force at the position's expectation value become similar. This is made possible if the extent of the wavefunction is small compared to the force's spatial variations. By considering just one dimension, for the sake of simplicity, assume that the wave packet, hence the probability density, has a sharp peak around position x_0 with a characteristic width δx. This means that there is a maximal probability of finding the particle in the immediate vicinity of x_0. The expectation value $\langle x \rangle$ is then not very different from x_0. Suppose further that the force hardly varies in this region where there is a significant probability of presence. It can then be written:

$$\langle F(x) \rangle = \int \rho(x) F(x) dx \approx \int_{\delta x} \rho(x) F(x) dx \tag{7.6}$$

$$= F(x_0) \int_{\delta x} \rho(x) dx$$

$$\approx F(x_0) \approx F(\langle x \rangle) \tag{7.7}$$

Therefore, if the force hardly varies in the region where the particle can be found, the two types of force estimates are equivalent. Key quantities that need to be considered are the wave packet extension and the characteristic length over which the force significantly changes. This consideration suggests a first criterion to decide whether a quantum treatment of the problem is required or if a classical approach would be sufficient. Strictly speaking, classical mechanics would then be obtained from Ehrenfest's theorem in the limiting case of a probability density expressed as a Dirac function so that the particle would be perfectly localized. However, given Heisenberg's inequality, we know that it would create a momentum issue. Therefore, the classical limit needs to be further investigated from the wavefunction aspect. This is the aim of the next section.

Question 7.1: **From Ehrenfest to Hamilton.** Show that Ehrenfest's theorem applied to the position observable yields the quantum equivalent of (6.26a) when the potential only depends on the position.

Answer: We use the same procedure for $\widehat{x}$ as with the momentum observable

$$\begin{aligned}\frac{d}{dt}\left(\langle\psi(t)|\widehat{x}|\psi(t)\rangle\right) &= \langle\psi(t)|\frac{1}{i\hbar}[\widehat{x},\widehat{H}]|\psi(t)\rangle = \frac{1}{i\hbar}\langle\psi(t)|[\widehat{x},\frac{\widehat{p}^2}{2m}]|\psi(t)\rangle \\ &= \frac{1}{2im\hbar}\langle\psi(t)|\widehat{p}[\widehat{x},\widehat{p}]+[\widehat{x},\widehat{p}]\widehat{p}|\psi(t)\rangle \\ &= \frac{1}{m}\langle\psi(t)|\widehat{p}|\psi(t)\rangle\end{aligned}$$

Question 7.2: **Virial theorem.** In classical physics, the virial theorem provides a convenient connection between the kinetic and the potential energies of a given system. In the quantum world, these quantities are described by operators and such a connection should be sought for their expectation values in a quantum state.

In this question, a unique particle in a one-dimensional (1D) world is considered. The Hamiltonian is written as

$$\widehat{H} = \frac{\widehat{p}^2}{2m} + \widehat{V}(x)$$

where $\widehat{x}$ and $\widehat{p}$ are the position and momentum operators, respectively.

1. Consider an operator that is defined by $\widehat{xp}$. Show that the time derivative of the expectation value for $\widehat{xp}$, in a quantum state $|\psi\rangle$, is proportional to the expectation value of the commutator $[\widehat{xp},\widehat{H}]$ in the same quantum state. It is advisable to explicitly make use of $|\psi\rangle$.

2. Assume the potential can be written as $\widehat{V}(x) = \alpha\widehat{x}^n$, where α is a (real) constant and n a positive or negative integer. Show the virial quantum theorem
$$\frac{d}{dt}\langle\psi|\widehat{xp}|\psi\rangle = 2\langle\psi|\frac{\widehat{p}^2}{2m}|\psi\rangle - n\alpha\langle\psi|\widehat{x}^n|\psi\rangle$$

3. Consider any potential of the form $\widehat{V}(x) = \sum_{n=0}^{\infty}\alpha_n\widehat{x}^n$, where α_n are real numbers and show that, when the state is represented by an eigenket of the Hamiltonian $|\phi\rangle$:
$$2\langle\phi|\frac{\widehat{p}^2}{2m}|\phi\rangle = \langle\phi|\widehat{x}\frac{\partial\widehat{V}}{\partial x}|\phi\rangle$$
The virial theorem is well known in classical physics and has a wide range of applications from thermodynamics to astrophysics. Applications of this relation to quantum and statistical physics will be given in Volume 2.

Answer:

1. Using Ehrenfest's theorem:

$$\frac{d}{dt}\langle\psi|\widehat{x}\widehat{p}|\psi\rangle = \frac{1}{i\hbar}\langle\psi|[\widehat{x}\widehat{p}, \widehat{H}]|\psi\rangle + \langle\psi|\frac{\partial(\widehat{x}\widehat{p})}{\partial t}|\psi\rangle$$

Since $\widehat{x}\widehat{p}$ does not explicitly depend on time, it becomes

$$\frac{d}{dt}\langle\psi|\widehat{x}\widehat{p}|\psi\rangle = \frac{1}{i\hbar}\langle\psi|[\widehat{x}\widehat{p}, \widehat{H}]|\psi\rangle$$

2. Consider the commutator $[\widehat{x}\widehat{p}, \widehat{H}] = [\widehat{x}\widehat{p}, \frac{\widehat{p}^2}{2m}] + [\widehat{x}\widehat{p}, \alpha\widehat{x}^n]$.
Once the Hamiltonian has been expressed in terms of the kinetic and potential operators

$$[\widehat{x}\widehat{p}, \widehat{H}] = \left[\widehat{x}, \frac{\widehat{p}^2}{2m}\right]\widehat{p} + \widehat{x}\,[\widehat{p}, \alpha\widehat{x}^n] = \frac{1}{2m}[\widehat{x}, \widehat{p}^2]\widehat{p} + \frac{\hbar}{i}n\alpha\widehat{x}^n$$

Since $[\widehat{x}, \widehat{p}^2] = [\widehat{x}, \widehat{p}]\widehat{p} + \widehat{p}[\widehat{x}, \widehat{p}] = 2i\hbar\widehat{p}$, it is then found $[\widehat{x}\widehat{p}, \widehat{H}] = 2i\hbar\frac{1}{2m}\widehat{p}^2 + \frac{\hbar}{i}n\alpha\widehat{x}^n$.
It immediately gives $\frac{d}{dt}\langle\psi|\widehat{x}\widehat{p}|\psi\rangle = 2\langle\psi|\frac{\widehat{p}^2}{2m}|\psi\rangle - n\alpha\langle\psi|\widehat{x}^n|\psi\rangle$.

3. This is a two-stage process. Firstly, examine what happens if the potential is expressed as a power series. All operations are linear and the previous question operator can be written:

$$[\widehat{x}\widehat{p}, \widehat{H}] = \left[\widehat{x}, \frac{\widehat{p}^2}{2m}\right]\widehat{p} + \sum_n \alpha_n\widehat{x}\,[\widehat{p}, x^n] = \frac{1}{2m}[\widehat{x}, \widehat{p}^2]\widehat{p} + \frac{\hbar}{i}\sum_n n\alpha_n\widehat{x}^n$$

$$= 2i\hbar\frac{1}{2m}\widehat{p}^2 + \frac{\hbar}{i}\widehat{x}\frac{\partial\widehat{V}(x)}{\partial x}$$

The quantum virial theorem is then found as

$$\frac{d}{dt}\langle\psi|\widehat{x}\widehat{p}|\psi\rangle = 2\langle\psi|\frac{\widehat{p}^2}{2m}|\psi\rangle - \langle\psi|\widehat{x}\frac{\partial\widehat{V}(x)}{\partial x}|\psi\rangle$$

Since the Hamiltonian eigenstates' time evolution depends on the associated energy:

$$|\phi_n(t)\rangle = e^{-i\varepsilon_n t/\hbar}|\phi_n(0)\rangle$$

Operator's expectation values do not explicitly depend on time and in the particular case of operator $\widehat{x}\widehat{p}$, we have

$$\frac{\partial}{\partial t}\langle\phi_n|\widehat{x}\widehat{p}|\phi_n\rangle = \frac{\partial}{\partial t}\langle\phi_n(0)|\widehat{x}\widehat{p}|\phi_n(0)\rangle = 0$$

Consequently, it yields $0 = 2\langle\phi|\frac{\widehat{p}^2}{2m}|\phi\rangle - \langle\phi|\widehat{x}\frac{\partial\widehat{V}(x)}{\partial x}|\phi\rangle$.
A relation between the kinetic and potential energies expectation values (in an eigenstate of $\widehat{H}$) is then found: $2\langle\phi|\frac{\widehat{p}^2}{2m}|\phi\rangle = \langle\phi|\widehat{x}\frac{\partial\widehat{V}(x)}{\partial x}|\phi\rangle$.

Question 7.3: **Hellmann–Feynman theorem.** Assume that an arbitrary Hamiltonian is a function of a parameter γ. Obviously, its eigenspectrum is expected to also vary with γ. Let $|\phi_i(\gamma)\rangle$ be any of its eigenkets and $\varepsilon_i(\gamma)$ the associated eigenenergy. Show that (if the differential exists): $\frac{\partial\varepsilon_i}{\partial\gamma} = \langle\phi_i|\frac{\partial\widehat{H}(\gamma)}{\partial\gamma}|\phi_i\rangle$.

Answer: The eigenenergy can be differentiated according to

$$\frac{\partial \varepsilon_i}{\partial \gamma} = \frac{\partial}{\partial \gamma}\langle \phi_i|\widehat{H}(\gamma)|\phi_i\rangle = \left(\frac{\partial}{\partial \gamma}\langle \phi_i|\right)\widehat{H}(\gamma)|\phi_i\rangle + \langle \phi_i|\widehat{H}(\gamma)\left(\frac{\partial}{\partial \gamma}|\phi_i\rangle\right) + \langle \phi_i|\frac{\partial \widehat{H}(\gamma)}{\partial \gamma}|\phi_i\rangle$$

$$= \varepsilon_i \frac{\partial \langle \phi_i|\phi_i\rangle}{\partial \gamma} + \langle \phi_i|\frac{\partial \widehat{H}(\gamma)}{\partial \gamma}|\phi_i\rangle$$

Obviously, since all eigenkets are normalized $\langle \phi_i|\phi_i\rangle = 1$, the first term of the last line vanishes. This result, now known as the "Hellmann–Feynman" theorem was rediscovered several times. One early demonstration dates back to 1922 and is attributed to Wolfgang Pauli. As mentioned in the 1939 paper Feynman (Fig. 7.3) wrote (upon J. C. Slater's suggestion), when he was still an undergraduate student [36], this result is useful for computing forces on nuclei in a molecule from the mere knowledge of the electron distribution, in much the same way as in a pure classical electrostatic problem. A more mathematical use will be given in Volume 2.

7.2. Transition from Quantum to Classical Hamilton–Jacobi's Equation

Planck's law for black-body radiation was obtained by introducing the $\hbar$ constant. It was clear from the beginning that Rayleigh–Jeans' power spectrum was to be recovered from Planck's result in the continuous limit, thus by setting $\hbar \approx 0$. We will study to what extent Planck's constant is enough to define the perimeter of action for quantum theory. The conditions are thus evaluated for Schrödinger's equation to reconnect to a trajectory in phase space. Under such conditions, can we infer what mathematical object should progressively substitute for the wavefunction when quantum effects can be neglected? Obviously, if a classical trajectory becomes a pertinent concept, the extension of probability density in position space should narrow down to a perfect localization. Therefore, it becomes essential to carefully follow the density's behaviour as the system approaches the classical world. In that respect, $\psi(\boldsymbol{r}, t)$ being a complex-valued function, it can be written from its associated probability density $\rho(\boldsymbol{r}, t)$ and a phase factor as[1]

$$\psi(\boldsymbol{r}, t) = \sqrt{\rho(\boldsymbol{r}, t)}\; e^{iS(\boldsymbol{r},t)/\hbar} \tag{7.8}$$

[1] The line of thought which is developed here extends that which was used to detail the mechanism of WKB approximation in Section 4.8.

where $S(\boldsymbol{r}, t)$ is a real-valued function. In what follows, for the sake of simplicity, we omit to specify the spatial and temporal dependencies of ρ and S.

Since $-\frac{\hbar^2}{2m}\nabla_r^2$ is the kinetic energy operator (see Section 3.3.4), the kinetic energy density can be written as

$$\begin{aligned}
&\psi^*(\boldsymbol{r},t)\left(-\frac{\hbar^2}{2m}\nabla_r^2\psi(\boldsymbol{r},t)\right)\\
&\quad = -\frac{\hbar^2}{2m}\sqrt{\rho}\, e^{-iS/\hbar}\vec{\nabla}_r \cdot \left(\frac{1}{2}\frac{\vec{\nabla}_r\rho}{\sqrt{\rho}}\, e^{iS/\hbar} + \frac{i}{\hbar}\sqrt{\rho}\vec{\nabla}_r S\, e^{iS/\hbar}\right)\\
&\quad = -\frac{\hbar^2}{2m}\left(\frac{1}{2}\nabla^2\rho - \frac{1}{4}\frac{(\vec{\nabla}_r\rho)^2}{\rho} + \frac{i}{\hbar}(\rho\nabla_r^2 S + \vec{\nabla}\rho\cdot\vec{\nabla}_r S) - \frac{1}{\hbar^2}\rho(\vec{\nabla}_r S)^2\right)
\end{aligned} \tag{7.9}$$

The total energy density is

$$\psi^*(\boldsymbol{r},t)\left(i\hbar\frac{\partial}{\partial t}\psi(\boldsymbol{r},t)\right) = \frac{i\hbar}{2}\frac{\partial\rho}{\partial t} - \rho\frac{\partial S}{\partial t} \tag{7.10}$$

or, introducing $U(\boldsymbol{r})$ as the potential, with the contribution of the kinetic energy density (7.9):

$$-\frac{\hbar^2}{2m}\left(\frac{1}{2}\nabla_r^2\rho - \frac{1}{4}\frac{(\vec{\nabla}_r\rho)^2}{\rho} + \frac{i}{\hbar}(\rho\nabla^2 S + \vec{\nabla}_r\rho\cdot\vec{\nabla}S) - \frac{1}{\hbar^2}\rho(\vec{\nabla}_r S)^2\right) + U(\boldsymbol{r})\rho \tag{7.11}$$

The total energy density bears two equal expressions:

$$\psi^*(\boldsymbol{r},t)\left[\left(\frac{-\hbar^2}{2m}\right)\nabla_r^2\psi(\boldsymbol{r},t) + U(\boldsymbol{r})\psi(\boldsymbol{r},t)\right] = \psi^*(\boldsymbol{r},t)\left(i\hbar\frac{\partial}{\partial t}\psi(\boldsymbol{r},t)\right) \tag{7.12}$$

but also, according to expressions (7.10) and (7.11)

$$\begin{aligned}
&\frac{i\hbar}{2}\frac{\partial\rho}{\partial t} - \rho\frac{\partial S}{\partial t}\\
&\quad = -\frac{\hbar^2}{2m}\left(\frac{1}{2}\nabla_r^2\rho - \frac{1}{4}\frac{(\vec{\nabla}_r\rho)^2}{\rho} + \frac{i}{\hbar}(\rho\nabla^2 S + \vec{\nabla}_r\rho\cdot\vec{\nabla}S) - \frac{1}{\hbar^2}\rho(\vec{\nabla}_r S)^2\right)\\
&\qquad + U(\boldsymbol{r})\rho
\end{aligned}$$

If the real and imaginary components are separated, two equalities are obtained[2]:

$$\frac{\partial \rho}{\partial t} = -\vec{\nabla}_r \cdot \left(\rho \frac{\vec{\nabla}_r S}{m}\right) \tag{7.13a}$$

$$\frac{\partial S}{\partial t} = \frac{\hbar^2}{2m}\frac{\nabla_r^2 \sqrt{\rho}}{\sqrt{\rho}} - \frac{1}{2}m\left(\frac{\vec{\nabla}_r S}{m}\right)^2 - U \tag{7.13b}$$

If the current probability density is set to $\boldsymbol{j}(\boldsymbol{r}, t) = \rho \frac{\vec{\nabla}_r S}{m}$, Eq. (7.13a) establishes its continuity.[3] It should be noted that this equation does not invoke Planck's constant. Consequently, setting $\hbar \to 0$ will not fully guarantee a classical description for the system unless the density is mechanically subjected to shrink to a single point. The description would then be a mere probability-based classical description, not a classical and deterministic system in Newton's sense.

Question 7.4: **From action to momentum.** Check that $\boldsymbol{j}(\boldsymbol{r}, t) = \rho \frac{\vec{\nabla}_r S}{m}$ is indeed equivalent to (3.48). In the case of a free particle, give a physical interpretation of $\vec{\nabla}_r S$.

Answer: Starting from (3.48) and expressing the wavefunction by (7.8)

$$\begin{aligned}\boldsymbol{j}(\boldsymbol{r}, t) &= \frac{1}{m}\mathrm{Re}\left\{\sqrt{\rho}\, e^{-iS/\hbar}\frac{\hbar}{i}\vec{\nabla}_r[\sqrt{\rho}\, e^{iS/\hbar}]\right\} \\ &= \frac{1}{m}\mathrm{Re}\left\{\sqrt{\rho}\, e^{-iS/\hbar}\frac{\hbar}{i}e^{iS/\hbar}[\,\vec{\nabla}_r\sqrt{\rho} + \frac{i}{\hbar}\sqrt{\rho}\vec{\nabla}_r S]\right\}\end{aligned}$$

The only real-valued term between the braces does indeed yield $\boldsymbol{j}(\boldsymbol{r}, t) = \frac{1}{m}\rho\vec{\nabla}_r S(\boldsymbol{r}, t)$.
When a free particle is considered, the phase of the plane wave is given by

$$iS(\boldsymbol{r}, t)/\hbar = i\boldsymbol{k}\cdot\boldsymbol{r} - \omega t$$

so that $S(\boldsymbol{r}, t) = \boldsymbol{p}\cdot\boldsymbol{r} - \varepsilon t$.
Consequently, $\vec{\nabla}_r S = \boldsymbol{p}$. Note that this equality only holds for a free particle.

Question 7.5: Give the expression for the electric current of a charged particle q in the presence of an electromagnetic field represented by the vector potential $\boldsymbol{A}$.

[2]The potential is an observable. Therefore, it is real valued.

[3]Since only $\vec{\nabla}_r S$ intervenes, it is understood that the wavefunction phase is determined as a constant.

Answer: The answer to the previous question is repeated, replacing the previous calculation $\frac{\hbar}{i}\vec{\nabla}_r$ by $\frac{\hbar}{i}\vec{\nabla}_r - q\widehat{\boldsymbol{A}}$.

Equation (7.13b), obtained after equating the real parts of (7.12), is now considered. If $\hbar$ is unconditionally set to 0, it becomes

$$\frac{\partial S}{\partial t} = -\frac{1}{2}m\left(\frac{\vec{\nabla}_r S}{m}\right)^2 - U(\boldsymbol{r}) \tag{7.14}$$

Undeniably, this expression bears a close resemblance to (6.39). However, since $S/\hbar$ represents the phase of the wavefunction, it depends on the position (and possibly the time). Therefore, unless the particle is totally free, $\vec{\nabla}_r S$ represents a field of possible momentum, i.e. the momentum depends on the position, $\vec{\nabla}_r S = \boldsymbol{p}(\boldsymbol{r})$. Taking into consideration the fact that there is still a continuity equation that involves a probability density, it is not enough to make $\hbar$ disappear from Schrödinger's equation to fully recover Newton's description; while quantization has vanished from the problem, the probabilistic description persists (see [63]). It simply becomes possible to express both expectation values using the density as $\langle \boldsymbol{r}\rangle = \int \boldsymbol{r}\rho(\boldsymbol{r})d^3\boldsymbol{r}$ and $\langle \boldsymbol{p}\rangle = \int(\vec{\nabla}_r S)\rho(\boldsymbol{r})d^3\boldsymbol{r}$. A classical deterministic picture is recovered on condition that the density takes a form arbitrarily close to $\rho(\boldsymbol{r}) = \delta(\boldsymbol{r} - \boldsymbol{r}_c(t))$ where $\boldsymbol{r}_c(t)$ corresponds to the classical position at time t.

From a pragmatic point of view, $\hbar$ is a constant and will always be present in Eq. (7.13b). The first term, the only one involving $\hbar$ explicitly, is usually known as the *quantum potential.* If one wishes to abandon a quantum description, it becomes necessary to evaluate under what condition it is justified to neglect this quantum potential, i.e. when it is legitimate to write:

$$\frac{\hbar^2}{2m}\frac{\nabla_r^2\sqrt{\rho}}{\sqrt{\rho}} \ll \frac{1}{2}m\left(\frac{\vec{\nabla}_r S}{m}\right)^2$$

It is under this condition that quantum effects can be discarded, simply because they would not be measurable. It is then useful to establish, from characteristic orders of magnitude for the quantities at stake, i.e. ρ and S, when

$$\hbar^2 \ll (\vec{\nabla}_r S)^2 \frac{\sqrt{\rho}}{\nabla_r^2 \sqrt{\rho}}$$

The quantity on the right obviously depends on the position. However, if in a finite region of space this relationship is not observed, the behaviour of the system will not be properly described by the classical Hamilton–Jacobi equation (7.14). An estimate in terms of orders of magnitude is often sufficient to form an opinion. For example, for a given system, $\nabla^2\sqrt{\rho}/\sqrt{\rho}$ is even larger when the local curvature of the probability density is high. This is locally a measurement of the spread of this density. The width of the distribution, which can be regarded as a characteristic size of the system, is then given by $\sqrt{\rho}/(\nabla_r^2\sqrt{\rho})$. Thus, the classical description is justified if $\vec{\nabla}_r S\sqrt{\sqrt{\rho}/(\nabla_r^2\sqrt{\rho})}$, which is the product of a momentum by a characteristic size, is large compared to $\hbar$.

Thus, we see that the transition between classical and quantum descriptions is not merely carried out by setting $\hbar \to 0$, but by assuming that any characteristic quantity of the system expressed in the same units (an "action", [E][T] or $[M][L]^2[T]^{-1}$) will be significantly larger than $\hbar$. This process of description change can be compared to that in which the speed of light tends to infinity to ensure the passage from electromagnetic optics to geometrical optics.

So we have here an important criterion for deciding whether the use of quantum physics is relevant, rather than a more classical approach for a given physical problem. If the characteristic action of the system is large compared to $\hbar$ then a classical treatment may suffice. *If the action is comparable to $\hbar$, the use of quantum rules must prevail.*

7.3. Particle Trajectories or Wave Interference?

7.3.1. *Large quantum numbers and Bohr's correspondence principle*

It will become apparent in many practical cases of Volume 2 that classical behaviour is often recovered at the limit of large quantum numbers. It is already possible to observe such a phenomenon if a classical ball bouncing between two perfectly rigid walls separated by a is considered from the infinite quantum well perspective. Whatever the energy stored in the system, a classical physics description always gives a uniform $1/a$ probability distribution for the ball. If the spatial resolution of the instrument used to observe the particle is δ, the probability for the particle to be captured is δ/a. On the other hand, the quantum solution is constructed from two counter-propagating plane waves with identical modulus wave-vectors $k_n = n\frac{\pi}{a}$, where n is a positive integer. The probability density results from the interference of these waves and yields $|\sqrt{2/a}\sin n(\frac{\pi}{a}x)|^2$. Obviously, this is not a uniform density and again the interference pattern is a clear sign of quantum behaviour. The probability of detection with a resolution δ is $\int_{-\delta/2}^{\delta/2}(2/a)\sin^2(nx\pi/a)dx = \delta/a - 2/(n\pi)\sin(n\pi\delta/a)$. As the quantum number n increases, the spatial period of oscillations reduces, and at the limit of large quantum numbers, the oscillating term is averaged out and the probability reaches the classical value δ/a. The fact that a classical behaviour is usually recovered from the quantum values at the large quantum numbers limit has been generalized since its introduction by Niels Bohr and is now quoted as the *correspondence principle* [9]. While this principle is still a matter of scientific (and philosophical) debate it is clear that the large quantum numbers criterion is not sufficient to shift the problem from the quantum to the classical world. As emphasized by Liboff [69] the uniformity of probability density is not enough to ensure a secured junction with the classical picture. In the quantum well problem, the energy spectrum, hence the electromagnetic absorption or emission spectrum, is driven by the value of $\hbar^2/(2ma^2)$. For large quantum numbers, the energy separation between consecutive levels is then given by $n\pi^2\hbar^2/(ma^2)$ which obviously does not match any continuous law. Therefore, it is difficult to blindly rely on the large quantum numbers limit to claim the disappearance of quantum effects and a safe recourse to a classical formalism.

7.3.2. *The noticeable interferences criterion*

Despite the failure of the continuous probability criterion seen above, which essentially refers to bound states, it might be advisable to consider what happens to scattering cases, i.e. those for which the wavefunction does not vanish at infinities. Recall that one obvious manifestation of quantum behaviour is the appearance of an interference pattern in the Young's double slit experiment for massive particles. A wave physics approach explains that for a spacing of d between the slits, the angular separation between fringes[4] is about $\Delta\theta \approx \pi/(k\times d)$ so, in terms of momentum, $\Delta\theta = \pi\hbar/(p\times d)$ (see Fig. 7.2).

Now consider the case of a ping-pong ball. Assume an approximate mass of one gram and a speed of about $10\,\mathrm{m\,s^{-1}}$. This yields a momentum of the order of $10^{-2}\,\mathrm{kg\,m\,s^{-1}}$. Given the size of the ball, we can hardly take a slit spacing smaller than $d \approx 1\,\mathrm{cm}$. We then find $\Delta\theta \approx 10^{-30}$. It would thus require positioning the screen at an astronomical distance to observe any interference effect! Another way to look at it is to estimate what speed (or momentum) would reduce the action $p \times d$ significantly enough to make it comparable to $\hbar$. The speed should then be lower than $10^{-28}\,\mathrm{m\,s^{-1}}$ which roughly corresponds to a cooling of $10^{-30}\,\mathrm{K}$. Not to mention the serious difficulties inherent to the duration of such an experiment when it takes more than 10^{20} years for the ball to travel one meter. Obviously, quantum effects due to the table tennis wave-ball duality are not readily observable and a conventional treatment is amply justified.

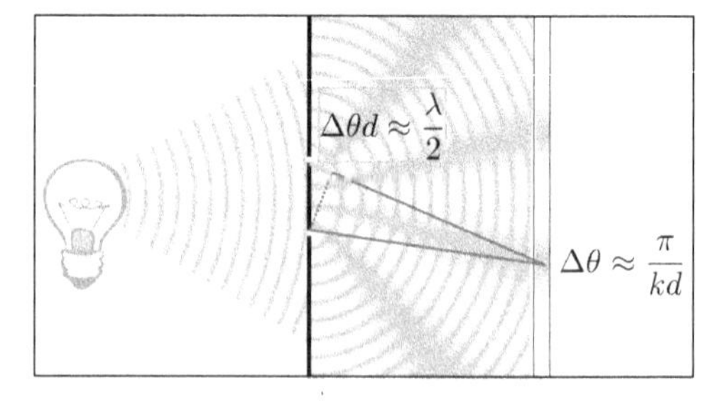

Fig. 7.2. Interference condition: for an interference effect to be observable, it is necessary that the distance separating the point of maximal intensity and the first cancellation (or lower intensity) point is experimentally resolved. The angular distance is given by the condition on the difference of optical paths: $d\Delta\theta \approx \lambda/2$.

[4]For a screen placed at distance L, the fringes separation is $L\Delta\theta$.

Question 7.6: **Gaussian distribution for a (very) light particle.** Consider a quantum particle whose mass is of the order of a hundredth of a microgram and whose position probability density can roughly be represented by the Gaussian:

$$\rho(x, t = 0) = Ae^{-\alpha x^2}$$

1. Find the mean square position. Derive a characteristic size of the wave packet and compare it to $-\sqrt{\rho}/\nabla^2\sqrt{\rho}$ in a mean position.
2. It is assumed that $\alpha \approx 10^{-6}\,\text{Å}^{-2}$. Show that the first term in (7.13b) is negligible if the spatial variations of S are of the order of $10^{-20}\,\text{kg}\,\text{m}\,\text{s}^{-1}$. Give a characteristic speed of the system in this case.

Answer: The wavefunction is written:

$$\psi(x, t) = \sqrt{\rho(x, t)}e^{iS(x,t)/\hbar}$$

It needs to be normalized:

$$\int \rho(x, t)dx = 1 \text{ if } A = \sqrt{\frac{\alpha}{\pi}}$$

1. The mean square position is then

$$\langle x^2 \rangle = \sqrt{\frac{\alpha}{\pi}} \int e^{-\alpha x^2} x^2 dx = \frac{1}{2\alpha}$$

Since the mean position is $x = 0$, the extent of the density is of the order of $\sqrt{1/(2\alpha)}$. In $x = 0$:

$$-\frac{\sqrt{\rho}}{\nabla_r^2\sqrt{\rho}} = [e^{-\alpha x^2/2}(\alpha e^{-\alpha x^2/2}(1 - x^2\alpha))^{-1}]_{x=0} = \frac{1}{\alpha}$$

Thus, in the Gaussian case $-\frac{\sqrt{\rho}}{\nabla_r^2\sqrt{\rho}}$ also measures the extent of the density. The importance of the first term in (7.13b) is conditioned by the spatial dimension of the position probability density.

2. If $\alpha \approx 10^{-6}\,\text{Å}^{-2}$, then the spatial extent is about $10^3\,\text{Å}$. The ratio between the first term of (7.13b) and the second term is then estimated by

$$\hbar^{-2}(\nabla S)^2 \frac{\sqrt{\rho}}{\nabla_r^2\sqrt{\rho}} \approx \hbar^{-2}(\nabla_r S)^2 \langle x^2 \rangle \approx \hbar^{-2}(\nabla S)^2 \frac{1}{\alpha}$$

Expression (7.13b) becomes a classical physics equation if $(\nabla_r S)^2 \langle x^2 \rangle \gg \hbar^2$. The system can thus be considered from a classical physics perspective if

$$\nabla_r S \gg \frac{\hbar}{\sqrt{\langle x^2 \rangle}} \approx \sqrt{\alpha} 10^{-34} \approx 10^{-37}\,\text{kg}\,\text{m}\,\text{s}^{-1}$$

Obviously, the 10^{-20} kg.m.s^{-1} momentum belongs to the classical region. Since, in classical physics $\nabla_r S$ is the momentum, and here $m \approx 10^{-11}$ kg, the limit speed is $v \approx 10^{-9}\,\text{m}\,\text{s}^{-1}$. This justifies the wide application range of classical physics and the fact that quantum effects only appear when much lighter particles are considered or when very low temperatures can be reached.

7.3.3. *The propagator and the multiple paths of a quantum particle*

If the state of a particle is represented by $|\psi(t_1)\rangle$ and subjected to the Hamiltonian $\widehat{H}$, the evolution operator $\widehat{U}(t_1, t_2) = e^{-i\widehat{H}(t_2-t_1)}$ (see Question *5.21*) provides a way to predict what the state will be at a later time t_2: $|\psi(t_2)\rangle = \widehat{U}(t_1, t_2)|\psi(t_1)\rangle$. This relation applies to any state under any circumstance, providing that $\widehat{H}$ exists. To consider the problem from a classical physics point of view, it is then possible to study how this relation applies to the position space representation. If the particle is in an eigenstate of the position operator at time t_1, say $|\psi(t_1)\rangle = |x_1, t_1\rangle$, then the probability amplitude to find it at another position x_2 at time t_2 is: $K(x_2, t_2; x_1, t_1) = \langle x_2, t_2|\widehat{U}(t_1, t_2)|x_1, t_1\rangle$. Since this is an amplitude of probability for the particle to change its position with time, $K(x_2, t_2; x_1, t_1)$ is called a *propagator*.

When it comes to finding connections between the quantum and classical worlds, it is useful to study not only a jump between two remote positions in a given time span but also to build some kind of trajectory. This is done by breaking down each time step into very short durations. If $\Delta t = t_2 - t_1$ is partitioned into a large number N of short time increments $\delta t = \Delta t/N$, the propagator is now written:

$$
\begin{aligned}
K(x_2, t_2; x_1, t_1) &= \langle x_2, t_2|\widehat{U}\left(t_1 + (N-1)\delta t, t_2\right) \ldots \widehat{U}(t_1, t_1 + \delta t)|x_1, t_1\rangle \\
&= \langle x_2, t_2| \prod_{j=1}^{N} \widehat{U}(t_1 + (j-1)\delta t, t_1 + j\delta t)|x_1, t_1\rangle
\end{aligned}
$$

Each elemental evolution operator changes the particle to a different state δt later. If a trajectory is pursued, it is then necessary to envisage what the position should be at each of these changes. This is done by introducing the closure[5] $\mathbb{1} = \int |x^{(j)}\rangle\langle x^{(j)}|dx^{(j)}$ so that:

$$
\begin{aligned}
&K(x_2, t_2; x_1, t_1) \\
&\quad = \langle x_2, t_2| \prod_{j=1}^{N} \int |x^{(j)}\rangle\langle x^{(j)}|dx^{(j)} \widehat{U}(t_1 + (j-1)\delta t, t_1 + j\delta t)|x_1, t_1\rangle
\end{aligned}
\tag{7.15}
$$

[5]Here $x^{(j)}$ is a variable and refers to all possible position eigenstates while x_1 and x_2 are fixed values for the initial and final positions, respectively.

The transition amplitude between the two extreme positions at times t_1 and t_2 can then be seen as a succession of tiny propagations:

$$K(x_2, t_2; x_1, t_1)$$
$$= \int \cdots \int K(x_2, t_2; x^{(N)}, t_2 - \delta t) \ldots K(x^{(1)}, t_1 + \delta t; x_1, t_1) dx^{(1)} \ldots dx^{(N)}$$

This could very well be seen as the particle jumping from one position to another at each time increment. However, there is a dramatic difference with a classical trajectory. At each step here, the particle is offered the possibility to be anywhere in the position space. Hence, every possibility is explored. This means that during its journey from x_1 to x_2, the quantum particle probes every possible trajectory and the resulting probability amplitude is the sum of the probability amplitudes associated with all the possible paths: $K(x_2, t_2; x_1, t_1) = \sum_\Gamma K_\Gamma(x_2, t_2; x_1, t_1)$ where Γ is a possible path. Of course, if obstacles are positioned on its way, such as the two-slit mask in Young's interference experiment, the number of possible paths will be significantly reduced. However, the final result, given by the measurement probability $|K(x_2, t_2; x_1, t_1)|^2$, will be obtained from the sum over the allowed trajectories. In actual facts, time cannot be broken down in the way we proceeded and the correct way to carry out this calculation is using an integral over the paths. Nevertheless, the connection with a classical particle behaviour may then raise the following question: "How does this universal exploration by the quantum particle reduce to the single trajectory that is observed in classical physics?". This interrogation prompted R. P. Feynman [35] (Fig. 7.3) to propose his own *sum over paths* formulation of quantum mechanics (see [37, 68]).

Feynman postulated that a propagator exists for each path and can be written as $K_\Gamma(x_2, t_2; x_1, t_1) = Ae^{iS_\Gamma/\hbar}$ where A is the normalization constant and S_Γ is equivalent to $\mathcal{A}_\Gamma$, the *classical* action computed along the path under consideration: $S_\Gamma \equiv \mathcal{A}_\Gamma = \int_{t_1 \atop \text{path } \Gamma}^{t_2} \mathcal{L}(x, \dot{x}, t) dt$. The ratio between the classical action and $\hbar$ gives the phase angle contribution to the global propagator.

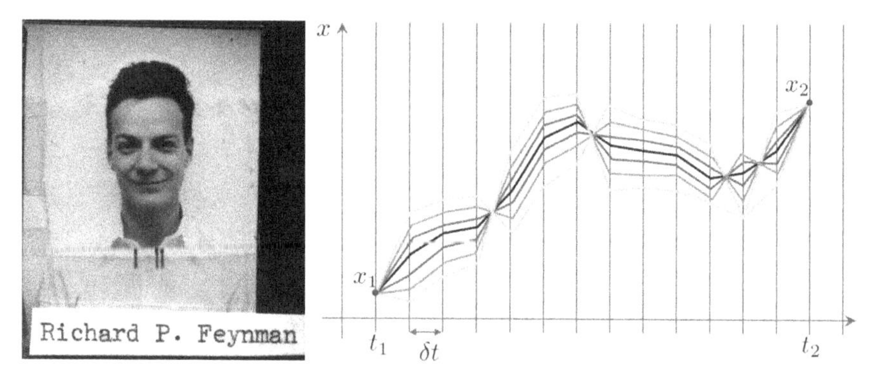

Fig. 7.3. Left: Richard Feynman won the Nobel Prize in 1965, jointly with J. Schwinger and S.-I. Tomonaga, for their work on quantum electrodynamics. This picture was taken for his security clearance badge while working (1942–1945) on the Manhattan Project at Los Alamos (Courtesy of Los Alamos National Laboratory). In 1986, as a member of the Rogers Commission, Feynman also proved that the Challenger shuttle explosion was created by the lack of resilience of the O-rings at low temperature. Right: There is a large variety of quantum paths that have to be summed upon. Each segment corresponds to a propagator value. The overall propagator for one path is characterized by its classical action. The lower the action (darker line), the closer to the classical trajectory. Photo credits: public domain.

Such a picture sheds new light on how a multiple trajectory scenario shrinks to a unique classical path. The action of a classical object is always a very large number compared to $\hbar$. Therefore, the phase angle in the sum over paths $K(x_2, t_2; x_1, t_1) = \sum_\Gamma Ae^{iS_\Gamma/\hbar}$ changes very rapidly from one path to another. The global result is then the sum of the phase factors that uniformly cover the full angle range between 0 and 2π and finally yields 0. The only contributions for which the action does not change significantly from one path to the next are those close to the path which bears the least action value. This is indeed the definition of the minimal value, the path around which the variation of action is zero to the first order. Therefore, in the classical case for which the action is large compared to $\hbar$, the only significant contributions are those infinitely close to the classical trajectory given by the least action principle.

Chapter 7: Nuts & Bolts

- *If the characteristic* **action** *of a system is not much greater than $\hbar$, the system must usually be described in quantum terms.*
- ***Ehrenfest's theorem*** *concerns the time evolution of an observable mean quantum value. If $\widehat{H}$ is the Hamiltonian, it states that $\frac{d}{dt}(\langle\widehat{A}\rangle_\psi) = \langle\frac{1}{i\hbar}[\widehat{A},\widehat{H}]\rangle_\psi$.*

Bibliography

[1] M. Abolins, R. L. Lander, W. A. W. Mehlhop, N. H. Xuong and P. M. Yager, Production of multimeson resonances by $\pi+$ p interaction and evidence for a π ω resonance, *Phys. Rev. Lett.* **11**(8) (1963) 381.

[2] S. K. Allison, Arthur Holly Compton 1892–1962, *Biographical Memoirs*, National Academy of Sciences, Vol. **38** (1965) pp. 81–110.

[3] R. A. Alpher, H. Bethe and G. Gamow, The origin of chemical elements, *Phys. Rev.* **73** (1948) 803.

[4] G. Arfken, H. Weber and F. E. Harris, *Mathematical Methods for Physicists*, 7th edn. (Elsevier, 2012).

[5] M. Arndt, O. Nairz, J. Vos-Andreae, C. Keller, G. van der Zouw and A. Zeilinger, Wave particle duality of C60 molecules, *Nature* **401** (1999) 680–682.

[6] L. Ballentine, *Quantum Mechanics: A Modern Development* (World Scientific Publishing Company, 1998).

[7] C. H. Bennett and G. Brassard, Quantum cryptography: Public key distribution and coin tossing, in *Proceedings of IEEE International Conference on Computers, Systems and Signal Processing*, New York (1984), pp. 175–179.

[8] G. R. Blumenthal and R. J. Gould, Bremsstrahlung, synchrotron radiation, and Compton scattering of high-energy electrons traversing dilute gases, *Rev. Mod. Phys.* **42**(2) (1970) 237–270.

[9] N. Bohr, *Niels Bohr — Nobel Lecture: The Structure of the Atom*, Nobel Lectures, Physics 1922–1941 (The Nobel Foundation, 1922).

[10] N. Bohr and J. A. Wheeler, The mechanism of nuclear fission, *Phys. Rev.* **56** (1939) 426–450.

[11] M. Born, H. Born and A. Einstein, *The Born–Einstein Letters. Correspondence between Albert Einstein and Max and Hedwig Born from 1916 to 1955 with Commentaries by Max Born* (MacMillan Press, 1971).

[12] H. Brandhorst, J. Hickey, H. Curtis and E. Ralph, Interim solar cell testing procedures for terrestrial applications, *Nasa Technical Memorandum*, NASA TM X-71771:1–19 (1975).
[13] B. H. Bransden and C. J. Joachain, *Quantum Mechanics* (Prentice-Hall, 2000).
[14] D. Branson, Continuity conditions on Schrödinger wave functions at discontinuities of the potential, *Amer. J. Phys.* **47**(11) (1979) 1000–1003.
[15] D. Bull, A. Krekeler, M. Alfeld, J. Dick and K. Janssens. An intrusive portrait by Goya, *Burlington Mag.* **CLIII** (2011) 668–673.
[16] P. Busch, On the energy-time uncertainty relation. Part I: Dynamical time and time indeterminacy, *Found. Phys.* **20**(1) (1990) 1–32.
[17] O. Carnal and J. Mlynek, Young's double-slit experiment with atoms: A simple atom interferometer, *Phys. Rev. Lett.* **66**(21) (1991) 2689–2692.
[18] N. Chaika, N. N. Orlova, V. N. Semenov, E. Yu. Postnova, S. A. Krasnikov, M. G. Lazarev, S. V. Chekmazov, V. Yu. Aristov, V. G. Glebovsky, S. I. Bozhko and I. V. Shvets, Fabrication of [001]-oriented tungsten tips for high resolution scanning tunneling microscopy, *Sci. Rep.* **4** (2014) 3742; DOI:10.1038/srep03742.
[19] C. E. Cleeton and N. H. Williams, Electromagnetic waves of 1.1 cm wavelength and the absorption spectrum of ammonia, *Phys. Rev.* **45** (1934) 234–237.
[20] A. H. Compton, A quantum theory of the scattering of X-rays by light elements, *Phys. Rev. B* **21** (1923) 483.
[21] C. Davisson and L. H. Germer. Diffraction of electrons by a crystal of nickel, *Phys. Rev.* **30**(6) (1927) 705–741.
[22] C. J. Davisson. Are electrons waves? *J. Franklin Inst.* **206**(5) (1928) 597–623.
[23] J. H. de Boer, Ueber die natur der farbzentren in alkalihalogenid-kristallen, *Recl. Trav. Chim. Pays-Bas* **56**(3) (1937) 301–309.
[24] L. de Broglie, Recherches sur la théorie des quanta, *Ann. Phys.* **10**(3) (1925) 22–128.
[25] P. Debye, Zerstreuung von rontgenstrahlen und quantentheorie, *Phys. Z.* **24** (1923) 161–166.
[26] D. M. Dennison and J. D. Hardy, The parallel type absorption bands of ammonia, *Phys. Rev.* **39**(6) (1932) 938–947.
[27] R. H. Dicke, P. J. E. Peebles and P. G. Rolland and D. T. Wilkinson, Cosmic black body radiation, *Astrophys. J.* **142** (1965) 414–419.
[28] G. F. D. Duff and D. Naylor, *Differential Equations of Applied Mathematics* (Wiley, 1966).
[29] P. Ewald (Ed.), *Fifty Years of X-ray Diffraction* (Springer, 1962).
[30] P. Ehrenfest, Bemerkung über die angenaherte Gültigkeit der klassischen Mechanik innerhalb der Quantenmechanik **45** (1927) 455–457.
[31] P. Ehrenfest and T. Ehrenfest, *The Conceptual Foundations of the Statistical Approach in Mechanics* (Dover Publications, 2015).
[32] A. Einstein, Zur elektrodynamik bewegter korper, *Ann. Phys.*, **10** (1905) 891–921.

[33] J. K. Feathers, Luminescence dating and modern human origins, *Evol. Anthr.* **5**(1) (1996) 25–36.

[34] R. M. Feenstra, J. A. Stroscio, J. Tersoff and A. P. Fein, Atom-selective imaging of the GaAs(110) surface, *Phys. Rev. Lett.* **58**(12) (1987) 1192–1195.

[35] R. P. Feynman, Space-time approach to non-relativistic quantum mechanics, *Rev. Mod. Phys.* **20** (1948) 367–387.

[36] R. P. Feynman, Forces in molecules, *Phys. Rev.* **56**(4) (1939) 340.

[37] R. P. Feynman and A. R. Hibbs, *Quantum Mechanics and Path Integrals* (Dover Publications, 2010).

[38] D. J. Fixen, E. S. Cheng, J. M. Gales, J. C. Mather, R. A. Shafer and E. L. Wright, The cosmic microwave background spectrum from the full COBE FIRAS data set, *Astrophys. J.* **473** (1996) 576–587.

[39] S. Fölsch, J. Martínez-Blanco, J. Yang, K. Kanisawa and S. C. Erwin, Quantum dots with single-atom precision, *Nature Nanotechnol.* **9**(7) (2014) 505–508.

[40] J. Franck and G. Hertz, Über zusammenstöße zwischen elektronen und den molekülen des quecksilberdampfes und die ionisierungsspannung desselben, *Verh. Deutsche Phys. Ges.* **16** (1914) 457–467.

[41] D. L. Freimund, K. Aflatooni and H. Batelaan, Observation of the Kapitza–Dirac effect, *Nature* **413** (2001) 142–143.

[42] A. Friedman, Über die krümmung des raumes, *Z. Phys.* **10**(1) (1922) 337–386. Translated in "General Relativity and Gravitation" **31**(12) (1999).

[43] G. Gamow, Zur quantentheorie des atomkernes, *Z. Phys.* **51**(3) (1928) 204–212.

[44] H. T. Goldstein and C. P. Poople, *Classical Mechanics* (Pearson, 2014).

[45] J. P. Gordon, H. J. Zeiger and C. H. Townes, The maser: New type of microwave amplifier, frequency standard, and spectrometer, *Phys. Rev.* **99**(4) (1955) 1264–1274.

[46] R. W. Gurney and E. U. Condon, Quantum mechanics and radioactive disintegration, *Phys. Rev.* **33** (1929) 127–140.

[47] R. W. Gurney and N. F. Mott, Conduction in polar crystals. iii. on the colour centres in alkali-halide crystals, *Trans. Faraday Soc.* **34** (1938) 506–511.

[48] W. Hallwachs, Ueber den einfluss des lichtes auf electrostatisch geladene körper, *Ann. Phys.* **33** (1888) 301.

[49] M. J. Hardcastle, Jets, hotspots and lobes: what X-ray observations tell us about extragalactic radio sources, *Philos. Trans. Roy. Soc. London A: Math. Phys. Eng. Sci.* **363**(1837) (2005) 2711–2727.

[50] L. Harding, Alexander Litvinenko: the man who solved his own murder, *The Guardian* (2016).

[51] W. Heisenberg, Über den anschaulichen inhalt der quantentheoretischen kinematik und mechanik, *Z. Phys.* **43**(3) (1927) 172–198.

[52] C. J. Hensley, J. Yang and M. Centurion, Imaging of isolated molecules with ultrafast electron pulses, *Phys. Rev. Lett.* **109** (2012) 133202.

[53] H. Hertz, Ueber einen einfluss des ultravioletten lichtes auf die electrische entladung, *Ann. Phys.* **31** (1887) 983.

[54] B. J. Hillman, B. Ertl-Wagner and B. C. Wagner, *The Man who Stalked Einstein: How Nazi Scientist Philipp Lenard Changed the Course of History* (Rowman & Littlefield, 2015).

[55] S.-W. Hla, G. Meyer and K.-H. Rieder, Selective bond breaking of single iodobenzene molecules with a scanning tunneling microscope tip, *Chem. Phys. Lett.* **370**(3) (2003) 431–436.

[56] E. Hubble, A relation between distance and radial velocity among extra-galactic nebulae, *Proc. Natl. Acad. Sci. USA* **15** (1929) 168–173.

[57] E. D. Isaacs, A. Shukla, P. M. Platzman, D. R. Hamann, B. Barbiellini and C. A. Tulk, Covalency of hydrogen bond in ice: A direct X-ray measurement, *Phys. Rev. Lett.* **82**(3) (1999) 600–603.

[58] C. J. Isham. *Lectures on Quantum Theory* (Allied Publishers, 2001).

[59] I. James. *Remarkable Physicists: From Galileo to Yukawa* (Cambridge University Press, 2011).

[60] K. J. Jansky, Electrical disturbances apparently of extraterrestrial origin, *Proc. IRE* **21**(10) (1933) 1387–1398.

[61] K. Janssens, M. Alfeld, G. Van der Snickt, W. De Nolf, F. Vanmeert, M. Radepont, L. Monico, J. Dik, M. Cotte, G. Falkenberg, C. Milian and B. G. Brunetti, The use of synchrotron radiation for the characterization of artists pigments and paintings, *Ann. Rev. Anal. Chem.* **6** (2013) 399–425.

[62] P. L. Kapitza and P. A. M. Dirac, The reflection of electrons from standing light waves, *Math. Proc. Cambridge Philos. Soc.* **29**(02) (1933) 297.

[63] U. Klein, What is the limit of h goes to zero? *Amer. J. Phys.* **80**(11) (2012) 1009–1016.

[64] S. Koyama, K. Onozawa, K. Tanaka, S. Saito, S. M. Kourkouss and Y. Kato, Multiocular image sensor with on-chip beam-splitter and inner meta-micro-lens for single-main-lens stereo camera. *Opt. Express*, **24**(16) (2016) 18035–18048.

[65] W. Kuhn, Scattering of thorium "C" γ-radiation by radium "G" and ordinary lead, *Philos. Mag. J. Sci.* **8**(52) (1929) 625–636.

[66] P. Lenard, Ueber die lichtelektrische wirkung, *Ann. Phys.* **5** (1902) 149–198.

[67] C. Leonard, S. Carter and N. C. Handy, The barrier to inversion of ammonia, *Chem. Phys. Lett.* **370** (2003) 360–365.

[68] J.-M. Levy-Leblond and F. Balibar, *Quantique — Rudiments.* (Intereditions, 1984).

[69] R. L. Liboff, The correspondence principle revisited, *Phys. Today* **37**(2) (1984) 50.

[70] L. I. Mandelstam and I. E. Tamm, The uncertainty relation between energy and time in nonrelativistic quantum mechanics, *J. Phys.* (*USSR*) **9** (1945) 249–254.

[71] M. F. Manning, Energy levels of a symmetrical double minima problem with applications to the NH3 and ND3 molecules, *J. Chem. Phys.* **3** (1935) 136–138.

[72] P. B. Moon, Resonant nuclear scattering of gamma-rays: Theory and preliminary experiments, *Proc. Phys. Soc. A* **64**(1) (1951) 76.
[73] A. A. Penzias and R. W. Wilson, A measurement of excess antenna temperature at 4080 mc/s, *Astrophys. J.* **142** (1965) 419–421.
[74] C. Qi, A. N. Andreyev, M. Huyse, R. J. Liotta, P. Van Duppen and R. Wyss, On the validity of the Geiger-Nuttall alpha-decay law and its microscopic basis, *Phys. Lett. B* **734** (2014) 203–206.
[75] C. F. Quate, Vacuum tunneling: A new technique for microscopy, *Phys. Today* **39**(8) (1986) 26.
[76] J. C. Ranuarez, M. J. Deen and C.-H. Chen, A review of gate tunneling current in MOS devices, *Microelectron. Reliab.* **46**(12) (2006) 1939–1956.
[77] J. J. Rennilson and D. R. Criswell, Surveyor observations of lunar horizon-glow, *The Moon* **10** (1974) 121–142.
[78] W. Rueckner and J. Peidle, Young's double-slit experiment with single photons and quantum eraser, *Amer. J. Phys.* **81** (2013) 951–958.
[79] P. Shaw, *The Philosophical Works of the Honourable Robert Boyle*, Vol. 3 (W. and J. Innys, 1725).
[80] C. G. Shull, Single-slit diffraction of neutrons, *Phys. Rev.* **179**(3) (1969) 752–754.
[81] A. A. Sickafoose, J. E. Colwell, M. Horányi and S. Robertson, Photoelectric charging of dust particles in vacuum, *Phys. Rev. Lett.* **84** (2000) 6034.
[82] V. Spirko and W. P. Kraemer, Anharmonic potential function and effective geometries for the NH3 molecule, *J. Mol. Spectrosc.* **133**(2) (1989) 331–344.
[83] G. L. Squires, *Introduction to the Theory of Thermal Neutron Scattering* (Cambridge University Press, 1978).
[84] D. Stiévenard, *Nanoscience: Nanotechnologies and Nanophysics* (Springer, 2007).
[85] M. H. Stone, On one-parameter unitary groups in Hilbert space, *Ann. Math.* **33**(3) (1932) 643–648.
[86] J. T. Tate, The low potential discharge spectrum of mercury vapor in relation to iononization potentials, *Phys. Rev.* **7**(6) (1916) 686–687.
[87] M. Tegmark and J. A. Wheeler, 100 years of Quantum Mysteries, *Sci. American* **284**(2) (2001) 68–75.
[88] J. J. Thomson, Cathode rays, *Philos. Mag. Ser.* (5) **44** (1897) 293–316.
[89] A. Tonomura, J. Endo, T. Matsuda, T. Kawasaki and H. Ezawa, Demonstration of single-electron buildup of an interference pattern, *Am. J. Phys.* **57**(2) (1989) 117–120.
[90] J. T. Tate, The passage of low speed electrons through mercury vapor and the ionizing potential of mercury vapor, *Phys. Rev.* **10**(1) (1917) 81–83.
[91] H. Valladas, J. L. Joron, G. Valladas, O. Bar-Yosef, B. Arensburg, A. Belfer-Cohen, P. Goldberg, H. Laville, L. Meignen, Y. Rak, E. Tchernov, A. M. Tillier and B. Vandermeersch, Thermoluminescence dates for the Neanderthal burial site at Kebara in Israel, *Nature* **330** (1987) 159–160.
[92] E. Wigner, The unreasonable effectiveness of mathematics in the natural sciences, *Commun. Pure Appl. Math.* **13**(1) (1960) 1–14.

[93] R. W. Wood, The fluorescence of sodium vapour and the resonance radiation of electrons, *Philos. Mag. Ser.* 6 **10**(59) (1905) 1.
[94] R. W. Wood, Lecture demonstration of resonance radiation of sodium, *Phys. Rev.* **56** (1939) 1172.
[95] N. Wright and H. M. Randall, The far infrared absorption spectra of ammonia and phosphine gases under high resolving power, *Phys. Rev.* **44**(5) (1933) 391–398.

Index

CPSIA information can be obtained
at www.ICGtesting.com
Printed in the USA
FFHW010013051118
49232347-53483FF